S0-ANP-662

ORDINARY DIFFERENTIAL EQUATIONS WITH APPLICATIONS

EDWARD L. REISS

Courant Institute,
New York University

ANDREW J. CALLEGARI

State University of New York,
College at Purchase

DALJIT S. AHLUWALIA

Courant Institute,
New York University
and the University of South Florida

HOLT, RINEHART AND WINSTON

New York Chicago San Francisco Atlanta
Dallas Montreal Toronto London Sydney

This book was set in Times Roman by Mono of Maryland Incorporated

Editor: Holly Massey
Designer: Katrine Stevens
Printer and Binder: The Maple Press Company
Drawings: Eric G. Hieber Associates
Cover: Katrine Stevens

Library of Congress Cataloging in Publication Data

Reiss, Edward L.
Ordinary differential equations with applications.
1. Differential equations. I. Callegari, Andrew John, joint author. II. Ahluwalia, Daljit S., joint author. III. Title.
QA372.A37 515'.352 74-15606

ISBN 0-03-085872-0

Printed in the United States of America
6 7 8 9 038 9 8 7 6 5 4 3 2 1

PREFACE

This book is intended for a one-semester or one-quarter post-calculus course on ordinary differential equations. It is written in the spirit of applied mathematics. Thus it is a blend of technique, application, and theory.

The development is at an elementary level appropriate for students with a conventional calculus background. The students in our classes are typically mixed. They consist of engineering, mathematics, and science majors and students with interests in the social and economic sciences. Consequently we have attempted to appeal to a wide audience.

Applications are used to motivate and illustrate solution techniques, basic theory, and qualitative behavior of solutions. A wide selection of "modern" and traditional applications is included. Although they vary in difficulty and in depth, each application is essentially self-contained. The origins of the differential equations are explained and in some cases they are derived explicitly. The first in a group of applications usually requires little or no background in the specific area. They are drawn from population dynamics, ecology, biology, and so on. The later applications in each group are taken from traditional elementary problems in mechanics and electric circuit theory. It is not intended that all the applications in a specific group should be treated in class. In our own lectures we usually consider one or two in a specific group, depending on the students' interests and backgrounds. Sufficient detail is included so that the others can be assigned as exercises.

New ideas, techniques, and theoretical results are motivated by appealing to simple examples, known results from calculus, and sometimes by using simple applications. The qualitative behavior of solutions has been emphasized. It is a basic feature of the text. Elementary aspects of boundary value problems and eigenvalue problems are treated as soon as possible. They are not relegated to separate chapters at the end of the text.

The text contains the material that is usually treated in an elementary differential equations course. However, there are some significant differences in the arrangement and the emphasis. The study of nonlinear equations is delayed until Chapter 14. Our experience indicates that a firm understanding of linear, first-order, and second-order differential equations is an important requisite for the study of nonlinear equations. Without this background it is difficult for students to appreciate the distinctions in the solution techniques and the qualitative behavior of solutions of linear and nonlinear differential equations, and the need for analyzing special nonlinear equations with special techniques. As a consequence, we have emphasized the linear first-order equations more than is done in most books.

Second-order equations with constant coefficients are analyzed extensively because the techniques developed for these equations are fundamental for the study of higher-order linear equations, systems of linear equations, and for nonlinear second-order equations. Moreover, the second-order equations occur often in applications. We feel that a major goal of the course should be a firm understanding of the techniques, elementary theory, and the qualitative behavior of the solutions of this equation.

In preparing this text, we have assumed that students, such as our own, have little or no background in linear algebra and matrices. This is unfortunate, since matrices are a natural and efficient tool for the analysis of linear systems of ordinary differential equations. However, in the brief chapter on systems we have attempted to present the essential ideas in an elementary way by using only two simultaneous equations.

The book contains more than enough material for a one-semester course. Additional topics are included to provide the instructor with options in the planning of a syllabus. A typical course would cover Chapters 1–12 and Chapter 14, where, as we have remarked previously, only a selection of the applications need be discussed in class. If the students have had an introduction to differential equations in their calculus sequence, then Chapters 1–3 and the first few sections of Chapters 4 and 14 can be treated more rapidly. Selected material chosen from Chapters 13, 15, and 16 can then be used.

The homework exercises at the end of each section are usually graded in difficulty. Some are of routine nature, while others are intended to extend the ideas introduced in the text or to introduce new ideas and directions. In general, answers are provided at the end of the text for odd-numbered problems and for some of the more difficult even-numbered problems.

An asterisk on a section or exercise indicates either more difficult or more advanced material. However, no new knowledge is required for these sections.

Finally we would like to express our appreciation to our colleagues and students for their helpful and constructive criticisms. We are grateful to Professors Garret Etgen, University of Houston; Jack Goldberg, University of Michigan; Paul Gordon, Drexel; Lee Kaminetzky, City College of New York; Gerald Ludden, Michigan State University; and David Ullrich, North Carolina State University, for reading various versions of the manuscript. Ms. Jane Clifford, Mr. David Stampf, and Mr. Wayne Barrett read the manuscript and provided valuable assistance with the homework exercises. Finally, we wish to thank Ms. Connie Engel and Ms. Anita Costadasi for typing a major portion of the manuscript, and Ms. Jane Ross, Holly Massey, and the editorial staff of Holt, Rinehart and Winston for their cooperation and patience.

Edward L. Reiss
Andrew J. Callegari
Daljit S. Ahluwalia

New York City
September 1975

CONTENTS

chapter 1

INTRODUCTION

1.1 ORDINARY DIFFERENTIAL EQUATIONS

Since the time of Newton, physical problems have been investigated by formulating them mathematically as differential equations. Many important problems in the social and life sciences can also be expressed in terms of differential equations. *A differential equation is an equation for an unknown function, which contains at least one derivative of this function.*

In this book we shall be concerned primarily with differential equations for functions $y(x)$ of a single independent variable x. Such equations are called *ordinary differential equations.*

Some examples of ordinary differential equations are:

$$y' - (1 - y)y = 0, \tag{1}$$

$$y'' - y' + \tfrac{1}{3}(y')^3 + y = 0, \tag{2}$$

$$y'' + y' + y = \sin x, \tag{3}$$

where we have denoted the derivative of $y(x)$ by $y' = dy/dx$ and the second derivative by $y'' = d^2y/dx^2$. This notation and an analogous notation for higher-order derivatives will be used throughout the book. Equation (1) is a special case of the logistic equation. The logistic equation arises in mathematical studies in ecology, biology, and learning theory. The transmission of signals from the human eye to the brain is believed to be described by Eq. (2). It is called van der Pol's equation. Finally, Eq. (3) describes the damped oscillations of a pendulum and other vibrating mechanical and electrical systems.

The adjective "ordinary" is employed to distinguish these equations from partial differential equations. The latter are equations relating the partial derivatives of an unknown function of several variables. Since partial differential equations are *not* treated to any extent in this book, the term ordinary is omitted and we refer merely to a differential equation. We abbreviate differential equation by DE. The plural of DE will be denoted by DEs.

One way in which DEs are classified is according to their *order*.

definition

The order of a DE is the order of the highest derivative of the unknown function that appears in the equation.

Thus,

$$y' + xy = x^3$$

is a first-order DE since y' is the highest-order derivative of y. The DE

$$(y')^3 = \sin x$$

is also of first order even though the highest derivative term y' is raised to the third power. Similarly the DEs

$$y'' - (1 - y^2)y' + y = 0 \qquad \text{and} \qquad (y')^3 \sin(y'') + x^2 + y' + 16e^y = 0$$

are both second-order equations. The DE

$$y''' + 3y'' + y = \sin x$$

is a third-order DE.

Exercises

1. Determine which of the following equations are ordinary differential equations and give their order.

(a) $(y'')^2 + 2yy' = \sin x$ (b) $\dfrac{\partial^2 y}{\partial x^2} + \dfrac{\partial^2 y}{\partial z^2} = \sin(xz)$

(c) $y' + 3y'' + 6y = 4x$ (d) $(y'')^3 + y''' + 3y = 0$

(e) $y' + \sqrt{y'' + 4} = 7x$

(f) $\dfrac{\partial^2 y}{\partial x^2} + 6xz\dfrac{\partial y}{\partial z} = \cos z$

(g) $y'' + \dfrac{2}{x}y' + e^{-y} = 0$

(h) $y''' - \dfrac{1}{\sqrt{x}}y^{3/2} = 0$

1.2 SOLUTIONS OF A DE

In this section, a solution of a DE is defined, and we discuss several examples of solutions to specific DEs. In the present discussion and in the remainder of the book the open interval $a < x < b$ will be denoted frequently by (a, b), and the closed interval $a \leq x \leq b$ by $[a, b]$. We allow the possibilities that $a = -\infty$ and/or $b = +\infty$. Thus if $y(x)$ is defined on $(-\infty, \infty)$ this means that it is defined for all values of x. Furthermore, we shall use the notation $y(x) \equiv 0$ to mean that $y(x) = 0$ for all values of x.

definition 1

A solution of a DE on an interval (a, b) is a function that when substituted into the DE reduces it to an identity for all x in the interval.

We observe that a solution to an nth-order equation must have n derivatives.

EXAMPLE 1

Show that $y = ce^{-x}$ is a solution of the first-order DE

$$y' + y = 0$$

for all values of x. Furthermore, show that it is a solution for every value of the constant c.

solution. To substitute $y = ce^{-x}$ into the DE we must compute y':

$$y' = (ce^{-x})' = c(e^{-x})' = -ce^{-x}.$$

Then we have

$$y' + y = -ce^{-x} + ce^{-x} = c[-e^{-x} + e^{-x}] = 0.$$

Thus $y = ce^{-x}$ reduces the DE to the identity $0 = 0$ for all values of x. Since the identity is valid for any c, ce^{-x} is a solution for all values of the constant c. *Thus the DE has an infinite number of solutions, one for each value of c.* Graphs of the solution for typical values of c are sketched in Figure 1.1. We observe that for each value of c we have one curve in the xy-plane.

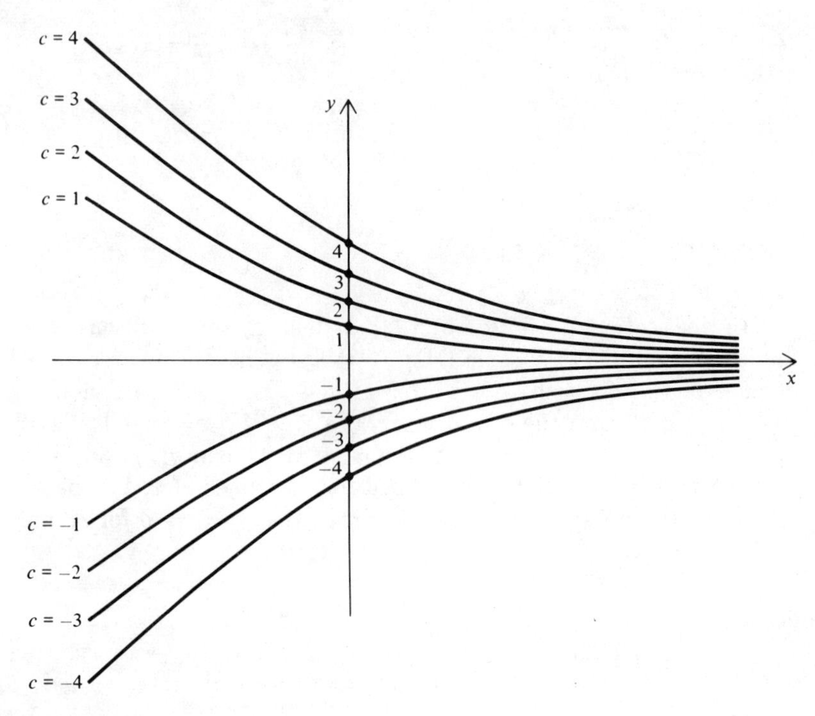

Figure 1.1

EXAMPLE 2

Show that for $-\infty < x < \infty$ the functions $y = c_1 \cos x$ and $y = c_2 \sin x$ are solutions of the DE

$$y'' + y = 0$$

for all values of the constants c_1 and c_2. Then show that

$$y = c_1 \cos x + c_2 \sin x \tag{1}$$

is a solution of the DE for each pair of values of c_1 and c_2.

solution. To show that $y = c_1 \cos x$ is a solution we must calculate y''. Thus we have

$$y' = (c_1 \cos x)' = c_1(\cos x)' = -c_1 \sin x$$

$$y'' = -(c_1 \sin x)' = -c_1 \cos x.$$

By substituting y and y'' into the DE we get

$$y'' + y = -c_1 \cos x + c_1 \cos x = c_1(-\cos x + \cos x) = 0.$$

Thus $y = c_1 \cos x$ reduces the DE to the identity $0 = 0$ for all values of x. Furthermore, $y = c_1 \cos x$ is a solution of the DE for any value of c_1. In a similar manner we can show that $y = c_2 \sin x$ is a solution for any value of c_2.

We now consider the function given in Eq. (1). We compute y'',

$$y' = (c_1 \cos x + c_2 \sin x)' = -c_1 \sin x + c_2 \cos x$$

$$y'' = (-c_1 \sin x + c_2 \cos x)' = -c_1 \cos x - c_2 \sin x.$$

Therefore we have

$$y'' + y = -c_1 \cos x - c_2 \sin x + c_1 \cos x + c_2 \sin x = 0.$$

Thus Eq. (1) is a solution of the DE for all values of c_1 and c_2.

The sum shown in Eq. (1), where c_1 and c_2 are any two constants, is a special case of the general concept of a linear combination of functions.

definition 2

Let $y_1(x)$ and $y_2(x)$ be two functions defined on an interval (a, b). Then the expression

$$y(x) = c_1 y_1(x) + c_2 y_2(x),$$

where c_1 and c_2 are any constants, is called a linear combination of $y_1(x)$ and $y_2(x)$. It is defined on (a, b). More generally, if $y_1(x), y_2(x), \ldots, y_n(x)$ are any n functions defined on an interval (a, b) and $c_1, c_2, \ldots, c_n$ are n constants then

$$y = c_1 y_1(x) + c_2 y_2(x) + \cdots + c_n y_n(x)$$

is a linear combination of $y_1(x), y_2(x), \ldots, y_n(x)$ and it is defined on (a, b).

The function $y(x)$ given in Eq. (1) is a linear combination of the two functions $y_1(x) = \cos x$ and $y_2(x) = \sin x$. We have shown in Example 2 that $y_1(x)$ and $y_2(x)$ are solutions of the DE and that y is a solution for *all* values of c_1 and c_2. We can restate this result as follows: Any linear combination of the two solutions y_1 and y_2 is also a solution. As we shall see in the next section there is a special class of DEs that possess the property that *any* linear combination of solutions is a solution. They are called linear homogeneous DEs. In general, DEs do not have this property, as we show in Exercise 10.

Exercises

1. Show that the given DEs have the indicated solutions. Give the values of x for which each solution is valid.
   (a) $y' + y = 9, \quad y = 9 + e^{-x}$
   (b) $y'' + 2y' + y = 0, \quad y = e^{-x} + xe^{-x}$
   (c) $y' - y^2 = 0, \quad y = -\dfrac{1}{x+4}$
   (d) $y' = -\dfrac{x}{\sin y}, \quad y = \cos^{-1}\left(\dfrac{x^2}{2} - 2\right)$

2. Show that each of the following DEs has the indicated solution. c, c_1, and c_2 are constants. Find the values of x and the values of the constants for which each solution is valid.
   (a) $y' + y = 9, \quad y = ce^{-x} + 9$
   (b) $y'' + 3y' + 2y = 4, \quad y = c_1 e^{-x} + c_2 e^{-2x} + 2$
   (c) $y'' = 0, \quad y = c_1 + c_2 x$ (d) $yy' + x = 0, \quad y = \pm\sqrt{c - x^2}$
   (e) $y' + \frac{2}{x} y = 4x, \quad y = x^2 + \frac{c}{x^2}$ (f) $(y')^2 - 4y = 0, \quad y = (x + c)^2$

3. Sketch the given solutions for typical values of the constant c in Exercises (2a), (2c), (2d), and (2f).

4. Find the values of the constant m for which $y = e^{mx}$ is a solution of the following DEs.
   (a) $y'' - y = 0$
   (b) $y' + ky = 0$, k is a prescribed constant
   (c) $y''' - y'' - 4y' + 4y = 0$

5. Find the values of the constant n so that $y = x^n$ is a solution of $x^2y'' + 3xy' - 3y = 0$. This above equation is called a Cauchy-Euler, or equidimensional, equation. It is discussed in Section 11.4.

6. Find the values of A and n so that $y = Ax^n$ is a solution of the DE $y'' - x^2y^3 = 0$.

7. The DE in Exercise 6 is a special case ($p = 2$, $q = 3$) of the Emden equation $y'' - x^p y^q = 0$. This equation arises in various astrophysical problems, including the study of the density of stars. Assume that $q \neq 1$. Find the values of $A \neq 0$ and n so that $y = Ax^n$ is a solution of the DE. Observe that for each pair of values (p, q) there is only one pair of values (A, n). What happens if $q = 1$?

8. For the following DEs verify that the given functions are solutions and that any linear combination of these solutions is also a solution.
   (a) $y'' + 3y' - 4y = 0, \quad e^x, e^{-4x}$
   (b) $y'' + 2y' + 2y = 0, \quad e^{-x} \sin x, e^{-x} \cos x$
   (c) $y''' - 4y'' + y' + 6y = 0, \quad e^{-x}, e^{2x}, e^{3x}$

9. For each of the following DEs show that the given functions y_1 and y_2 are solutions but that the specific linear combination $y_1 + y_2$ is not a solution.
   (a) $y'' + y = e^x, \quad y_1 = \cos x + \frac{1}{2}e^x, \; y_2 = \sin x + \frac{1}{2}e^x$
   (b) $yy' - x = 0 \quad y_1 = -x, \; y_2 = x$
   (c) $y' + y^3 = 0, \quad y_1 = \frac{1}{\sqrt{2x}}, \; y_2 = \frac{1}{\sqrt{2(x + 1)}}$

10. For each of the following sets of functions write one function of the set as a linear combination of the remaining functions.
   (a) $1 + e^x, e^x, 2$
   (b) $e^{-x}, 3e^x, e^x + e^{-x}$
   (c) $(x + 1), (x - 1), x$

*11. In Example 1 we saw that $y = ce^{-x}$ was a solution of $y' + y = 0$ for all values of c, and that for each value of c the solution was a curve in the xy-plane.
(a) Show that if c_1 and c_2 are two distinct values of c, then the two curves $y = c_1e^{-x}$ and $y = c_2e^{-x}$ never intersect.
(b) Show that if $P_0 = (x_0, y_0)$ is a point in the xy-plane, then there is only one value of c for which $y = ce^{-x}$ passes through P_0.

1.3 LINEAR DEs

In the first part of this book we shall be concerned mainly with linear DEs and the techniques for solving them. Linear DEs play a central role in the application of DEs to physical and other problems.

definition 1

A linear DE of order n is any DE that can be written as

$$y^{(n)} + b_{n-1}(x)y^{(n-1)} + b_{n-2}(x)y^{(n-2)} + \cdots + b_0(x)y = r(x) \tag{1}$$

where $b_0(x)$, $b_1(x)$, . . ., $b_{n-1}(x)$, and $r(x)$ are specified functions. The functions $b_0(x)$, $b_1(x)$, . . ., $b_{n-1}(x)$ are called the coefficients of the DE.

In Eq. (1) we have used the notation $y^{(k)} = d^ky/dx^k$ for the kth derivative of y, where k is any positive integer. The DE is of nth order because the order of the highest derivative is n. We observe that the manner in which the independent variable x occurs in Eq. (1) does not affect the definition of linearity.

definition 2

A DE that is not linear is called a nonlinear DE.

EXAMPLE 1

The first-order DE

$$y' + q(x)y = r(x) \tag{2}$$

where $r(x)$ and $q(x)$ are specified functions is a linear DE because if we set $n = 1$ and $b_0(x) = q(x)$ in Eq. (1), then Eqs. (1) and (2) are identical.

EXAMPLE 2

The first-order DE

$$y' + y^2 = r(x) \tag{3}$$

is nonlinear, since for $n = 1$ no term proportional to y^2 appears in Eq. (1).

EXAMPLE 3

The second-order DE

$$y'' + p(x)y' + q(x)y = r(x) \tag{4}$$

is linear. This is shown by setting $n = 2$, $b_1(x) = p(x)$, and $b_0(x) = q(x)$ in Eq. (1). Then Eqs. (1) and (4) are identical.

definition 3

If $r(x) = 0$ for all values of x, then the DE

$$y^{(n)} + b_{n-1}(x)y^{(n-1)} + \cdots + b_0(x)y = 0 \tag{5}$$

is a linear homogeneous DE of order n.

For example, $y' + y = 0$ and $y'' + y = 0$ are homogeneous DEs. The DE $y' + y = \sin x$ is not a homogeneous DE because $r(x) = \sin x$ is not zero for all values of x. *If $r(x) \neq 0$ for at least one value of x, then Eq. (1) is called an inhomogeneous DE and $r(x)$ is called the inhomogeneous term.*

All linear homogeneous DEs possess an important property that shall be used to construct solutions. It is stated in the following theorem.

theorem

(Superposition Property)

If $y_1, y_2, \ldots, y_m$ are m solutions of a linear homogeneous DE, then the linear combination of these solutions

$$y = c_1y_1 + c_2y_2 + \cdots + c_my_m$$

is a solution for all choices of the constants $c_1, c_2, \ldots, c_m$.

Example 3 of Section 1.2 is an illustration of this theorem, because there we have shown that the linear combination $y = c_1 \cos x + c_2 \sin x$ of the solutions $y_1 = \cos x$ and $y_2 = \sin x$ is a solution of the DE $y'' + y = 0$. A proof of this theorem is indicated in Exercises 9 and 10. Here we shall establish a special case of the theorem: Every linear combination of two solutions of the second-order linear homogeneous DE

$$y'' + b_1(x)y' + b_0(x)y = 0 \tag{6}$$

is a solution. Thus if y_1 and y_2 are two solutions of the DE Eq. (6), then we must show that the linear combination

$$y = c_1y_1 + c_2y_2 \tag{7}$$

is a solution for all values of the constants c_1 and c_2. We shall show that Eq. (7) is a solution by substituting it into Eq. (6). First we calculate y' and y''. We use the differentiation formulas, $(f(x) + g(x))' = f'(x) + g'(x)$ and $(cf(x))' = cf'(x)$, where c is a constant. Thus we obtain

$$y' = (c_1y_1 + c_2y_2)' = c_1y_1' + c_2y_2',$$

$$y'' = (c_1y_1' + c_2y_2')' = c_1y_1'' + c_2y_2''.$$

Substituting these expressions for y, y', and y'' into Eq. (6), we get

$$\begin{aligned} &y'' + b_1(x)y' + b_0(x)y \\ &\quad = c_1y_1'' + c_2y_2'' + b_1(x)(c_1y_1' + c_2y_2') + b_0(x)(c_1y_1 + c_2y_2) \\ &\quad = c_1[y_1'' + b_1(x)y_1' + b_0(x)y_1] + c_2[y_2'' + b_1(x)y_2' + b_0(x)y_2]. \qquad (8) \end{aligned}$$

Since y_1 and y_2 are solutions of Eq. (6)—that is, $y_1'' + b_1(x)y_1' + b_0(x)y_1 = 0$ and $y_2'' + b_1(x)y_2' + b_0(x)y_2 = 0$—the bracketed terms on the right side of Eq. (8) are equal to zero. Thus

$$y'' + b_1(x)y' + b_0(x)y = 0$$

and we conclude that $y = c_1y_1 + c_2y_2$ is a solution of the DE for all values of c_1 and c_2.

A special case of the theorem occurs when we consider a single solution y_1 of a linear homogeneous DE. Then we deduce the following corollary from the theorem with $m = 0$.

corollary

If y_1 is a solution of a linear homogeneous DE, then cy_1 is a solution for all values of the constant c.

The theorem, which is called the *superposition property* of solutions of *linear homogeneous* DEs, is a basic result that is used to construct solutions of linear DEs. The restrictions that the DE be linear and homogeneous are essential, as we demonstrate in Example 4 and Exercises 7 and 8.

EXAMPLE 4

Verify that $y = x^2$ and $y = (x + 1)^2$ are solutions of the nonlinear first-order DE $(y')^2 - 4y = 0$. Show, however, that the specific linear combination

$$y = x^2 + (x + 1)^2$$

is not a solution of the DE.

solution. If $y = x^2$, then $y' = 2x$, and substituting these expressions into the DE we get

$$(2x)^2 - 4(x^2) = 4x^2 - 4x^2 = 0.$$

Similarly, the reader should show that $y = (x + 1)^2$ is a solution. We now consider the linear combination ($c_1 = c_2 = 1$), $y = x^2 + (x + 1)^2$. Substituting this expression and its derivative $y' = 4x + 2$ into the DE we find that

$$(4x + 2)^2 - 4(2x^2 + 2x + 1) = 8x^2 + 8x \neq 0.$$

Hence $y = x^2 + (x + 1)^2$ is not a solution of the DE.

Exercises

1. Determine which of the following DEs are linear. For each of the linear DEs indicate which are homogeneous.
   (a) $xy'' + x^2y' + (\sin x)y = x$
   (b) $xy'' + x^2y' + (\sin x)y = y$
   (c) $xy' + 2e^xy = 0$
   (d) $y'' + y' + \sin y = 0$
   (e) $y'' + y' + \sin x = 0$
   (f) $y''' + y'' + x^2y' + x^3y = 0$
   (g) $y'' + x^2y' + (\sin x)y = y^2$
   (h) $yy' + y' + 2y = x$

2. Consider the linear homogeneous first-order DE $y' - xy = 0$. Show that $y = e^{x^2/2}$ is a solution and that $y = ce^{x^2/2}$ is a solution for all values of c.

3. Show that $y = x$ is a solution of the nonlinear DE $yy' - x = 0$. Verify that $y = cx$, $c^2 \neq 1$, is not a solution.

4. Suppose that the nonlinear DE (van der Pol's equation)

$$y'' - y' + \tfrac{1}{3}(y')^3 + y = 0$$

   has a solution $y_1(x)$. Show that $y = cy_1$ is not a solution unless $c = 0$ or $c^2 = 1$.

5. (a) Show that if y_1 is a solution of $y' + q(x)y = r(x)$, then $y = cy_1$, $c \neq 1$, is not a solution unless $r(x) \equiv 0$.
   (b) Repeat (a) for the DE $y'' + b_1(x)y' + b_0(x)y = r(x)$.

6. Show that $y_1 = e^{-x} + 2$ and $y_2 = e^x + 2$ are solutions of the inhomogeneous DE, $y'' - y = -2$. Verify that the linear combination $y = c_1y_1 + c_2y_2$ is not a solution unless $c_1 + c_2 = 1$.

7. Let y_1 and y_2 be two solutions of the DE $y' + q(x)y = r(x)$. Show by direct substitution into the DE that $y = y_1 + y_2$ is not a solution unless $r(x) = 0$ for all values of x.

8. Repeat Exercise 7 for $y'' + b_1(x)y' + b_0(x)y = r(x)$.

   The next two exercises suggest a proof of the theorem of this section.

9. Let $y_1, y_2, \ldots, y_m$ be m solutions of $y' + q(x)y = 0$. Show by substituting the linear combination of solutions, $y = c_1y_1 + \cdots + c_my_m$, into the DE that it is a solution.

*10. Repeat Exercise 9 for the general linear, homogeneous nth-order DE.

1.4 ORIGIN OF DIFFERENTIAL EQUATIONS

Most situations that occur in nature are too complicated to precisely formulate mathematically. Therefore, mathematical models are created by making certain simplifying assumptions concerning the phenomena and by neglecting certain of its aspects that are thought to be unessential. It is hoped that the resulting mathematical model, which is often expressed in terms of differential equations, is reasonable. The mathematical properties of the model are then studied and, where possible, solutions are obtained. A comparison of the solutions with the corresponding experimentally measured quantities usually determines whether the model is "good" or should be discarded. Newton's laws of motion, or Newtonian mechanics, which is a mathematical model for the motion of objects, dominated physics for many years. Many complicated physical phenomena were accurately described by these laws. They are still used today to study a great variety of problems. However, it was found that certain astronomical and atomic phenomena could not be adequately predicted by them. This does not mean that Newtonian mechanics is "wrong," but merely that as a mathematical model it has certain limits of applicability. New mathematical models, such as quantum mechanics and the theory of relativity, were created to obtain more accurate descriptions of nature. They too must have certain limits of applicability since they are only models.

In Sections 1.4-2 and in Appendix I we shall consider the mathematical formulation of some typical physical and biological problems. Other problems will be formulated in later chapters.

1.4-1 Initial-value problem; boundary-value problem

In addition to a differential equation, a mathematical model contains subsidiary conditions that solutions of the DE must satisfy. In the most common subsidiary condition, the value of the unknown function or the value of one of its derivatives is prescribed at a specified point $x = x_0$. If all such subsidiary conditions are prescribed at the same point x_0, then they are called *initial conditions*. The point x_0 is called the *initial point* and the prescribed values are called the *initial values*.

An *initial-value problem*, which we abbreviate as IVP, consists of determining the solutions of a DE that satisfy given initial conditions. The physical phenomenon being modeled usually determines the precise number of initial conditions, as we shall see in the following sections. An example of an IVP is: Find a solution of the DE $y'' + y = 0$ that satisfies the initial conditions $y(1) = 2$, $y'(1) = 0$. Here $x_0 = 1$ is the initial point, and the constants 2 and 0 are the initial values.

An extensive mathematical theory has been developed to determine the appropriate initial conditions in order that an IVP will have exactly one solution. The results of this theory show that, for a wide class of DEs, the appropriate initial conditions for an nth-order DE are to specify the unknown function and its first $n - 1$ derivatives at a given point x_0. Thus, for a first-order DE, $y(x_0)$ is given. For a second-order DE, $y(x_0)$ and

$y'(x_0)$ are given, and so on. This mathematical prescription of the initial conditions is usually in accord with the initial conditions deduced from physical considerations, as we illustrate in the following sections.

We discuss the actual solution of IVPs in subsequent chapters. In Exercises 2, 3, and 4 we indicate how subsidiary conditions are satisfied if solutions of the DE are known.

If subsidiary conditions are specified at more than one point, then they are called *boundary conditions*. Typical boundary conditions consist of specifying the unknown function or its derivatives at two points, x_0 and x_1. A *boundary-value problem*, which we abbreviate as BVP, consists of determining a solution of a DE in an interval (x_0, x_1) where the solution must also satisfy the given boundary conditions at x_0 and x_1. Two typical BVPs are:

1. Determine solutions of the DE

$$y'' - y = r(x)$$

on the interval $(0, 1)$ that satisfy the boundary conditions

$$y(0) = 1, \qquad y'(1) = 0.$$

This BVP has been used as a simple model to study heat conduction in the human body in connection with the early detection of breast cancer.

2. Determine solutions of the DE

$$y'' + \frac{3}{x}y' = 1 - \frac{1}{y^2}$$

in the interval $(0, 1)$ that satisfy the boundary conditions

$$y'(0) = 0, \qquad y(1) = 1.$$

This BVP occurs in studying the deformations of curved elastic membranes and in the study of dielectric liquids. As we shall see in Chapter 6, BVPs may have no solutions, one solution, or many solutions. The mathematical theory of BVPs is more complicated, and less well-developed, than the corresponding theory of IVPs.

Exercises

1. Classify each of the following as IVPs, BVPs, or neither.
   (a) $y' + 3y = 6, \qquad y(0) = 1$
   (b) $y' + 3y = 6, \qquad \int_0^1 y(x)\,dx = 2$

(c) $y'' + 3y' + 6y = 0, \quad y(0) = 1, y'(1) = 2$
(d) $y'' + \sin y = 0, \quad y(0) = 1, y'(0) = 0$
(e) $y'' + (y')^2 = 0, \quad y'(0) = 0, y(1) + 3y'(1) = 2$

(f) $y'' + 2y = 0, \quad y(0) = 1, \int_0^1 y'(x)\, dx \leq 3$

(g) $y''' + yy'' + y' = 0, \quad y(1) = 1, y'(1) = 0, y''(1) = 3$

2. We have seen in Example 1 of Section 1.2 that the DE $y' + y = 0$ has solutions $y = ce^{-x}$ for all values of c.
   (a) In this set of solutions find one solution that satisfies the condition $y(0) = 2$. (*Hint:* Choose c so that $y = 2$ when $x = 0$.)
   (b) Repeat (a) for the condition $y(3) = 2$.

3. We have shown in Example 2 of Section 1.2 that the DE $y'' + y = 0$ has solutions $y = c_1 \cos x + c_2 \sin x$ for all values of c_1 and c_2. (a) Find one solution of the DE that satisfies the initial conditions $y(0) = 0$, $y'(0) = 1$. (b) Repeat (a) for $y(0) = 1$, $y'(0) = 0$ and also for $y(0) = 2$, $y'(0) = 3$. (c) How many solutions can you find that satisfy the condition $y(0) = 0$?

4. Given that $y = c_1e^x + c_2e^{-x}$ is a solution of the DE $y'' - y = 0$ for all values of c_1 and c_2, solve the BVP

$$y'' - y = 0, \qquad y(0) = 1, y(1) = 0.$$

1.4-2 Population dynamics

Mathematical models are used to predict the growth and decay of populations. The population may be composed of people, of bacteria, or of some other species. The species may be isolated or there may be several interacting populations of different species. Mathematical ecology is the scientific discipline in which mathematical models of interacting populations are derived and analyzed.

In the simplest example of population dynamics the growth of a single, isolated species, such as a single species of bacteria in a laboratory culture, is studied. The number of members of the population at any time t, $N(t)$, is always an integer. As t varies, $N(t)$ changes suddenly as new members of the population are born or die (see Figure 1.2). We shall not discuss the theory of population dynamics that accounts for such integer variations. However, if the population is large, then the birth or death of a single member is insignificant compared to the total population size. Thus we assume that the discontinuous graph of Figure 1.2 can be approximated by a continuous curve. This is equivalent to approximating the actual, integer-valued population, $N(t)$ by a continuously varying function $y(t)$. Henceforth we shall refer to $y(t)$ as the population size at time t.

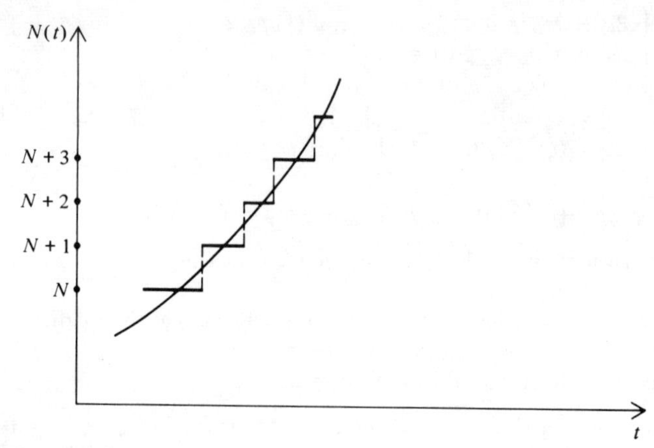

Figure 1.2

To derive a simple, population dynamics model, we assume that the quantity $y'(t)/y(t)$, which is the rate of change of the population per unit of population at time t, depends only upon the population that is present at time t. It does not depend on the value of t explicitly or on the population size at an earlier time. That is, $y'/y = F(y)$ or

$$y' = F(y)y, \tag{1}$$

where $F(y)$ is a specified function of y.

In the simplest model we take $F = k =$ constant. Then

$$y' = ky, \tag{2}$$

which is called the Malthusian law. It is a linear DE. To derive this DE we proceed as follows. Let m and n denote constant birth and death rates respectively for the members of the population. That is, m is the number of births per person per unit time and n is the number of deaths per person per unit time. Then the number of births in the interval Δt is given approximately by $my(t)\ \Delta t$, and the number of deaths is $ny(t)\ \Delta t$. If the population changes are due only to births and deaths, then the total change in the population, Δy, in the interval Δt is given by

$$\Delta y = \text{number of births} - \text{number of deaths} = my\ \Delta t - ny\ \Delta t.$$

Then, by dividing both sides of this equation by Δt, we have

$$\frac{\Delta y}{\Delta t} = (m - n)y. \tag{3}$$

Since

$$y' = \lim_{\Delta t \to 0} \frac{\Delta y}{\Delta t},$$

we obtain the DE in Eq. (2) by taking the limit of both sides of Eq. (3) as $\Delta t \to 0$. Then the constant k is given by $k = m - n$.

Since the initial size $a_0 > 0$ of the population is known, the population dynamics IVP is to find a solution of the DE in Eq. (1) that satisfies the initial condition

$$y(t_0) = a_0, \tag{4}$$

where t_0 is a given initial time.

Generalizations of the Malthusian law, Eq. (2), are obtained by assuming that the function $F(y)$ in Eq. (1) can be expanded in a power series; that is,

$$F(y) = k_0 + k_1 y + k_2 y^2 + \cdots,$$

where $k_0, k_1, \ldots$ are specified constants. In the special case in which $F(y) = k_0 + k_1 y$, Eq. (1) is reduced to

$$y' = (k_0 + k_1 y)y. \tag{5}$$

This nonlinear DE is called the logistic equation. It was originally derived by P. F. Verhulst, a Belgian demographer and mathematician. One of the first applications of this theory was to predict accurately the growth of the population of the United States for relatively long time periods. Since that time it has been applied in a variety of fields including ecology, sociology, psychology, and epidemiology. We discuss this DE in Section 14.6-1.

If the population is not isolated, and immigration into or emigration out of the population by members of the same species is permitted, then we must modify Eq. (1). If we denote the rate of change of the population due to these effects by $I(t)$, then Eq. (1) is modified to

$$y' = F(y)y + I(t).$$

Here $I(t)$ is a specified function, which is positive for immigration and negative for emigration. In particular, the Malthusian law with immigration is

$$y' = ky + I.$$

This is an inhomogeneous, first-order, linear DE.

Exercises

1. The rate of increase of a given population per unit of population is equal to 1/10. Given that the initial population size is 10,000 write an IVP that describes the growth of the population. Assume no immigration or emigration.

2. Suppose in Exercise 1 that certain seasonal factors cause a net immigration rate of $100 \cos 2t$. Write down the relevant IVP.

3. (a) A given population changes at a rate proportional to the number present. Write down the appropriate DE. (b) What can be said about the population if the proportionality constant is positive? negative? (c) Can such a population ever be increasing for a time and then decrease, or vice versa?

4. (a) A given population decreases at a rate proportional to the square of the population. Assuming that there is no immigration or emigration, and given that the initial population size is a_0 write down the IVP describing the population decay. (b) What is the sign of the proportionality constant?

5. In many circumstances the birth and/or death rate may depend explicitly on time. For example, bacteria in a laboratory culture may undergo division at a faster rate if the temperature is raised and at a decreased rate if the temperature is lowered. Assume that the temperature is externally varied with time. Then the birth rate m will depend explicitly on time t. If the death rate is constant and $m = m(t)$, write down a DE that describes the population changes.

The problem of population dynamics is a typical example of growth and decay problems. Other examples are discussed in the following exercises.

6. It is observed experimentally that the rate at which a radioactive substance, such as radium, disintegrates is proportional to the amount of the substance which is present. Write an IVP which describes the decay of 20 grams of radium. What is the sign of the proportionality constant?

7. A man has an ambitious investment scheme that will enable him to increase his wealth at a rate proportional to the cube of his present wealth. What is the DE that describes the increase in his wealth?

8. (a) Suppose that the supply S of and the demand D for a certain commodity depend on the price P only. Assuming that the price P changes at a rate proportional to the difference $S - D$, what is the DE for $P(t)$? What is the sign of the proportionality constant if we assume that prices rise when a commodity is scarce?
   (b) Extend the above model to include the possibility that S and D depend explicitly on time. This could account for seasonal factors in supply and demand.

9. An earthquake or hurricane warning is given in a city of 1 million population. People rush to leave the city. The rate at which they leave at any time t is proportional to the number of people still in the city. Set up the IVP that determines the number of people left in the city at any time.

chapter 2

THE HOMOGENEOUS FIRST-ORDER EQUATION

2.1 INTRODUCTION

The initial-value problem (IVP)

$$y' + q(x)y = 0, \qquad y(x_0) = a_0$$

occurs frequently in mathematical models of real problems. Here the coefficient $q(x)$ is a specified function. For simplicity of presentation, we shall assume that it is a continuous function on an interval $\alpha < x < \beta$. The reason for this assumption will become apparent later in the section. Initial value problems whose coefficients are discontinuous functions are discussed in the exercises of Section 2.2. The number x_0 is the given initial point in this interval and a_0 is the given initial value. We have already shown that (1) is a linear DE; see Example 1 in Section 1.3.

2.2 THE HOMOGENEOUS EQUATION AND THE INITIAL VALUE PROBLEM

The IVP

$$y' + q(x)y = 0 \tag{1}$$

$$y(x_0) = a_0 \tag{2}$$

for the homogeneous DE is to find a function $y(x)$ with a continuous first derivative that is a solution of DE (1) and has the given value a_0 at the initial point $x = x_0$. We illustrate the techniques for solving the IVP in the following example.

EXAMPLE

Solve the IVP,

$$y' + 2y = 0, \qquad y(0) = 4.$$

solution. We assume that the IVP has a solution $y(x)$. Since $y(0) = 4$ is positive, it follows from the continuity of the solution that $y(x)$ must be positive in some interval about the initial point $x = 0$. Then it is permissible in this interval to divide both sides of the DE by y. This gives

$$\frac{y'}{y} = -2. \tag{3}$$

Since y is positive we have from the differentiation formula for $\ln y$ that $y'/y = d/dx(\ln y) = (\ln y)'$. Then Eq. (3) is equivalent to

$$(\ln y)' = -2.$$

We integrate both sides of this equation and get,

$$\ln y = c_1 - 2x, \tag{4}$$

where c_1 is a constant of integration.

Equation (4) is of the form $\ln A = B$. From calculus we know that $\ln A = B$ implies that $A = e^B$. Therefore we solve Eq. (4) for y and get

$$y = e^{c_1 - 2x}.$$

Now we use the property, $e^{a+b} = e^a e^b$ of the exponential function to rewrite y as

$$y = e^{c_1}e^{-2x} = ce^{-2x}, \tag{5}$$

where $c = e^{c_1}$ is an arbitrary positive constant. It is positive because $e^{c_1} > 0$ for any c_1.

Thus we have shown that if the DE has a positive solution, then it is given by Eq. (5). This does not prove that Eq. (5) is a solution or even that the DE has a solution; it merely gives a candidate for a solution. To verify that Eq. (5) is a solution we shall substitute it directly into the DE. Thus we first calculate y',

$$y' = (ce^{-2x})' = -2ce^{-2x}.$$

By substituting these expressions for y and y' into the DE we get

$$y' + 2y = -2ce^{-2x} + 2ce^{-2x} = c(-2e^{-2x} + 2e^{-2x}) = 0. \tag{6}$$

This shows that Eq. (5) solves the DE for all values of x. Furthermore, it shows that Eq. (5) is a solution for all values of the constant c. These values can be positive or negative—not just positive, as the derivation of Eq. (5) required. *Thus c is an arbitrary constant.* We observe from Eq. (5) that the solution is of one sign for all values of x: It is positive if $c > 0$ and it is negative if $c < 0$.

The solution given by Eq. (5) can be interpreted geometrically as a family or collection of curves depending on the constant c. For each value of c we get one curve of this family; see Figure 2.1. To solve the IVP, we wish to select c so that Eq. (5) satisfies the initial condition. Geometrically, this means that we choose the curve in Figure 2.1 that passes through the point ($x = 0$, $y = 4$). Thus by inserting $y = 4$ and $x = 0$ into Eq. (5) we get

$$y(0) = c = 4.$$

The initial condition determines only one value of c. By substituting this value into Eq. (5) we get a solution of the IVP as

$$y = 4e^{-2x}.$$

(See Figure 2.1.)

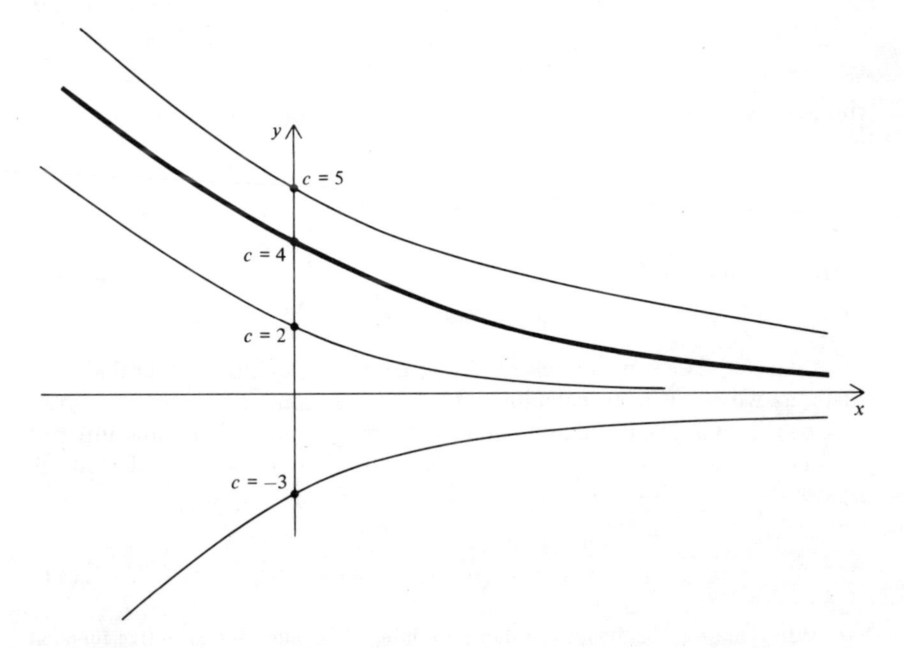

Figure 2.1

We observe that as x increases from its initial value $x = 0$, the solution decreases monotonically and approaches zero as $x \to \infty$. As x decreases from $x = 0$, the solution increases monotonically and

$$\lim_{x \to -\infty} y = \infty.$$

The solution is positive for all values of x; that is, it has the same sign as the initial value $y(0) = 4$.

The analysis given in the Example suggests a procedure for determining a solution of the IVP, Eqs. (1) and (2), where $q(x)$ is any given continuous function and x_0 and a_0 are any given numbers. We assume that the DE has a solution of one sign; i.e., the solution is not zero for any value of x. We divide both sides of DE (1) by y to obtain

$$\frac{y'}{y} + q(x) = 0. \tag{7}$$

Since $y'/y = (\ln |y|)'$, Eq. (7) is equivalent to

$$(\ln |y|)' = -q(x). \tag{8}$$

We must use the absolute value signs in the argument of the logarithm because we allow for positive or negative solutions. We integrate Eq. (8) and get

$$\ln |y| = c_1 - Q(x), \tag{9}$$

where c_1 is a constant of integration, and the function $Q(x)$ is any indefinite integral[1] of q; that is,

$$Q(x) = \int^x q(z)\, dz. \tag{10}$$

In the Example we have

$$q(x) = 2 \qquad \text{and} \qquad Q(x) = 2x.$$

If $q(x)$ is a continuous function, then its indefinite integral always exists, as we recall from calculus. This is the reason why we assume $q(x)$ to be continuous. If q is a discontinuous function, then its indefinite integral may not exist. The solution of IVPs with discontinuous coefficients is discussed in Exercises 6 and 7.

We solve Eq. (9) for $|y|$ and obtain

$$|y| = e^{c_1 - Q(x)} = e^{c_1} e^{-Q(x)} = ce^{-Q(x)}, \tag{11}$$

[1] In some calculus books the indefinite integral is called the primitive function or the antiderivative. We use the notation $\int^x f(z)\, dz$ for an indefinite integral of $f(x)$. The notation $\int f(x)\, dx$ is also used in calculus texts for indefinite integrals.

where $c = e^{c_1}$ is a *positive* constant. If $y > 0$, then we can remove the absolute value sign in Eq. (11) and obtain

$$y = ce^{-Q(x)}. \tag{12}$$

If $y < 0$, then we substitute $|y| = -y$ in Eq. (11) and we get $y = -ce^{-Q(x)}$. However, $-c$ is an arbitrary *negative* constant. Thus we can include in Eq. (12) all solutions that are either positive for all x or negative for all x by allowing c to be an arbitrary constant.

We have shown that if DE (1) has a solution of one sign, then it is given by Eq. (12). We now verify that Eq. (12) is indeed a solution of the DE by substituting it into Eq. (1). We use the differentiation formulas

$$[e^{f(x)}]' = f'(x)e^{f(x)} \qquad \text{and} \qquad Q'(x) = \left[\int^x q(z)\,dz\right]' = q(x).$$

Then we have

$$y' = [ce^{-Q(x)}]' = c(-Q')e^{-Q} = -cqe^{-Q}. \tag{13}$$

Then by substituting Eqs. (12) and (13) into Eq. (1) we get

$$y' + qy = -cqe^{-Q} + qce^{-Q} = c(-qe^{-Q} + qe^{-Q}) = 0.$$

Thus Eq. (12) is a solution of Eq. (1) for all values of c. Furthermore, it is of one sign because $e^{-Q(x)} > 0$. The sign of y is the same as the sign of c.

We could use Eq. (12) to solve the IVP, Eqs. (1) and (2), in the same way as we did in the example. Thus we could pick the constant c in Eq. (12) to satisfy the initial condition. A useful formula for a solution of the IVP can also be derived by observing that $Q(x)$ in Eq. (12) is any indefinite integral of $q(x)$. If we choose the specific integral

$$Q(x) = \int_{x_0}^{x} q(z)\,dz,$$

where the initial point x_0 is taken as the lower limit, then Eq. (12) can be written as

$$y = c\exp\left[-\int_{x_0}^{x} q(z)\,dz\right]. \tag{14}$$

To satisfy the initial condition of Eq. (2) we require that $y = a_0$ at the initial point $x = x_0$. Setting $x = x_0$ and $y = a_0$ in Eq. (14), and using

$$\int_{x_0}^{x_0} q(z)\,dz = 0,$$

we find that $c = a_0$. Then Eq. (14), with $c = a_0$, yields the solution of the

IVP

$$y = a_0 \exp\left[-\int_{x_0}^{x} q(z)\, dz\right]. \tag{15}$$

Thus, for any choice of a_0, x_0 and the coefficient $q(x)$, Eq. (15) provides a solution of the IVP. This shows that the IVP, Eqs. (1) and (2), always has at least one solution. In Section 2.3-1 we shall demonstrate that this is the only solution of the IVP.

Exercises

1. Solve the following DEs and sketch solutions for typical values of the constant of integration:
   (a) $y' + xy = 0$ (b) $y' + (\sin x)y = 0$
   (c) $y' - xy = 0$ (d) $y' + (1 - x)y = 0$

2. Solve the following IVPs, and give the interval for which the solution is valid.
   (a) $y' + (\cos x)y = 0, \quad y(0) = 2$ (b) $y' + 4x^3y = 0, \quad y(1) = 1$
   (c) $y' + e^x y = 0, \quad y(0) = 3$

3. Show that the DE $y' + (q_1(x) + q_2(x))y = 0$ has the solution $y = y_1y_2$, where y_1 and y_2 are solutions of the DEs $y_1' + q_1y_1 = 0$ and $y_2' + q_2y_2 = 0$, respectively.

4. Solve the following IVPs, using Eq. (15).
   (a) $y' + 3\,(\sin x)y = 0, \quad y(0) = 1/2$
   (b) $y' + (6x^2 + 2x)y = 0, \quad y(1) = 1$

5. (a) Show that $y = a_0e^{Q(x_0)-Q(x)}$ is a solution of the IVP, Eqs. (1) and (2), for any choice of indefinite integral $Q(x)$.
   (b) Show that if we use the indefinite integral

$$Q(x) = \int_{x_0}^{x} q(z)\, dz$$

   then the formula in (a) reduces to Eq. (15).

*6. In many problems $q(x)$ is not a continuous function for all values of x. We recall from calculus that a function may fail to be continuous for several reasons. Two simple types of discontinuities are: (1) A jump discontinuity; for example, the function

$$f(x) = \begin{cases} x, & 0 \le x \le 1 \\ 2, & x > 1 \end{cases}$$

has a jump discontinuity at $x = 1$. (2) Unbounded discontinuity; for example, the functions $1/x$, $1/x^2$ are not continuous at $x = 0$ because they are not bounded as $x \to 0$. In each of the following exercises give the points of discontinuity of $q(x)$. Solve the given IVP using the method in this section. Determine the largest interval containing the initial point for which the solution is a continuous function.

(a) $y' - (\cot x)y = 0, \qquad y(\pi/2) = 1$

(b) $y' - \dfrac{1}{x}y = 0, \qquad y(1) = 1$

(c) $y' + \dfrac{1}{x^2}y = 0, \qquad y(-1) = 2$

(d) $y' + (\csc x)y = 0, \qquad y(\pi/2) = 4$

(e) $y' + \dfrac{1}{1+x}y = 0, \qquad y(0) = 1$

*7. Consider the DE $y' + q(x)y = 0$.

(a) If $q(x)$ is a continuous function on an interval $\alpha < x < \beta$, then show that all solutions of the DE given by Eq. (14) are continuous and have continuous first derivatives in (α, β).

(b) If $q(x)$ has a point of discontinuity, then the solutions of the DE may or may not be continuous at this point. For example, show that the solutions of the DEs in Exercise 6(a) and (b) are continuous at the points of discontinuity of $q(x)$, and that the solutions in Exercise 6(c), and (e) are discontinuous.

2.3 THE INTEGRATING FACTOR

In the preceding section we have shown that every solution of one sign of the DE

$$y' + q(x)y = 0 \tag{1}$$

is given by

$$y = ce^{-Q(x)} = c \exp\left[-\int^x q(z)\,dz\right]. \tag{2}$$

However, there may be other solutions that are not given by Eq. (2). In particular, there may be solutions that change sign. We shall use the integrating factor method to show that every solution of the DE is given by Eq. (2). The integrating factor will also provide another technique for solving the DE. In fact it will be used in Section 3.1 to solve the inhomogeneous DE.

The basic idea of the method is contained in the following observation about the solution given by Eq. (2). We multiply both sides of Eq. (2) by e^Q and obtain

$$ye^Q = c = \text{constant}. \tag{3}$$

Therefore, by differentiating Eq. (3) we get $(ye^Q)' = 0$ or, equivalently,

$$(ye^Q)' = y'e^Q + yQ'e^Q = e^Q(y' + qy) = 0. \tag{4}$$

In Eq. (4) we have used the result $Q' = q$. The last term in Eq. (4) is merely DE (1) multiplied by the nonzero factor e^Q.

The integrating factor method consists of reversing the above steps. That is, we consider any solution y of the DE and we multiply both sides of Eq. (1) by e^Q, where Q is any indefinite integral of q. Then, as we see by reversing the steps in Eq. (4), the resulting expression is

$$(ye^Q)' = 0. \tag{5}$$

Since Eq. (5) is easily integrated, e^Q is called an integrating factor. We now integrate Eq. (5) and obtain

$$ye^Q = c = \text{constant}. \tag{6}$$

Finally, by multiplying both sides of Eq. (6) by e^{-Q} we obtain the solution given in Eq. (2). Thus we have shown that *every solution of DE* (1) *is given by* Eq. (2). *Each solution is obtained by appropriately choosing the value of c in the solution given in Eq.* (2). We refer to Eq. (2) as a *general solution* of the DE (1). The solution given in (2) is zero only if $c = 0$, and it is then zero for all values of x. If $c \neq 0$, then the solution is not zero at any point because $e^{-Q} > 0$.

We shall use these observations to prove that the IVP for Eq. (1) has only one solution. However, we first illustrate how to use the integrating factor to find general solutions.

EXAMPLE

Find a general solution of

$$y' + 2xy = 0,$$

using the integrating factor method.

solution. Since $q(x) = 2x$, a specific choice for Q is

$$Q = \int^x 2x\, dz = x^2.$$

A corresponding integrating factor is $e^Q = e^{x^2}$. We multiply both sides of the DE by e^{x^2} and obtain

$$e^{x^2}y' + 2xe^{x^2}y = 0.$$

Since $(e^{x^2}y)' = e^{x^2}y' + 2xe^{x^2}y$, we can rewrite this equation as $(e^{x^2}y)' = 0$. Therefore $e^{x^2}y = c$ or

$$y = ce^{-x^2}$$

is a general solution of the DE.

*2.3-1 A uniqueness theorem

We have shown in the previous section that the DE

$$y' + q(x)y = 0$$

always has at least one solution. It is given by Eq. (2) of Section 2.3. Furthermore, in the example of Section 2.2, we saw that the IVP had exactly one solution. In the following theorem we shall show that if $q(x)$ is a continuous function on an interval (α, β), then the corresponding IVP has exactly one solution. The continuity requirement on $q(x)$ is important; because as we indicate in Exercises 3 and 4, IVPs with discontinuous coefficients may have no solutions, one solution, or even an infinite number of solutions.

theorem

If $q(x)$ is a continuous function on the interval (α, β) and x_0 is a given point in this interval, then the IVP

$$y' + q(x) = 0, \qquad y(x_0) = a_0$$

where a_0 is the given initial value, has at most one solution.

proof. Every solution of the DE is given by $y = ce^{-Q(x)}$, where $Q(x)$ is any indefinite integral of $q(x)$. Since $q(x)$ is a continuous function in (α, β), its indefinite integrals and hence the solutious are defined and continuous functions in (α, β). In particular, if we choose for Q the integral

$$Q(x) = \int_{x_0}^{x} q(z)\, dz,$$

then every solution of the DE is given by

$$y = c \exp\left(-\int_{x_0}^{x} q(z)\, dz\right). \tag{1}$$

That is, to every solution of the DE there is a corresponding value of c. If we set $x = x_0$ and use the initial condition, we get $a_0 = c$. This value of c is the only value that satisfies the initial condition and hence the corresponding solution is the only solution of the DE that satisfies the initial condition. Thus the IVP has exactly one solution and the theorem is proved.

Exercises

1. Use an integrating factor to find a general solution of the following DEs:
   (a) $y' + xy = 0$, (b) $y' + (1 + x^2)y = 0$,
   (c) $y' + (x \sin x)y = 0$, (d) $y' + x^n y = 0$, $n \geq 1$.

*2. Use an integrating factor to solve the following problems. In each case the coefficient $q(x)$ is not continuous for all values of x. Give the largest interval containing the initial point for which your solution is valid.

(a) $y' + (\cot x)y = 0, \quad y(\pi/2) = -1$

(b) $y' + (\ln x)y = 0, \quad y(1) = 2$ (c) $y' + \frac{1}{x}y = 0, \quad y(1) = 1$

(d) $y' + \frac{1}{x}y = 0, \quad y(-2) = 3$ (e) $y' + \frac{2x+1}{x}y = 0, \quad y(1) = 1$

3. (a) Consider the IVP

$$y' - \frac{1}{x}y = 0, \qquad y(0) = 0.$$

Here $q(x)$ is not continuous at the initial point $x = 0$. Show that there are an *infinite* number of solutions to this IVP. (b) Consider the IVP

$$y' - \frac{1}{x}y = 0, \qquad y(0) = 1.$$

Show that there is no solution.

*4. How many solutions can you find to each of the following IVPs?

(a) $y' - \frac{3}{\sqrt{x}}y = 0, \quad y(0) = 0$

(b) $y' - \frac{3}{x}y = 0, \quad y(0) = a_0, \quad a_0$ any constant

(c) $y' + (\tan x)y = 0, \quad y(\pi/2) = a_0, \quad a_0$ any constant

2.4 APPLICATIONS AND QUALITATIVE FEATURES OF THE SOLUTION

In this section we study the qualitative behavior of the solution of the homogeneous IVP for specific choices of the coefficient $q(x)$. We shall interpret the solutions in terms of some of the mathematical models that were derived in Section 1.4-2 and Appendix I as well as other models.

2.4-1 Population dynamics

We consider a colony of a species of organisms. For example, the species may be bacteria, or viruses, or people. Let $y(t)$ denote the number of members in the population at time t. The simplest mathematical model for the growth or decay of populations is to assume (see Section 1.4-2) that the rate of change of the population at t is proportional to the size of the population at t. That is,

$$y' = ky, \tag{1}$$

where k is the proportionality constant. It is defined by $k \equiv m - n$, where m and n are the birth and death rates of the population respectively. If $k < 0$, the death rate exceeds the birth rate, and conversely if $k > 0$.

Equation (1) is an example of the *law of mass action*, which is used as a model for many reactions and processes in biology and chemistry; see Section 2.4-2.

The population a_0 at an initial time t_0 is a prescribed positive number. For simplicity we take $t_0 = 0$. Thus

$$y(0) = a_0 > 0. \tag{2}$$

We wish to determine the subsequent growth or decay of the population. Hence we must solve the IVP given in Eqs. (1) and (2).

Since $a_0 > 0$, we know from Section 2.3 that the solution is positive for all $t \geq 0$. Hence dividing Eq. (1) by y, we obtain

$$\frac{y'}{y} = (\ln y)' = k. \tag{3}$$

We integrate (3) and get $\ln y = c_1 + kt$. Solving for y gives $y = e^{c_1+kt} = e^{c_1}e^{kt}$, and finally

$$y = ce^{kt}, \tag{4}$$

where $c = e^{c_1}$ is an arbitrary constant. We evaluate c by substituting the initial condition given by Eq. (2) into Eq. (4). This gives the solution of the IVP as

$$y = a_0e^{kt}. \tag{5}$$

The behavior of the solution depends on the sign of k. If $k = 0$, the birth and death rates are equal, and then the solution is $y(t) = a_0 =$ constant. If $k < 0$ (that is, if the death rate exceeds the birth rate), then

$$\lim_{t\to\infty} y = a_0 \lim_{t\to\infty} e^{kt} = 0.$$

Hence $y(t)$ decreases to zero monotonically from its initial value a_0 as t increases. Specifically, the solution decays exponentially to zero as $t \to \infty$. The coefficient k is called the decay rate. A graph of the solution is sketched in Figure 2.2.

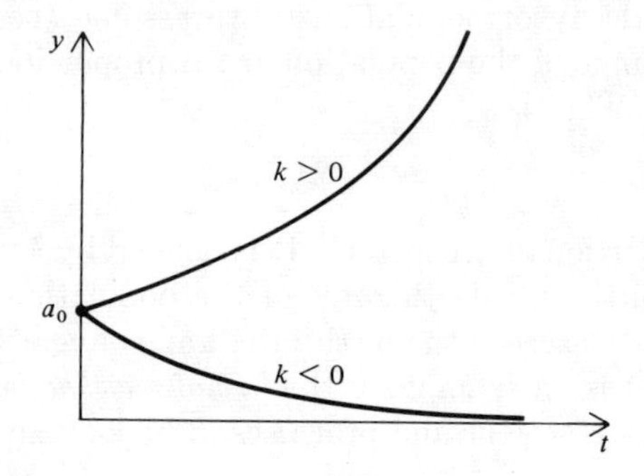

Figure 2.2

We shall now show that for $k < 0$ the larger the values of $|k|$, the faster the solution decays as t increases. Let $y_1(t)$ and $y_2(t)$ denote the solutions of Eqs. (1) and (2) corresponding to two different negative values $k_1 < k_2 < 0$ of k, and the same value of a_0. Then

$$y_1 = a_0 e^{k_1 t}, \qquad y_2 = a_0 e^{k_2 t}$$

and consequently

$$\frac{y_1}{y_2} = e^{(k_1 - k_2)t}. \tag{6}$$

Since $(k_1 - k_2)t < 0$, $e^{(k_1-k_2)t} < 1$ for $t > 0$. Applying this fact to Eq. (6) gives $y_1 < y_2$. Since $|k_1| > |k_2|$, this shows that the larger the value of $|k|$ for $k < 0$, the faster the solution decays as t increases.

If $k > 0$, then

$$\lim_{t \to \infty} e^{kt} = \infty,$$

and the solution given by Eq. (5) is said to grow exponentially; see Figure 2.2. The coefficient k is then called the growth rate. As $t \to \infty$, the size of the population $y(t)$ becomes unbounded. This is to be expected intuitively for this population model, since the birth rate exceeds the death rate when $k > 0$. Of course, unbounded populations have never been observed. In fact, comparison with experiments and with known population data indicates that the model gives reasonable results only for a short time after the initial time. Thus there must be defects in the mathematical model of population dynamics with m and n independent of t and y. Clearly, as the population increases, other effects such as overcrowding and competition for food will change the birth and death rates. These effects are neglected when k is constant. In Chapter 14 we shall discuss models of population dynamics that include these effects and lead to nonlinear DEs. The solutions of these equations are bounded as $t \to \infty$.

In the case of bacterial populations, it is of interest to determine the time required for the population to double itself. This is called the doubling time. Since this happens quite rapidly, we can use Eq. (5) with $k > 0$ to determine the value of $t = T$ such that $y = 2a_0$. Substituting this into Eq. (5) gives

$$2a_0 = a_0 e^{kT}. \tag{7}$$

We cancel the factor a_0 on both sides of Eq. (7) and then take the logarithms of both sides of the resulting equation. This gives

$$T = \frac{\ln 2}{k}$$

for the doubling time. It is called the *law of Malthus*.[2] The doubling time is independent of the initial time and the initial value a_0.

We can also interpret the IVP in Eqs. (1) and (2) in terms of the electric circuit problems in Appendix I.3 and other physical problems.

Exercises

1. The population in a country changes at a rate proportional to the number of people present. If the proportionality constant is 0.04 and the population in the year 1970 was 200 million, find the population in the year 1995.

2. According to the 1940 census the population of Pearl River was 5000 and in 1960 it was 10,000. If the population is increasing at a rate proportional to the current population, find in what year the population will be 20,000.

3. The population of certain bacteria in a pint of pond water is initially 1000. After 1 hour it is 1500. Find the growth rate k for the population, assuming that the model given by Eq. (1) applies. Also find after how many hours the population will be 2000 (the doubling time) and 3375.

   Exercise 4 illustrates that the simple model of population growth can produce reasonable answers for short periods of time beyond the initial time, but give inaccurate results for larger values of time.

4. The population of the United States was 3.9 million in 1790 and 5.3 million in 1800. Assume that the rate of change of population at time t is proportional to the number of people present at time t. Determine the population of the United States in 1830 and in 1960. (The actual populations were 17.1 million in 1840 and 179.3 million in 1960.)

[2] Thomas Malthus (1766–1834), an English economist, concluded from his famous studies of economic growth that "population, when unchecked, increases in a geometrical ratio. Subsistence increases only in an arithmetical ratio." He suggested that populations growing without checks would double every 25 years. If this is true for the human population, then $T = 25$ and hence $k = (\ln 2)/25 \approx .028$.

Exercises 5 and 6 indicate that the world's population is growing at a rate that is even *faster* than that predicted by the model given by Eq. (1).

5. The world population (in billions) was 0.8 in 1750 and 1.0 in 1800. Using the model given by Eq. (1) determine the 1950 population. The actual population was 2.5 billion. Thus the world has been growing at a rate that is faster than exponential growth.

6. Using the populations 1.7 billion for 1900 and 2.5 billion in 1950 repeat Exercise 5 to determine the population in 1960. (The actual population was approximately 3.0 billion. In fact, the doubling times of the world's population has decreased from 175 years in 1800 to 39 years in 1960. The model given by Eq. (1) predicts a *constant* doubling time. This greater than exponential growth has stimulated the concern with the problems of overpopulation.)

7. If a given species is introduced into a new environment and no natural predators or other ecological checks exist to limit their population growth, then dramatic growth rates can occur for a number of years. This can have either beneficial or detrimental effects on man. Typical examples of detrimental growth are the Japanese beetle in the United States and rabbits in Australia. An example of beneficial growth occurred in San Francisco Bay, where in 1879 and 1881 about 450 striped bass from the Atlantic Ocean were introduced. By 1889, commercial fishermen had caught approximately 1.25 million pounds of striped bass. Assuming that the number of striped bass increased at a rate proportional to the number present, that the average weight of a fish was 1 pound, that all 450 fish were introduced in 1881, and that 5 percent of the existing fish were caught in 1889, determine the growth rate k of the striped bass.

8. Assume that the growth or decay of a given population is described by the IVP given in Eqs. (1) and (2). Show that if the size, $y(t)$, of the population is known at two different times, t_1 and t_2, then the constant k in the DE given in Eq. (1) is

$$k = (t_2 - t_1)^{-1} \ln [y(t_2)/y(t_1)].$$

2.4-2 Chemical and biochemical kinetics

We consider a single chemical species confined in a closed volume. A reaction occurs in which this chemical is converted into one or more other chemicals. The concentration $y(t)$ of the species is the amount of the chemical per unit volume. The dimensions of y are, for example, pounds per gallon or moles per liter. We assume that the concentration has the same instantaneous value at each point of the volume, and thus y depends only on the time t. This is usually called a *well-stirred system*. For such systems experimental evidence indicates that the *law of mass action* is an accurate model. For a single reacting chemical the law of mass action states that the rate at which the reaction is proceeding at any instant,

$y'(t)$, is a function of the concentration $y(t)$ of the chemical that is present at that instant. We can express this as $y' = F(y)$. The simplest example of this law is the first-order reaction

$$y' = ky. \tag{1}$$

The constant k is the reaction rate. Its dimensions are $(\text{time})^{-1}$. We must have $k < 0$, because y is diminishing. In using the law of mass action, we have assumed that no chemicals are entering from outside of the volume, that the temperature is constant, and that the products of the reaction (the chemicals that are produced by the reaction) have a negligible influence on the reaction. Typical reactions that follow Eq. (1) are the decomposition of dinitrogen pentoxide into oxygen and nitrogen dioxide, which can be written as

$$N_2O_5 \rightarrow 2NO_2 + \tfrac{1}{2}O_2,$$

and the decomposition of dimethyl ether into methane, carbon monoxide, and hydrogen, which can be written as

$$(CH_3)_2O \rightarrow CH_4 + CO + H_2.$$

The IVP for Eq. (1) is completed by specifying the concentration at an initial time. For simplicity, we can take the initial time as $t = 0$. Thus we have

$$y(0) = a_0. \tag{2}$$

EXAMPLE

A single chemical is being converted into another chemical at a reaction rate of $k = -1/2$ per second. The initial concentration of the chemical is $a_0 = 10$ moles per liter. Determine the subsequent variation of the concentration $y(t)$ by assuming that the reaction is of the first order. Find the time at which the concentration has decreased to one half of its initial value. This time is called the *half-life*.

solution. Since we have a first-order reaction, the IVP that we must solve is

$$y' + \tfrac{1}{2}y = 0, \qquad y(0) = 10. \tag{3}$$

The general solution is found by dividing both sides of the DE by y and then integrating. This gives

$$y = ce^{-t/2},$$

where c is an arbitrary constant. We determine c by requiring this general solution to satisfy the initial condition. This yields $c = 10$, and hence the solution of the IVP is

$$y = 10e^{-t/2}. \tag{4}$$

To determine the half-life we set $y = a_0/2 = 5$ in the solution given in Eq. (4) and solve for the value of t. Then

$$5 = 10e^{-t/2}, \text{ or } \qquad t = 2 \ln 2.$$

The reader should verify that in general the half-life is equal to $(-\ln 2)/k$. It is independent of the initial concentration a_0.

Exercises

1. A chemical decomposes in a first-order reaction into several other chemicals. The initial concentration of the chemical is 10 moles per liter, and after 30 minutes the concentration is 4 moles per liter. (a) Find the concentration at the time t. (b) Find the half-life. (c) At what time will the concentration be reduced to 10 percent of the original concentration?
2. Express the reaction rate, in terms of the half-life, for a first-order reaction.
3. Repeat Exercise (1) if the initial concentration is 15 moles per liter and after 1 hour the concentration is 6 moles per liter.
4. A chemical decomposes in a first-order reaction. Measurements indicate that the concentration was reduced by 3 moles per liter in the third hour of the reaction and by 1 mole per liter in the fifth hour. Determine the concentration at time t. What is the initial concentration?
5. Consider the IVP $y' - ky = 0$, $y(0) = a_0$. (a) Sketch a graph of $\ln y$ vs. t for the solution of the IVP. (b) Give a geometric interpretation for the reaction rate k. In practice, to determine whether a given chemical reaction is a first-order reaction, experimental observations must be used. The concentration y is measured at several different times during the reaction and a graph of $\ln y$ vs. t is plotted. If the experimental values lie on a straight line (or nearly on a straight line), then the reaction is of the first order. If the graph is not a straight line, then higher-order reactions are involved and the governing DE is nonlinear (see Section 14.6-2).
6. The evaporation of ether proceeds at a rate that is proportional to the amount of ether present. Suppose that we have a container of volume V filled with ether and that 10 percent of the ether evaporates in 4 hours. Find the amount of ether left after 12 hours.
7. A tank contains 100 gallons of a solution of water and salt with concentration of 8 pounds of salt per gallon. In order to decrease the concentration of the salt solution, pure water flows into the tank at a rate of 4 gallons per minute. The mixture is kept uniform by stirring, and the mixture flows out of the tank at the same rate. Let $s(t)$ be the amount of salt in the tank at time t. Then ds/dt = rate in − rate out. Find how much salt is left in the tank at any time $t > 0$. In addition, find the amount of salt after 25 minutes.
8. As X rays travel through the body, they are absorbed. Lambert's law of absorption states that the rate of change of the intensity of the X rays with respect

to the distance traveled in the body is proportional to the intensity itself. The proportionality constant depends on the density of the body tissue. Let $I(x)$ be the intensity of the X rays at a distance x below the surface of the skin. Find the intensity at any distance x below the surface, assuming that the intensity of the X rays at the surface of the skin is I_0.

9. A dose of m_0 milligrams of a drug is injected into the bloodstream. The drug leaves the blood and enters the urine at a rate proportional to the amount of the drug present in the blood. Find the time at which the amount of the drug in the bloodstream will be 20 percent of the original amount. (*Hint*: To derive an appropriate DE, treat the bloodstream as a container, as in Exercise 7. This is often done to simplify complicated biological processes and is the basic idea of the subject of compartment analysis.)

*10. Suppose that, in addition to passing into the urine, the drug in Exercise 9 is absorbed by an organ of the body at a rate proportional to the amount of the drug in the bloodstream. Let the proportionality constant for this absorption be half the proportionality constant for passage into the urine. (a) Find the time at which the amount of the drug in the bloodstream is 20 percent of the original amount. Compare your answer with that of Exercise 9. (b) In order to maintain a drug level of no less than 50 percent of the original dosage in the blood at all times, how often should injections of dosage m_0 be given?

2.5 PERIODIC COEFFICIENTS; TIME-VARYING ELECTRIC CIRCUITS

In many applications the coefficient $q(x)$ in the IVP

$$y' + q(x)y = 0, \qquad y(x_0) = a_0 \tag{1}$$

is not a constant. We shall now study this IVP when $q(x)$ is a periodic function. Such IVPs occur in a great variety of applications. We first define a periodic function.

definition

$f(x)$ is a periodic function if there is a number $P > 0$ such that

$$f(x + P) = f(x) \tag{2}$$

for all x. The number P is called a period of $f(x)$. If there is a smallest positive value of $P = P_0$ for which Eq. (2) is true then P_0 is called the fundamental period.[3]

Thus a periodic function $f(x)$ repeats itself when its argument is increased by P. We recall that $\sin x$ and $\cos x$ are periodic functions of fundamental period 2π, and $\tan x$ is a periodic function of fundamental period π.

[3] A periodic function may not have a fundamental period. For example, the constant function is periodic and it has any value of P for a period.

In many applications it is of interest to determine, if the solution of the IVP given in Eq. (1) is a periodic function when $q(x)$ is a periodic function, and to determine the period of the solution. We illustrate this in Exercises 4–6.

We now consider the *RC* electric circuit, where the circuit parameters R and C vary with the time t. The case of constant circuit parameters is equivalent to the population dynamics problem with constant decay rate considered in Section 2.4-1. The DE for the *RC* circuit is (see Appendix I.3),

$$y' + q(t)y = 0, \tag{1}$$

where $q(t) \equiv [R(t)C(t)]^{-1}$ and $y(t)$ is the charge. The charge on the capacitor at $t = 0$ is specified as $y(0) = a_0$. We wish to study the subsequent decay of the charge.

Time-varying circuit parameters are observed experimentally. For example, the variation in the properties of the circuit elements may be caused by heat generated due to the passage of current. Alternatively, the time variation may be imposed by external means. For example, in the carbon-granule microphone, which is used in telephones, the resistance is a variable that depends on the pressure exerted on the carbon granules by a diaphragm. The movement of the diaphragm is caused by the pressure of the sound waves, which are generated by speaking into the telephone.

EXAMPLE

At $t = 0$ a unit charge is placed on the capacitor of a time-varying *RC* circuit. From experiments with the circuit it is observed that the resistance R varies so that the coefficient in DE (1) oscillates about an average value, which we take as unity. The capacitance C is constant. Thus we assume that

$$q(t) = 1 + b \cos t$$

where b is a given real number such that $|b| < 1$. We do not use a specific value for b because we wish to study the subsequent ($t > 0$) response of the circuit for different values of b. $q(t)$ is a periodic function of fundamental period 2π.

solution. The IVP that we must solve is

$$y' + (1 + b \cos t)y = 0, \qquad y(0) = 1.$$

The solution is

$$y = e^{-t}e^{-b \sin t}. \tag{3}$$

We now discuss certain features of this solution. We observe from Eq. (3) that y is a positive function. It decays to zero as $t \to \infty$ because of the factor e^{-t}. From the DE we have

$$y' = -(1 + b \cos t)y. \tag{4}$$

Since $|b| < 1$, then the factor $1 + b \cos t$ in Eq. (4) is positive for all t. Consequently, $y' < 0$ for all t because $y(t)$ in Eq. (4) is positive (see Eq. (3)). This implies that y is a monotonically decreasing function.

It is important to observe that the oscillation of the resistance is periodic, but the solution is not periodic. Thus periodic coefficients do not always imply a periodic solution. The necessary and sufficient conditions on the coefficient of the DE to insure a periodic solution are given in Exercise 6.

Exercises

1. Suppose that the parameters of an RC circuit vary in such a way that $q(t) = 1 - \frac{1}{2}\sin t$. If $y(0) = 1$, find the charge at any time t. Show that $y(t)$ is not a periodic function.
2. When current passes through a resistor, the resistor heats up and its resistance R increases. Assume that $R(t) = R_0 + \alpha^2 t$, where R_0 and α are constants. Find the charge $y(t)$ at any time t given that the capacitance C is constant and that the initial charge is a_0.
3. The inductance in an RL circuit with constant resistance is given as $L = L_0 + \beta^2 t$, where L_0 and β are constants. Find the current $i(t)$ at any time t. Assume that the applied voltage $E = 0$ and that $i(0) = a_1$. (See Appendix I.3.)
4. (a) Solve the IVP $y' + (a + 3\sin x)y = 0$, $y(0) = 1$, where a is a constant. (b) Are there any values of a for which the solution is a periodic function? (c) The average value of $q(x)$ over one period is defined as,

$$A = \frac{1}{P}\int_x^{x+P} q(z)\,dz$$

Show that for $q(x) = a + 3\sin x$, $A = a$.

5. (a) Solve the IVP $y' + (\sin^2 x)y = 0$, $y(0) = 1$.
   (b) Is the solution periodic? What is the period of $q(x)$?
   (c) What is the average value of $q(x) = \sin^2 x$ over one period?
   (d) If $q(x) = \sin^2 x - \frac{1}{2}$ in (a), is the solution periodic? Observe that the average value of $q(x)$ over one period is now zero.

The results of the previous two exercises and the Example suggest the following theorem:

theorem

The solution of the IVP given by Eq. (1) with $a_0 \neq 0$ is a periodic function with fundamental period P, if and only if $q(x)$ is a periodic function with fundamental period P and with zero average value.

*6. Prove this theorem.

7. The growth rate of bacteria is known to be sensitive to temperature. Suppose that the temperature of a bacteria culture is varied periodically and that this causes the birth and death rates of the bacteria to oscillate so that k = birth rate − death rate = $\sin 2x$. Determine the number of bacteria at any time t given that the initial number is 1000. Graph your solution.

chapter 3

THE FIRST-ORDER INHOMOGENEOUS LINEAR EQUATION

3.1 THE INHOMOGENEOUS EQUATION AND THE INITIAL VALUE PROBLEM

The IVP for the inhomogeneous DE is to determine a function $y(x)$ that satisfies

$$y' + q(x)y = r(x) \tag{1}$$

and the initial condition

$$y(x_0) = a_0. \tag{2}$$

Here the prescribed function $r(x)$ is the inhomogeneous term, and the prescribed numbers x_0 and a_0 are the initial point and the initial value of the solution, respectively. In physical problems the inhomogeneous term corresponds to an externally applied force or stimulus. It is called the forcing function. For example, in the population dynamics model in Section 1.4-2, it represents the immigration rate into the population.

We illustrate the techniques of solving Eqs. (1) and (2) by considering a specific initial-value problem in the following example.

EXAMPLE

Solve the IVP

$$y' + 2y = 1, \qquad y(0) = 0.$$

solution. We solve the IVP by using the integrating factor e^Q, where Q is given by

$$Q(x) = \int^x q(z)\,dz = \int^x 2\,dx = 2x.$$

We multiply both sides of the DE by the integrating factor e^{2x}. This gives

$$(y' + 2y)e^{2x} = e^{2x}.$$

Since $(ye^{2x})' = (y' + 2y)e^{2x}$, this equation can be written as

$$(ye^{2x})' = e^{2x}.$$

We integrate both sides of this equation and get

$$ye^{2x} = \tfrac{1}{2}e^{2x} + c$$

where c is an integration constant. By multiplying both sides of this equation by e^{-2x} we get

$$y = ce^{-2x} + \tfrac{1}{2}. \tag{3}$$

In deriving Eq. (3) we have tacitly assumed that the DE has a solution. Thus we have shown that if the DE has a solution, it is given by Eq. (3). It is easy to prove by direct substitution that Eq. (3) is indeed a solution of the DE.

As in the case of the homogeneous DE, Eq. (3) is a one-parameter family of solutions, one solution for each value of the constant c.

To solve the IVP we require that the solution given by Eq. (3) satisfies the initial condition $y(0) = 0$. This yields

$$y(0) = c + \tfrac{1}{2} = 0 \qquad \text{or} \qquad c = -\tfrac{1}{2}.$$

A solution of the IVP is then obtained by substituting this value of c into Eq. (3). This gives

$$y = \tfrac{1}{2} - \tfrac{1}{2}e^{-2x} = \tfrac{1}{2}(1 - e^{-2x}).$$

The solution is a monotonically increasing function. As $x \to \infty$ the solution $y(x) \to \frac{1}{2}$, because $e^{-2x} \to 0$ as $x \to \infty$.

The disadvantage in solving the IVP by using the integrating factor, as we did in the preceding example, is that we must remember that e^Q is an integrating factor. However, it is unnecessary to explicitly remember the integrating factor as we illustrate in Exercise 4.

The analysis given in this example suggests a procedure for solving the IVP given in Eqs. (1) and (2), where $q(x)$ and $r(x)$ are any given continuous functions and x_0 and a_0 are any given numbers. We assume that

DE (1) has a solution $y(x)$. We multiply both sides of the DE by the integrating factor e^Q, where

$$Q(x) = \int^x q(z)\, dz.$$

This gives

$$e^Q(y' + qy) = e^Q r. \tag{4}$$

The basic property of the integrating factor is that the left side of Eq. (4) can be written as $e^Q(y' + qy) = (e^Q y)'$. Thus Eq. (4) is equivalent to

$$(e^Q y)' = e^Q r.$$

We integrate both sides of this equation and get

$$e^Q y = c + \int^x e^{Q(z)} r(z)\, dz$$

where c is the arbitrary constant of integration. By multiplying both sides of this equation by $e^{-Q(x)}$ we finally obtain

$$y(x) = ce^{-Q(x)} + e^{-Q(x)} \int^x e^{Q(z)} r(z)\, dz. \tag{5}$$

Thus we have shown that if Eq. (1) has a solution, then it is given by Eq. (5). We leave it as an exercise for the reader to show by direct substitution that Eq. (5) is a solution of the DE for all values of c. The calculation is similar to the one given in Section 2.2 for the homogeneous equation. Thus every solution of DE (1) is given by Eq. (5).

To solve the IVP given in Eqs. (1) and (2), we require the solution, Eq. (5), to satisfy the initial condition given in Eq. (2). This gives an equation that can be solved for c as we did in the example. However, this calculation can be simplified by recalling that e^Q is an integrating factor for any choice of the indefinite integral Q. We now select the specific integral

$$Q(x) = Q_0(x) = \int_{x_0}^x q(z_1)\, dz_1,$$

where the lower limit x_0 is the initial point. This integral, which was used in Section 2.2, has the important property that

$$Q_0(x_0) = \int_{x_0}^{x_0} q(z_1)\, dz_1 = 0.$$

Similarly, for the indefinite integral

$$\int^x e^{Q(z)} r(z)\, dz,$$

we use the specific integral

$$\int_{x_0}^{x} e^{Q_0(z)} r(z)\, dz.$$

Then the solution given by Eq. (5) can be written as

$$y(x) = ce^{-Q_0(x)} + e^{-Q_0(x)} \int_{x_0}^{x} e^{Q_0(z)} r(z)\, dz. \tag{6}$$

We now require that this solution satisfies the initial condition $y(x_0) = a_0$. Thus by substituting $x = x_0$ and $y = a_0$ in Eq. (6), we get

$$a_0 = ce^{-Q_0(x_0)} + e^{-Q_0(x_0)} \int_{x_0}^{x_0} e^{Q_0(z)} r(z)\, dz.$$

Since $e^{-Q_0(x_0)} = e^0 = 1$, we conclude from this equation that $c = a_0$. By substituting this value of c into Eq. (6) we get

$$y(x) = a_0 e^{-Q_0(x)} + e^{-Q_0(x)} \int_{x_0}^{x} e^{Q_0(z)} r(z)\, dz \tag{7}$$

as a solution of the IVP.

Equation (7) gives a formula for the solution of the IVP. It is not advisable to memorize this formula. The procedure used in the example should be used in order to solve specific IVPs.

Exercises

1. Solve the following IVPs and sketch the solutions.
   (a) $y' + y = e^{-x}, \quad y(0) = 1$ (b) $y' - 3y = -3x, \quad y(0) = 0$
   (c) $y' + xy = x + 1, \quad y(1) = 3$ (d) $y' + y = xe^{3x}, \quad y(0) = 2$

2. Verify by direct substitution that y given by Eqs. (6) or (7) is a solution of DE (1).

3. (a) Solve the DE $y' - by = e^{st}$, where b and s are constants. Be sure to solve the DE for $s = b$ as well as for $s \neq b$.
   (b) Sketch some typical solutions when $s = b$ and $s \neq b$.

4. (a) Find an integrating factor $f(x)$ for the DE $y' + 6y = r(x)$, where $r(x)$ is a specified function by proceeding as follows. Multiply the DE by an unknown function f. Determine f so that the left side of the resulting DE is the exact derivative

   $$f(y' + 6y) = (fy)'.$$

   Show that f satisfies the DE $f' - 6f = 0$ and hence $f = e^{6x}$ is an integrating factor. (b) Find an integrating factor for the DE $y' + x^3 y = x^2$. (c) Derive a formula for the integrating factor for the DE $y' + q(x)y = r(x)$ by proceeding as in (a).

5. Consider the IVP

$$y' = r(x), \qquad y(x_0) = a_0,$$

where $r(x)$ is a continuous function for all values of x. (a) Show, by integrating both sides of the DE between the limits x_0 and x, that the solution of the IVP is

$$y(x) = a_0 + \int_{x_0}^{x} r(z)\, dz.$$

(b) Observe that if $r(x)$ is continuous for all values of $x \geq x_0$, then the solution given in (a) is valid for the same values of x.

*6. In the following DEs either $r(x)$ or $q(x)$, or both, are discontinuous for some value(s) of x. Determine the points of discontinuity and then solve each IVP. What is the largest interval (α, β) for which your solution is valid?

(a) $y' + \dfrac{3}{x}\, y = 1, \qquad y(1) = \dfrac{1}{4}$ (b) $y' + \dfrac{3}{x}\, y = 1, \qquad y(-1) = \dfrac{1}{4}$

(c) $y' - y \cot x = \sin x, \qquad y\left(\dfrac{\pi}{2}\right) = 0$

(d) $y' - \dfrac{2}{x}\, y = x^2 \sin 3x, \qquad y\left(\dfrac{\pi}{6}\right) = \dfrac{2}{\pi}$

(e) $y' + \dfrac{2}{x}\, y = \dfrac{1}{x^3}, \qquad y(-1) = 1$

*7. Assume that y_1 and y_2 are any two solutions of the IVP $y' + q(x)y = r(x)$, $y(x_0) = a_0$, with $q(x)$ and $r(x)$ continuous. Show that $z = y_1 - y_2$ satisfies a homogeneous DE with zero initial condition. Thus the uniqueness theorem of Section 2.3-1 implies that $z \equiv 0$, and hence $y_1 \equiv y_2$. This shows that the given IVP has no more than one solution.

3.2 PROPERTIES AND QUALITATIVE FEATURES OF THE SOLUTION

We shall now study properties of the solution

$$y(x) = ce^{-Q(x)} + e^{-Q(x)} \int^{x} e^{Q(z)} r(z)\, dz, \qquad Q = \int^{x} q(z_1)\, dz_1 \tag{1}$$

of the inhomogeneous DE

$$y' + q(x)y = r(x) \tag{2}$$

that are important for its interpretation in applications. We observe that

Eq. (1) is the sum of the functions $y_c(x)$ and $y_p(x)$, which are defined by

$$y_c(x) = ce^{-Q(x)} \tag{3}$$

and

$$y_p(x) = e^{-Q(x)} \int^x e^{Q(z)} r(z)\, dz. \tag{4}$$

It is clear that y_c is a general solution of the homogeneous DE (see Eq. (14) in Section 2.2). We refer to y_c as a *complementary solution* of Eq. (2). We now show that $y_p(x)$ is a solution of the inhomogeneous DE (2). It is called a *particular solution*. Thus we first compute y_p':

$$y_p' = \left[e^{-Q} \int^x e^Q r\, dz\right]' = e^{-Q} e^Q r + \left[-Q' e^{-Q} \int^x e^Q r\, dz\right].$$

Since $Q' = q$, we can rewrite this equation, using Eq. (4), as

$$y_p' = r - qe^{-Q} \int^x e^Q r\, dz = r - qy_p.$$

We conclude that

$$y_p' + qy_p = r, \tag{5}$$

or y_p is a solution of the inhomogeneous DE. Thus we have shown that *every solution of the inhomogeneous DE is the sum of a general solution of the homogeneous DE and a particular solution of the inhomogeneous DE. We refer to the solution given in Eq.* (1) *as a general solution of the inhomogeneous DE.*

The function

$$y = a_0 e^{-Q_0(x)} + e^{-Q_0(x)} \int_{x_0}^x e^{Q_0(z)} r(z)\, dz, \qquad Q_0(x) = \int_{x_0}^x q(z)\, dz, \tag{6}$$

is a solution of the IVP for DE (2), as we have shown in Eq. (7) of Section 3.1. It contains the effects of the initial condition a_0 and the forcing function $r(x)$. We shall now show that this solution is given by the superposition of these two effects. Equation (6) is the sum of the functions $H(x)$ and $F(x)$, where

$$H(x) = a_0 e^{-Q_0(x)} \tag{7}$$

and

$$F(x) = e^{-Q_0(x)} \int_{x_0}^x e^{Q_0(z)} r(z)\, dz. \tag{8}$$

Since $Q_0(x)$ is an indefinite integral of $q(x)$ and $H(x)$ is a_0 times $e^{-Q_0(x)}$, and in addition $Q_0(x_0) = 0$, we conclude that $H(x)$ is a solution of the IVP

$$H' + q(x)H = 0, \qquad H(x_0) = a_0. \tag{9}$$

Furthermore, by following the same steps that led to Eq. (5), we can show that $F(x)$ is a solution of the IVP

$$F' + q(x)F = r(x), \qquad F(x_0) = 0. \tag{10}$$

We shall not give the details of this calculation.

Thus the solution, Eq. (6), of the IVP is

$$y(x) = H(x) + F(x). \tag{11}$$

$H(x)$ represents the influence of the initial data on the solution. It is independent of $r(x)$. The function $F(x)$ gives the influence of the forcing function $r(x)$ on the solution. It is independent of a_0. It is called the *forced response.* The forced response is a particular solution of the DE since $F(x)$ satisfies Eq. (2); see Eq. (10). The forcing function is frequently called the input and the solution the output.

EXAMPLE 1

Determine the forced response for the IVP

$$y' + 2y = 1, \qquad y(0) = 0.$$

that was solved in the example in Section 3.1.

solution. The forced response $F(x)$ is the solution of the IVP

$$F' + 2F = 1, \qquad F(0) = 0.$$

It is

$$F = \tfrac{1}{2}(1 - e^{-2x}). \tag{12}$$

The function $H(x)$ is the solution of the homogeneous DE $H' + 2H = 0$ that satisfies the given initial condition, which for this problem is $H(0) = 0$. Since $H(0) = 0$, the solution is $H(x) \equiv 0$. Thus the solution of the IVP is given by $y = F(x)$.

The term $y_t = -\frac{1}{2}e^{-2x}$ in the solution given in Eq. (12) decays to zero as $x \to \infty$. It is called a transient, since its contribution to the solution is transitory. The term $y_s = \frac{1}{2}$ does not decay to zero as $x \to \infty$. It is called a steady-state response. This suggests the following general definition.

definition

A transient is a term in the solution of an IVP that approaches zero as $x \to \infty$. Let $y_t(x)$ denote the sum of all the transient terms. The steady-state response y_s is then $y_s = y - y_t$, where y is the solution of the IVP.

We observe that if $y(x) = y_t(x)$ for all values of x, then there is no steady-state response. The solution is a transient.

EXAMPLE 2

Solve the IVP

$$y' - ky = A, \qquad y(0) = 1,$$

where k and A are given constants. Determine the transient and steady-state responses.

solution. We shall solve this problem by using the integrating factor e^{-kx}. Thus the DE multiplied by e^{-kx} is

$$(e^{-kx}y)' = Ae^{-kx}. \tag{13}$$

We integrate Eq. (13) and then multiply both sides of the resulting equation by e^{-kx}. This gives the general solution

$$y = ce^{kx} - \frac{A}{k}. \tag{14}$$

We determine the constant c by requiring Eq. (14) to satisfy the initial condition. This gives

$$y(0) = c - \frac{A}{k} = 1.$$

Therefore $c = 1 + (A/k)$, and the solution of the IVP is

$$y = \left(1 + \frac{A}{k}\right)e^{kx} - \frac{A}{k}. \tag{15}$$

A sketch of the solution is given in Figure 3.1 for $k < 0$.

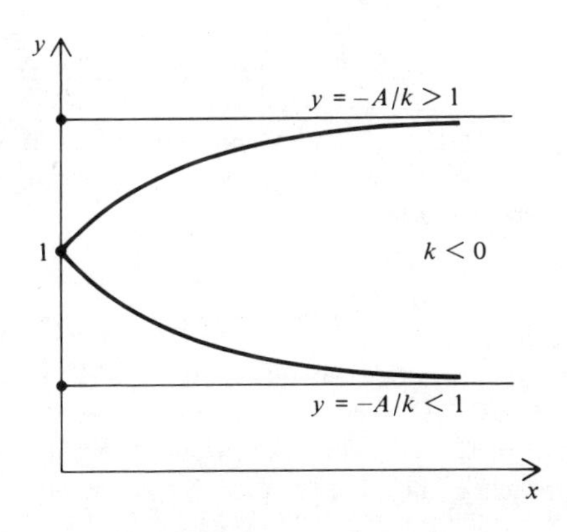

Figure 3.1

We can write the solution given in Eq. (15) as $y = H + F$ where

$$H = e^{kx}, \qquad F = -\frac{A}{k}(1 - e^{kx}). \tag{16}$$

Here H satisfies the homogeneous DE and the given initial condition, and F satisfies the inhomogeneous DE and the initial condition $F(0) = 0$. If $k < 0$, then H decays exponentially to zero as $x \to \infty$. We recall that H represents the response to the initial data and F is the forced response. We observe that the forcing term A appears in the forced response only.

For $k < 0$, the term $(1 + A/k)e^{kx}$ in Eq. (15) is a transient response since it decays to zero as $x \to \infty$. The remaining term $-A/k$ in Eq. (15) is the steady-state response because it does not approach zero as $x \to \infty$. For large values of x, the solution approaches the steady-state response. If $A = -k$, then the transient term vanishes for all x, and the solution is $y = -A/k = 1$ for all x.

If $k > 0$, then $(1 + A/k)e^{kx}$ grows exponentially as $x \to \infty$, if $A \neq -k$. This term dominates $-A/k$, and the solution is unbounded as $x \to \infty$.

Exercises

1. Find a complementary and a particular solution of the following DEs.
(a) $y' + 3y = 6$ (b) $y' + x^2y = 6x^2$
(c) $y' + y = \dfrac{1}{1 + e^{2x}}$

2. For each of the following IVPs determine the forced response $F(x)$, the response to the initial condition $H(x)$, and the transient and steady-state responses.
(a) $y' + y = e^{-x}, \quad y(1) = 2$ (b) $y' + 3y = 3x, \quad y(0) = 1$
(c) $y' - xy = 2x, \quad y(-1) = 1$ (d) $y' + (\sin x)y = \sin x, \quad y(0) = 1$
(e) $y' - ky = x, \quad y(0) = 0$, here k is a constant.

3. Solve the IVP $y' - ky = A$, $y(0) = a_0$. Compare your answer with Eq. (15), which gives the solution for $a_0 = 1$.

4. (a) Solve the IVP $y' + y = A \sin \omega x$, $y(0) = 1$, where A and ω are constant. In this case $r(x)$ is periodic, with period $2\pi/\omega$. A is called the amplitude of the forcing function. (b) Find the transient and steady-state responses. (c) Is the solution periodic? (d) What is the period of the steady-state response? (e) Sketch the steady-state solution for $\omega = 1$.

5. (a) Solve the IVP
$$y' - ky = A \sin \omega x, \qquad y(0) = a_0,$$
where k, A, ω, and a_0 are given constants. (b) Find the steady-state response, assuming that $k < 0$. (c) Observe that the steady-state response is periodic, with a period equal to the period of the forcing function. Thus as x gets large the solution approaches the steady-state response. (d) Is the full solution periodic?

*6. Show that the steady-state solution in Exercise 5 is equal to $A_0 \sin(\omega x - \phi)$, where

$$A_0 = \frac{A}{\sqrt{k^2 + \omega^2}} \qquad \text{and} \qquad \tan \phi = \frac{\omega}{k}.$$

A_0/A is called the amplification factor, and ϕ is called the phase angle.

7. Is the solution of each of the following IVPs periodic?
(a) $y' + (\sin x)y = \sin x, \quad y(0) = 1$
(b) $y' + (\cos^2 x)y = 1 + \cos 2x, \quad y(0) = 1$
In both cases the coefficients and inhomogeneous terms are periodic functions with a common period.

8. The DE $y' + q(x)y = r(x)$ has a complementary solution $y = ce^{-Q(x)}$. Find a particular solution by assuming that $y_p = u(x)e^{-Q(x)}$. Insert this into the DE and show that $u(x)$ satisfies the simple DE $u' = r(x)e^{Q(x)}$. Find $u(x)$ and thus y_p. This technique is called *variation of parameters* or *variation of the constant*, since the foregoing equation for y_p is of the same form as the complementary solution with the constant c replaced by a function $u(x)$.

9. Find a particular solution of the following DEs by variation of parameters.
(a) $y' + 6y = 4$ (b) $y' - y = 2x$
(c) $y' + (\sinh x)y = 3 \sinh x$

3.3 APPLICATIONS

In this section we shall discuss several applications of the IVP,

$$y' - ky = r(x), \qquad y(0) = a_0,$$

where k is a constant. This IVP arises frequently in many applications, with x usually representing time.

3.3-1 Mathematical models in psychology: learning theory

The process of learning has been extensively studied by psychologists. A simple mathematical model of a certain class of learning processes will now be derived. It has been used to describe memorization and the acquisition of skills such as typing and piano playing. For simplicity we shall consider the memorization process. We wish to study the relationship between the amount of material memorized and the time spent in memorizing. We refer to the amount of material memorized as the attainment. The time spent in memorization is denoted by t, and the attainment at time t is denoted by $y(t)$. The total amount of material to be memorized is M. The quantity M is frequently called the maximum attainment.

To determine $y(t)$ we assume that the rate of change in attainment at time t, $y'(t)$, is proportional to the amount of material that remains to be memorized; that is,

$$y' = A(M - y),$$

where $A > 0$ is a proportionality constant, which is a measure of the natural learning ability. We assume that at the beginning of the memorization process, $t = 0$, no material has been memorized. Thus the IVP for the memorization process is

$$y' + Ay = AM, \tag{1}$$

$$y(0) = 0. \tag{2}$$

As in Section 3.2, a solution is obtained by multiplying both sides of the DE by the integrating factor e^{At}. In this way we get the solution of the IVP as

$$y = M(1 - e^{-At}). \tag{3}$$

From Eq. (3) we get

$$y' = AMe^{-At}. \tag{4}$$

The solution given by Eq. (3) is sketched in Figure 3.2. The solution increases rapidly for small values of t and it approaches M as $t \to \infty$. This shows that most of the material is memorized in a relatively short time. As the maximum attainment is approached, the rate of attainment is small. This is confirmed by Eq. (4), which shows that the attainment rate is a monotonically decreasing function. This is an example of the so-called "law of diminishing returns."

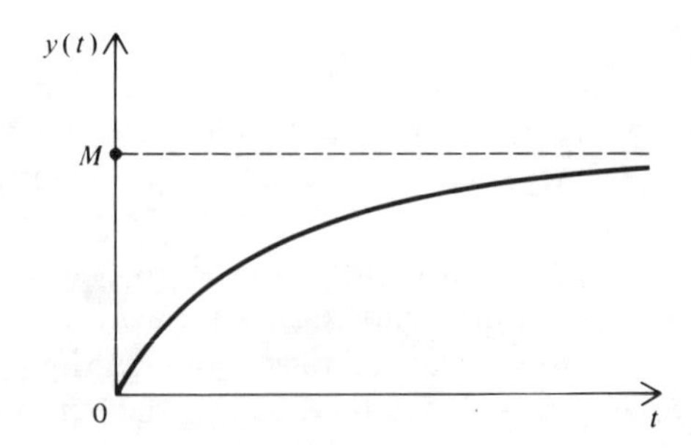

Figure 3.2

It is common experience in many learning situations that in the beginning of the process, learning is slow. Then after some minimal amount of proficiency has been achieved, the learning is rapid for some interval of time. Finally, the rate of learning decreases to zero as $t \to \infty$. More sophisticated models of learning, that have these features lead to non-linear DEs.

Exercises

1. In 10 minutes a person has memorized 15 words out of a list of 30 words. Determine the learning ability A of the person.
2. If a person can memorize 20 words out of a list of 50 words in 20 minutes, how long will it take to memorize 40 words?
3. Two people are given a list of 20 words to memorize. After 10 minutes, they have memorized 15 and 10 words respectively. (a) Find the learning ability A for each person. (b) Compare the learning abilities found in part (a) and observe whether the results make intuitive sense.
4. Determine the learning ability A if a person can achieve 50 percent of the maximum attainment after a time T.
5. The model for learning discussed in the text does not explicitly account for fatigue or forgetting. If the learning process is lengthy, one way to include the effect of fatigue is to assume that the constant A in the DE given in Eq. (1) is a function of time. Since fatigue would tend to slow learning, $A(t)$ should be taken as a positive monotone decreasing function.
   (a) Given that $A(t) = A_0/(1 + t)$, find the solution of the DE, Eq. (1), that satisfies the initial condition $y(0) = 0$.
   (b) Find the time T at which $y = M/2$.
   (c) Compare this result with the time at which $y = M/2$ when $A = A_0 =$ constant.
6. Find a formula for the general solution of the DE $y' = A(t)(M - y)$.
7. As a person learns, he tends to forget. To include the effect of forgetting in our learning model, we assume that the rate at which forgetting takes place is proportional to the amount already learned. Thus the IVP (1), (2) is replaced by $y' = A(M - y) - By$, $y(0) = 0$, where B is a positive constant. The term $-By$ accounts for the forgetting since it produces a decrease in attainment. (a) Solve this IVP. (b) Determine $\lim_{t\to\infty} y(t)$. Observe that it is not equal to M. Thus if M is the amount to be memorized, this model predicts that M is never actually attained. This would seem reasonable, for example, in memorizing a very long list of syllables (M large), then forgetting makes it impossible to successfully memorize the total list M. (c) Show that if B is small, then $\lim_{t\to\infty} y(t)$ is approximately equal to M.

3.3-2 Population dynamics with immigration and emigration

In Section 2.4-1 we studied the dynamics of populations according to the law of mass action,

$$y' = ky, \tag{1}$$

where $y(t)$ is the number of members in the population at time t, and k is the difference between the birth and death rates. In deriving Eq. (1) we assumed that the population was isolated; that is, that no new members were added or subtracted by external means. We shall now consider a generalization of Eq. (1) in which we allow immigration into or emigration out of the population. Thus we modify Eq. (1) to

$$y' - ky = r(t), \tag{2}$$

where $r(t)$ is the rate that members of the population are being added or subtracted from outside the system. If $r > 0$, then we have immigration and if $r < 0$, we have emigration. The size of the population is specified as a_0 at an initial time, which for simplicity we take as $t = 0$. Thus we have

$$y(0) = a_0. \tag{3}$$

EXAMPLE

Determine the population growth if $k = -3$ and the immigration law is the periodic function

$$r(t) = 1000(1 + b \sin t). \tag{4}$$

The initial population is specified as a_0 at $t = 0$. Graphs of the immigration law, Eq. (4), are given in Figure 3.3 when $|b| > 1$ and $|b| < 1$. If $|b| < 1$, then $r(t) > 0$ and there is only immigration. If $|b| > 1$, then $r(t)$ changes sign as t varies, and there is alternately immigration and emigration. Alternate immigration and emigration has been observed in most Western European countries. For example, a substantial emigration from Sweden occurred in 1865–1915 and a substantial immigration into Sweden which started about 1940 has continued growing until the present time.

solution. We must solve the IVP,

$$y' + 3y = 1000(1 + b \sin t), \qquad y(0) = a_0.$$

We use the integrating factor e^{3t} to obtain a general solution

$$y = ce^{-3t} + 1000e^{-3t} \int^t (1 + b \sin s)e^{3s}\, ds \tag{5}$$

where c is an integration constant. We evaluate the integral in Eq. (5) and obtain

$$y = ce^{-3t} + \frac{1000}{3} + 100b(3 \sin t - \cos t).$$

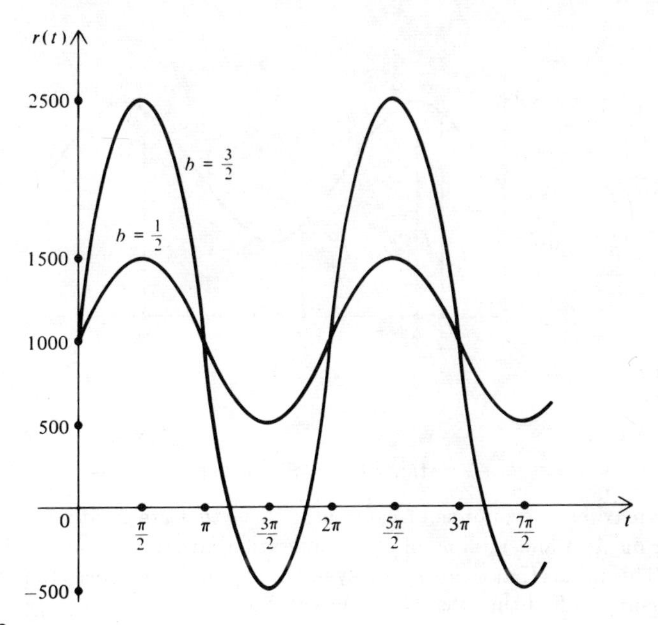

Figure 3.3

The integral of $(\sin s)e^{3s}$ was evaluated by using integration by parts twice. We determine the constant c by requiring that this general solution satisfy the initial condition. This gives the solution of the IVP as

$$y = \left(a_0 - \frac{1000}{3} + 100b\right)e^{-3t} + \frac{1000}{3} + 100b(3 \sin t - \cos t). \tag{6}$$

The term proportional to e^{-3t} in Eq. (6) is a transient. Thus, for sufficiently large t, the solution is approximated by

$$y \approx y_s(t) = \frac{1000}{3} + 100b(3 \sin t - \cos t) \tag{7}$$

which is the steady-state response. If $r(t) \equiv 0$, then the solution of the IVP is $y = a_0e^{-3t}$. Hence, if there were no immigration or emigration, the population would decay to zero.

We shall now study the steady-state response. It is the sum of a constant (1000/3) and a periodic function. They are particular solutions corresponding to the constant and periodic terms respectively in the forcing function (4). A sketch of y_s for $b = \frac{1}{2}$ is given in Figure 3.4. We observe from the figure that it is a periodic function oscillating about 1000/3. The steady-state population varies periodically in response to the periodic immigration law, Eq. (4). The maxima and minima of the immigration occur at $t = \pi/2, 3\pi/2, 5\pi/2, \ldots$. By using calculus, we find that the maxima and minima of y_s occur at $t = \alpha + n\pi$, $n = 1, 2, \ldots$, where α is defined

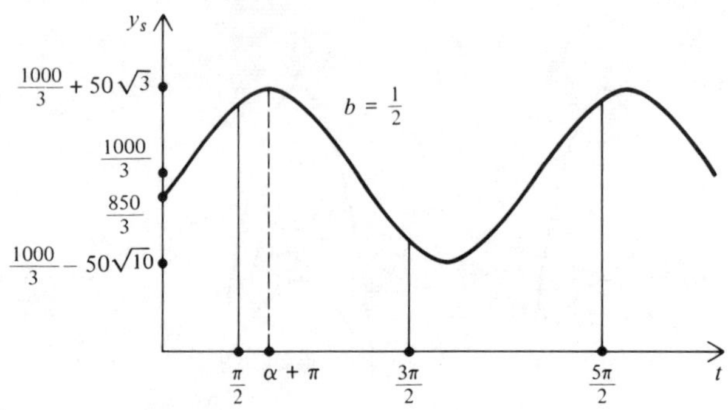

Figure 3.4

by

$$\alpha = \tan^{-1}(-3) \approx -1.25 \text{ rad}.$$

Thus the maxima and minima of y_s occur at times that are greater than the corresponding maxima and minima of the immigration law, as we show in Figures 3.3 and 3.4. This means that there is a delay in the population response to immigration. Furthermore, we find that the minimum value of y_s is

$$\min [y_s(t)] = \frac{1000}{3} - 100\sqrt{10}b. \tag{8}$$

We recall that $y > 0$ because it is the number of members of the population. Thus in order that $\min [y_s] > 0$, we conclude from Eq. (8) that

$$b < \frac{\sqrt{10}}{3}.$$

Thus b cannot be made too large; that is, large amounts of emigration would reduce the population to zero.

Exercises

1. Let $b = 0$ in the example in the text. (a) Determine the population at any time t. (b) What is the limiting population size as $t \to \infty$? (c) Show that if $a_0 > 1000/3$, then the population always decreases despite the immigration. (d) What happens to the population if $a_0 < 1000/3$?

2. Consider a population with $k = -.1$, $a_0 = 1000$, and an immigration rate $r(t) = 50$. (a) Find the population at any time t. (b) Show that the population decreases for all time. (c) Find the limiting population size. (d) Find the time at which the population is equal to 600.

3. The rate at which the population of a country is increasing due to immigration is equal to a constant A. Assume in addition that the difference between the

birth and death rates is a negative constant k, and that the initial population is a_0. (a) Find the population at any time $t > 0$. (b) What is the limiting population? (c) Show that the population decreases if $a_0 > -A/k$ and increases if $a_0 < -A/k$. (d) The government of the given country wants to encourage immigration (increase A) in order to keep the population above the initial value a_0 for all time t. Determine the smallest value of A for which this is possible.

4. If the immigration law is linear, i.e., $r(t) = r_0(1 + bt)$ and the population growth is determined by the IVP given by Eqs. (2) and (3), then find the population at time t.

5. In the Andaman Islands the population was 10,000 in 1970. The birth rate exceeds the death rate by two percent. Furthermore, a population of 1000 immigrates every year. Find the population in 1980. If the government of Andaman decides that the immigration would not be allowed once the population has reached 70,000, find the year in which the immigration would be discontinued.

6. Consider a population where the birth rate exceeds the death rate and the difference between them is .1. Suppose people immigrate at a constant rate $r(t) = A$. (a) Find the time at which the initial population will double. (b) Compare the doubling time in (a) with the result with no immigration.

7. The government of an overpopulated country, where the birth rate exceeds the death rate, wants to stabilize (or decrease) the population by encouraging emigration. Can this be done if a constant rate of emigration is maintained, and the DE (2) is used to model the population?

3.3-3 Linear motion of a particle

We consider a particle of mass m that moves in a vertical straight line. For example, a parachutist can be considered as a particle that is acted on by the force of gravity and a force due to resistance of the air. The descent of the parachutist occurs in a straight vertical line, if we neglect the effects of the earth's curvature and rotation and horizontal wind forces.

The displacement of the particle at time t measured from a fixed point 0 is denoted by $y(t)$. The velocity $v(t)$ is the rate of change of displacement with respect to time and thus it is given by $v = y'$.

Newton's second law of motion states that

$$F = mv' = ma,$$

where F is the total external force, m is the mass of the particle, and $a = v'$ is the acceleration of the particle in the direction of the force F. Here m is assumed to be constant.

There are two systems of dimensions which are commonly used in mechanics. In the metric system, the basic units are centimeters (cm), grams (gm), and second (sec) for length, mass, and time, respectively. Then the units of velocity and acceleration are cm/sec and cm/sec^2 respectively. The unit of force is called a dyne. It is the force necessary to give a mass of 1 gram an acceleration of 1 cm/sec^2. Since $F = ma$, the dimensions of a dyne are gm-cm/sec^2.

In the English system, the basic units are feet (ft), pounds (lb), and seconds (sec) for length, force, and time, respectively; a force of 1 pound moves a mass of 1 slug at an acceleration of 1 ft/sec^2. Since $F = ma$ and hence $m = F/a$, the dimensions of a slug are lb-sec^2/ft. The metric and English systems are also known as the CGS (centimeter-gram-second) and the FPS (foot-pound-second) systems, respectively.

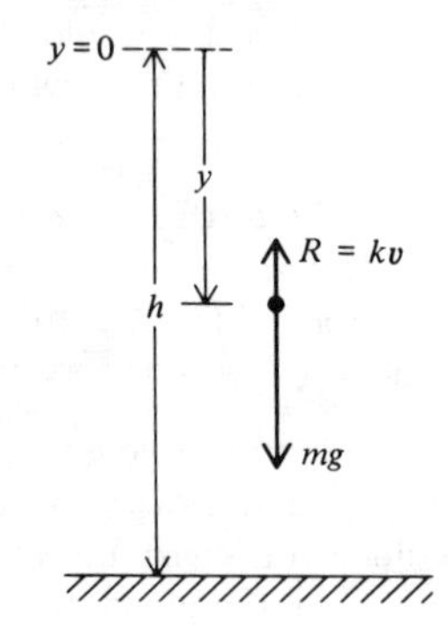

Figure 3.5

The forces acting on the particle as it falls (see Figure 3.5) are its weight mg, and the force of air resistance, which we approximate by kv. Here g is the gravitational constant (32 ft/sec^2 or 980 cm/sec^2) and k is a coefficient of friction for the air. In Appendix I.1 we have shown that Newton's second law implies that the velocity of the particle must satisfy the IVP

$$v' + \frac{k}{m}v = g, \qquad v(0) = v_0, \tag{1}$$

where v_0 is the initial velocity.

EXAMPLE

A stone weighing $\frac{1}{4}$ lb falls from rest from a tall building. It is determined from experiments that the air resistance is $R = v/160$ lb, where v is the velocity. Find the velocity of the stone at time t.

solution. The forces acting on the stone are the weight, which is equal to $\frac{1}{4}$ lb acting downward, and the air resistance, $R = v/16$ lb, acting upward. The total downward force is

$$F = \frac{1}{4} - \frac{v}{160}.$$

The mass of the particle is its weight divided by g, or $m = 1/(4g)$. From the IVP, Eqs. (1) and (2) (Newton's second law) with $g = 32$, we get

$$v' + \tfrac{4}{5}v = 32, \qquad v(0) = 0.$$

We solve this IVP by using the integrating factor $e^{4/5t}$. The solution is

$$v = 40(1 - e^{-4/5t}).$$

The term $-4e^{-(4/5)t}$ is a transient because it decays as $t \to \infty$. The solution approaches the steady-state response, which is the constant $v =$ 40 ft/sec. Thus the steady-state velocity is called the *limiting velocity*. For the general IVP given in Eq. (1), we can show (see Exercise 4) that the limiting velocity is then given by,

$$v = mg/k. \tag{2}$$

Thus the heavier the particle, the larger the limiting velocity; and the larger the air resistance, the smaller the limiting velocity.

Exercises

1. (a) A 150-lb parachutist opens his parachute when his downward velocity is 100 ft/sec. If the force of air resistance is $25v$, where v is the velocity, find the velocity of the parachutist at any time t. (b) At what time will his velocity be 20 ft/sec?

2. If the parachutist in Exercise 1 is carrying an additional weight of P lb, find his velocity for any time. When will the velocity be 20 ft/sec?

3. A 2-lb ball is thrown vertically upward with a velocity of 60 ft/sec. Use Newton's law to formulate an IVP. Find the velocity as a function of time. For what value of t will the ball stop rising? How high will it have gone? Neglect air resistance.

4. Solve the IVP given in Eq. (1). Derive Eq. (2) for the limiting velocity.

5. Suppose an object is thrown upward with velocity v_0. Assume that a resistance force $R = kv$ is acting on the body. (a) Find the time it takes the object to reach its maximum height. (b) Find the maximum height.

6. Show directly from the DE given in Eq. (1) that if the initial velocity of a body moving in a resisting medium is greater than the limiting velocity, then the body will decelerate for all t. What does the DE imply if $v_0 < mg/k$?

7. A ball of mass m falls under gravity into a barrel containing a viscous liquid, such as oil or tar. The motion of the ball is resisted by the viscous liquid which exerts a force μv on the ball. Here μ is the coefficient of viscosity. It is temperature-dependent, and thus if the liquid is heated as the ball falls, then μ will decrease with time. We take

$$\mu = \mu_0 \left(\frac{1}{1+t} \right).$$

(a) If the initial velocity is v_0, find the velocity for all subsequent times. (b) Is there a limiting velocity?

8. A stone is dropped from a height of 300 ft into a body of water. Assume that the air resists the motion of the stone with a force $k_1 v$ while the water exerts a force $k_2 v$. We assume that $k_2 > k_1$. (a) Write down an IVP for the velocity of the stone. (b) In this problem the coefficient of v changes abruptly from k_1 to k_2 as the stone hits the water. Thus this coefficient is discontinuous. Is the solution for v continuous? (c) Show that the acceleration v' is not continuous.

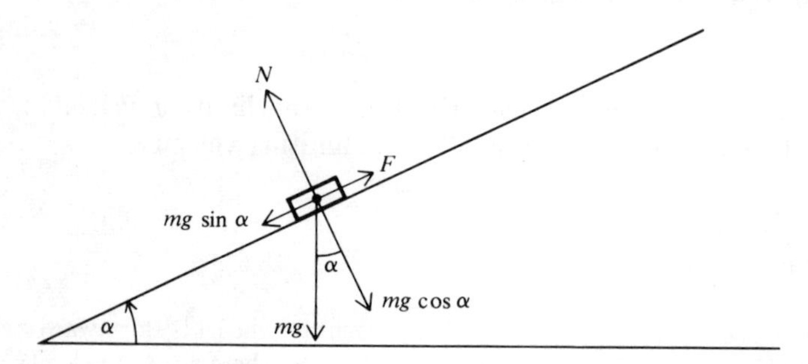

Figure 3.6

9. Consider a block of mass m moving on an inclined plane (see Figure 3.6). Neglect air resistance. The forces acting on the block are the force of gravity, the normal force N, and a frictional force exerted by the plane on the block. The frictional force F acting on the moving block is proportional to the normal force and thus we write $F = \mu mg \cos \alpha$, where α is the angle of inclination of the plane and μ is the proportionality constant (coefficient of friction). Since the block moves *along* the plane, we know that $N = mg \cos \alpha$. (a) Apply Newton's second law to derive an IVP for the velocity $v(t)$ of the block moving along the plane, using $v(0) = v_0$. (b) If $v_0 > 0$, find $v(t)$. (c) If the block starts at 100 ft from the bottom of the plane, how long will it take to reach the bottom? (d) Let $v_0 = 0$, and assume that the coefficient of friction is the same whether the block is sliding or stationary. Show that if $\mu \geq \tan \alpha$, then the block will not move.

chapter 4

THE SECOND-ORDER HOMOGENEOUS LINEAR EQUATION WITH CONSTANT COEFFICIENTS

4.1 INTRODUCTION

The second-order homogeneous linear DE,

$$y'' + 2by' + cy = 0, \tag{1}$$

where b and c are given real numbers, occurs in a variety of problems, as we have shown in Appendices I.2 and I.3. The methods we develop for solving this equation are easily generalized to solve linear DEs with constant coefficients of any order, as we shall see in subsequent chapters.

We observed in Section 1.4-1 that

$$y(x_0) = a_0, \qquad y'(x_0) = a_1 \tag{2}$$

are appropriate initial conditions for DE (1). Here a_0 and a_1 are specified constants, and x_0 is the given initial point. Thus the IVP for DE (1) is to determine solutions of the DE that satisfy the initial conditions given in Eq. (2).

The main goal in this chapter is to develop techniques to solve the IVP and to study properties of the solution.

4.2 EXPONENTIAL SOLUTIONS

In this section we show how to find solutions of

$$y'' + 2by' + cy = 0, \tag{1}$$

where b and c are given real constants. First we consider the special case $c = 0$, $b \neq 0$. Then Eq. (1) becomes

$$y'' + 2by' = 0. \tag{2}$$

By integrating both sides of the equation with respect to x, we obtain

$$y' + 2by = c_1 \tag{3}$$

where c_1 is an arbitrary constant of integration. This is a first-order inhomogeneous linear equation. It can be solved by using the integrating factor e^{2bx}. The solution of Eq. (3) is

$$y = c_2 e^{-2bx} + \frac{c_1}{2b}, \tag{4}$$

where c_2 is an arbitrary constant. It is a solution of Eq. (2) for all values of c_1 and c_2. In particular, if we take $c_2 = 1$ and $c_1 = 0$, we see that $y = e^{-2bx}$ is a solution of Eq. (2). In general, we call a solution of a DE in the form $y = e^{mx}$, $m =$ constant, an exponential solution. Hence DE (2), with $b \neq 0$, always has an exponential solution. We have already shown in Chapter 2 that the first-order homogeneous linear DE with a constant coefficient has an exponential solution. Of course, if a homogeneous linear DE has an exponential solution e^{mx}, then $y = Ae^{mx}$ is a solution for all values of the constant A. However, when we use the term *exponential solution*, we explicitly refer to the solution $y = e^{mx}$, where the constant A is set equal to one.

We shall now show that the DE (1) with $c \neq 0$ has exponential solutions. Thus we substitute $y = e^{mx}$, $m =$ constant, into Eq. (1) and use the differentiation formulas $(e^{mx})' = me^{mx}$ and $(e^{mx})'' = m^2e^{mx}$. This gives

$$m^2e^{mx} + 2bme^{mx} + ce^{mx} = 0.$$

Then by factoring out e^{mx} in this expression, we get

$$(m^2 + 2bm + c)e^{mx} = 0. \tag{5}$$

Since $e^{mx} \neq 0$ for any value of mx, we conclude from Eq. (5) that the DE has an exponential solution, e^{mx}, if m is a root of the quadratic equation

$$m^2 + 2bm + c = 0. \tag{6}$$

This is called the *characteristic equation* of DE (1).

EXAMPLE 1

Find the characteristic equation of the DE

$$y'' - 3y' + 2y = 0.$$

solution. We substitute $y = e^{mx}$, where m is constant, into the DE. We use the relations $(e^{mx})' = me^{mx}$ and $(e^{mx})'' = m^2e^{mx}$. This gives $(m^2 - 3m + 2)e^{mx} = 0$. Since $e^{mx} \neq 0$, we conclude from this equation that the term in parentheses must be zero. This gives the characteristic equation

$$m^2 - 3m + 2 = 0.$$

To determine the exponential solutions of DE (1), we must evaluate the roots m_1 and m_2 of the characteristic equation (6). The roots of the characteristic equation are called the *characteristic roots*. We use the solution formula for quadratic equations and find that the characteristic roots m_1 and m_2 of Eq. (6) are

$$m_1 = -b + \sqrt{b^2 - c}, \qquad m_2 = -b - \sqrt{b^2 - c}. \tag{7}$$

The nature of the characteristic roots depends on the sign of $b^2 - c$. There are three cases:

Case I. $b^2 - c > 0$; then m_1 and m_2 are real and unequal.

Case II. $b^2 - c < 0$; then m_1 and m_2 are complex conjugate roots, which are given by

$$m_1 = -b + i\sqrt{c - b^2}, \qquad m_2 = -b - i\sqrt{c - b^2}, \tag{8}$$

where $i = \sqrt{-1}$.

Case III. $b^2 - c = 0$; then both m_1 and m_2 are equal to $-b$.

Corresponding to each characteristic root, we have an exponential solution of the DE. Therefore in Case I we have two exponential solutions, which we call $y_1(x)$ and $y_2(x)$. They are given by

$$y_1 = e^{m_1x} = \exp\left[(-b + \sqrt{b^2 - c})x\right],$$

$$y_2 = e^{m_2x} = \exp\left[(-b - \sqrt{b^2 - c})x\right]. \tag{9}$$

In Case III we have only one exponential solution: $y_1(x)$. It is given by

$$y_1 = e^{m_1x} = e^{m_2x} = e^{-bx}. \tag{10}$$

In Case II there are no real roots, and hence there are no solutions of the form e^{mx}, with m a real constant. However, as we shall see in the next section, solutions of the DE can be constructed by using the complex conjugate roots given in Eq. (8). The technique for finding exponential solutions is illustrated in the following example.

EXAMPLE 2

Find the exponential solutions of

$$y'' - 3y' + 2y = 0.$$

solution. In Example 1 we found by substituting $y = e^{mx}$ into the DE that the characteristic equation is $m^2 - 3m + 2 = 0$. The roots of this quadratic equation are

$$m_1 = 2 \qquad \text{and} \qquad m_2 = 1.$$

Since these roots are real and distinct, this is an example of Case I. Thus we have found the exponential solutions

$$y_1 = e^{m_1 x} = e^{2x} \qquad \text{and} \qquad y_2 = e^{m_2 x} = e^x.$$

We have shown in Section 2.2 that $y = ce^{-Q(x)}$ is a solution of the first-order DE $y' + q(x)y = 0$, for all values of the constant c. By the appropriate selection of the constant c, it is possible for the general solution to satisfy the single initial condition $y(x_0) = a_0$ for any choice of a_0. That is why we called $y = ce^{-Q(x)}$ a general solution. Thus the IVP for the first-order DE can always be solved by using a general solution.

We wish to define a general solution for the second-order DE such that the IVP

$$y'' + 2by' + cy = 0, \tag{11}$$

$$y(x_0) = a_0, \qquad y'(x_0) = a_1 \tag{12}$$

can be solved for any choice of a_0 and a_1 by using a general solution. We observe that for the IVP given by Eqs. (11) and (12) there are two initial conditions that must be satisfied. This suggests that a general solution should contain two arbitrary constants. Then both initial conditions can be satisfied simultaneously by choosing the constants appropriately.

When the characteristic roots are real and unequal (Case I) DE (11) has the two exponential solutions y_1 and y_2 given in Eq. (9). Since Eq. (11) is a homogeneous linear DE, we know (see Section 1.3) that any linear combination

$$y = c_1 y_1 + c_2 y_2 \tag{13}$$

of two solutions is a solution for all choices of the constants c_1 and c_2. The solution given by Eq. (13) contains two arbitrary constants. Hence it is a candidate for a general solution of the DE.

If the solutions y_1 and y_2 of DE (11) are proportional—that is, if there is a constant $A \neq 0$ such that

$$y_1(x) = Ay_2(x)$$

for all values of x—then the solution given by Eq. (13) can be written as

$$y = c_1 y_1 + c_2 y_2 = c_1 A y_2 + c_2 y_2 = (c_1 A + c_2) y_2.$$

This solution contains only one arbitrary constant, $c_3 = c_1 A + c_2$. Thus it cannot qualify as a general solution because we have only one constant at our disposal to satisfy two initial conditions. Consequently, in forming a general solution the functions y_1 and y_2 cannot be proportional. We observe that in Case I, the two exponential solutions $y_1 = e^{m_1 x}$ and $y_2 = e^{m_2 x}$, where $m_1 \neq m_2$ are not proportional (show this).

In Section 4.5 we shall precisely define a general solution of DE (11) and give conditions on the solutions y_1 and y_2 that will insure that Eq. (13) is a general solution. For the present we shall be concerned with illustrating the techniques for constructing solutions of IVPs.

EXAMPLE 3

Solve the IVP

$$y'' - 3y' + 2y = 0, \tag{14}$$

$$y(0) = 1, \qquad y'(0) = 0. \tag{15}$$

solution. In Example 2 we showed that DE (14) has the exponential solutions $y_1 = e^{2x}$ and $y_2 = e^x$. Therefore, the linear combination

$$y = c_1 y_1(x) + c_2 y_2(x) = c_1 e^{2x} + c_2 e^x \tag{16}$$

is a solution of the DE for all values of the constants c_1 and c_2. Since $m_1 = 2 \neq m_2 = 1$, the solutions y_1 and y_2 are not proportional. In Section 4.5 we shall show that it is a general solution. We now wish to determine c_1 and c_2 so that Eq. (16) satisfies the initial conditions given in Eq. (15). We therefore substitute $y = 1$ and $x = 0$ in Eq. (16) and $y' = 0$ and $x = 0$ in $y' = 2c_1 e^{2x} + c_2 e^x$. This gives

$$y(0) = c_1 + c_2 = 1 \tag{17}$$

and

$$y'(0) = 2c_1 + c_2 = 0. \tag{18}$$

Equations (17) and (18) are a system of two algebraic equations for determining the constants c_1 and c_2 so that the solution given by Eq. (16) satisfies the initial conditions. To solve the algebraic equations we eliminate c_2 by subtracting Eq. (18) from Eq. (17). This gives $c_1 - 2c_1 = 1$ or $c_1 = -1$. By substituting $c_1 = -1$ into Eq. (18), we obtain $c_2 = 2$. We now use these values for c_1 and c_2 in Eq. (16) and get

$$y = -e^{2x} + 2e^x$$

as a solution of the IVP.

Exercises

1. Determine the characteristic equation and all the exponential solutions for each of the following DEs.
   (a) $y'' - 4y' + 3y = 0$ (b) $y'' - y = 0$
   (c) $y'' - 4y' + 4y = 0$ (d) $y'' + 2y' - 35y = 0$
   (e) $y'' - y' + 3y = 0$ (f) $y'' + 6y' = 0$

2. Solve the following IVPs.
   (a) $y'' - 6y' + 5y = 0, \quad y(0) = 1, \quad y'(0) = 0$
   (b) $y'' - 4y = 0, \quad y(7) = 0, \quad y'(7) = 2$
   (c) $y'' + 2y' - 35y = 0, \quad y(1) = 1, \quad y'(1) = 1$
   (d) $y'' - y = 0, \quad y(0) = 1, \quad y'(0) = 0$

3. If $m_1 \neq m_2$, show that e^{m_1x} and e^{m_2x} are not proportional.

4. Find all values of b for which the DE $y'' + 2by' + y = 0$ has two exponential solutions, with real exponent m.

5. Find all values of c for which the DE $y'' + cy = 0$ has exponential solutions, with m real.

6. (a) Show that $y'' - 4y = 0$ has solutions $\sinh 2x$ and $\cosh 2x$, and thus that $y = c_1 \sinh 2x + c_2 \cosh 2x$ is a solution for all values of c_1 and c_2.
   (b) Show that $y = b_1e^{2x} + b_2e^{-2x}$ is also a solution for all values of b_1 and b_2.
   (c) Verify that if $c_1 + c_2 = 2b_1$ and $c_2 - c_1 = 2b_2$, then the solutions given in (a) and (b) are equal.

7. Show that the DE $y'' + 2by' = 0$ always has two exponential solutions if $b \neq 0$. (*Hint:* $e^0 = 1$.)

8. Solve the IVP $y'' - y = 0$, $y(0) = a_0$, $y'(0) = 0$, where a_0 is any prescribed constant. Show that this solution is equal to a_0Y_1, where Y_1 is the solution of the IVP $y'' - y = 0$, $y(0) = 1$, $y'(0) = 0$.

9. Solve the IVP $y'' - 6y' + 8y = 0$, $y(0) = 0$, $y'(0) = a_1$. Show that this solution is equal to a_1Y_2, where Y_2 is the solution of the IVP $y'' - 6y' + 8y = 0$, $y(0) = 0$, $y'(0) = 1$.

*10. Prove without actually solving that a solution of the IVP

$$y'' + 2by' + cy = 0, \qquad y(x_0) = a_0,\ y'(x_0) = a_1,$$

can be written as $y = a_0Y_1 + a_1Y_2$, where Y_1 and Y_2 satisfy the given DE and the initial conditions $Y_1(x_0) = 1$, $Y_1'(x_0) = 0$ and $Y_2(x_0) = 0$, $Y_2'(x_0) = 1$. Since Y_1 and Y_2 are independent of a_0 and a_1, this result implies that y depends linearly on a_0 and a_1. Hence if we double the initial values, we double the solution. (*Hint:* Verify that $y = a_0Y_1 + a_1Y_2$ satisfies the given IVP.)

4.3 CASE II. COMPLEX CHARACTERISTIC ROOTS

If $b^2 - c < 0$ (Case II), the roots of the characteristic equation

$$m^2 + 2bm + c = 0 \tag{1}$$

are the complex conjugate numbers

$$m_1 = -b + i\sqrt{c - b^2} \quad \text{and} \quad m_2 = -b - i\sqrt{c - b^2}. \tag{2}$$

Since the corresponding exponentials are e^{mx}, with m a complex number, we must define these functions. We first consider the number e^C, where C is an arbitrary complex number. We recall that any complex number C can be written as

$$C = A + iB,$$

where A and B are real numbers and we use the letter $i = \sqrt{-1}$ to denote the imaginary unit. A and B are called the real and imaginary parts of C respectively. We wish to define e^C so that it has algebraic properties similar to the real-valued exponential and such that it reduces to the real-valued exponential when $B = 0$.

definition

Let $C = A + iB$. Then e^C is the complex number that is defined by

$$e^C = e^A(\cos B + i \sin B), \tag{3}$$

where e^A is the real-valued exponential.

The real and imaginary parts of e^C are therefore $e^A \cos B$ and $e^A \sin B$, respectively. If C is real, then $B = 0$ and Eq. (3) is reduced to the real exponential $e^C = e^A$. If $A = 0$, so that C is pure imaginary, then since $e^0 = 1$ we conclude from Eq. (3) that

$$e^C = e^{iB} = \cos B + i \sin B.$$

This identity is called Euler's formula.

We show in Appendix II by direct application of Eq. (3) that e^C has the following algebraic properties. Let C_1 and C_2 be any two complex numbers. Then

$$e^{C_1 + C_2} = e^{C_1}e^{C_2}$$

and

$$\frac{e^{C_1}}{e^{C_2}} = e^{C_1 - C_2} = e^{C_1}e^{-C_2}. \tag{4}$$

These properties are, of course, well known for the real-valued exponential.

The function e^{mx}, where x is a real variable and $m = \alpha + i\beta$ (α and β real numbers) is a complex number for each fixed x. It is defined by setting $C = mx$ or, equivalently, $A = \alpha x$ and $B = \beta x$ in Eq. (3). This gives

$$e^{mx} = e^{\alpha x}(\cos \beta x + i \sin \beta x) \tag{5}$$

as the definition of e^{mx}. If m is a real number ($\beta = 0$), then Eq. (5) implies that $e^{mx} = e^{\alpha x}$, since $\cos 0 = 1$ and $\sin 0 = 0$. Thus we recover the ordinary exponential function. As x varies, the complex number e^{mx} varies. Hence Eq. (5) defines a complex-valued function of the real variable x. In analogy with complex numbers, we refer to the real-valued functions $e^{\alpha x} \cos \beta x$ and $e^{\alpha x} \sin \beta x$ as the real and imaginary parts of the complex-valued function e^{mx}. If $\alpha = 0$, and thus $m = i\beta$ is a pure imaginary number, then Eq. (5) implies that

$$e^{i\beta x} = \cos \beta x + i \sin \beta x. \tag{6}$$

It follows by using Eq. (4) with $C_1 = m_1 x$ and $C_2 = m_2 x$ that the complex-valued exponential function has the following important property that is analogous to the real-valued exponential function:

$$e^{(m_1+m_2)x} = e^{m_1 x} \cdot e^{m_2 x}, \tag{7}$$

where m_1 and m_2 are any two complex numbers. In addition,

$$(e^{mx})' = me^{mx} \tag{8}$$

and

$$e^{mx} \neq 0 \tag{9}$$

for any complex number m, as we demonstrate in Appendix II.

We now return to our study of the DE

$$y'' + 2by' + cy = 0, \tag{10}$$

when $b^2 - c < 0$ (Case II). We insert the complex-valued function $y = e^{m_1 x}$, where m_1 is the complex root of the characteristic equation given by Eq. (2), into Eq. (10) using the differentiation formula given in Eq. (8). This gives, in the usual way,

$$y'' + 2by' + cy = (m_1^2 + 2bm_1 + c)e^{m_1 x}. \tag{11}$$

Since m_1 is a characteristic root, the right side of Eq. (11) is zero and $e^{m_1 x}$ is a *complex-valued solution* of the DE. Similarly $e^{m_2 x}$ is a complex-valued solution. If we use the definition given in Eq. (5) and the roots given in Eq. (2), these solutions are

$$\begin{aligned} y = z_1(x) &\equiv e^{m_1 x} = e^{-bx}[\cos(\sqrt{c - b^2}\,x) + i \sin(\sqrt{c - b^2}\,x)], \\ y = z_2(x) &\equiv e^{m_2 x} = e^{-bx}(\cos(\sqrt{c - b^2}\,x) - i \sin(\sqrt{c - b^2}\,x)], \end{aligned} \tag{12}$$

where we have used $\cos(-A) = \cos A$ and $\sin(-A) = -\sin A$, with $A = \sqrt{c - b^2}$ in obtaining $z_2(x)$.

EXAMPLE 1

Find the complex-valued exponential solutions of

$$y'' + 9y = 0.$$

solution. We look for a solution $y = e^{mx}$. Substituting this for y and the corresponding expression for y'' into the DE, we find that the characteristic equation is

$$m^2 + 9 = 0.$$

The roots of this equation are the pure imaginary numbers

$$m_1 = i3 \qquad \text{and} \qquad m_2 = -i3.$$

Thus the exponential solutions are $z_1 = e^{i3x}$ and $z_2 = e^{-i3x}$, which are complex-valued functions. Using Eq. (6) with $\beta = 3$, we obtain

$$z_1 = e^{3ix} = \cos 3x + i \sin 3x,$$

$$z_2 = e^{-3ix} = \cos(-3x) + i \sin(-3x)$$

$$= \cos 3x - i \sin 3x.$$

We have shown that in Case II (when $b^2 - c < 0$), DE (10) has the two complex-valued solutions given by Eq. (12). We shall now show how to combine these two complex-valued solutions to obtain two real-valued solutions of the DE. We recall that if y_1 and y_2 are two real-valued solutions of the DE, then every linear combination, $y = c_1y_1 + c_2y_2$, where c_1 and c_2 are arbitrary real-valued constants, is also a solution of the DE. The same result can be established for linear combinations of complex-valued solutions (see Exercise 7). Thus any linear combination of the two complex-valued solutions given in Eq. (12) is a solution.

We now show that the specific linear combinations

$$y_1(x) = \frac{1}{2} z_1 + \frac{1}{2} z_2$$

and

$$y_2(x) = -\frac{i}{2} z_1 + \frac{i}{2} z_2$$

are real-valued solutions of DE (10). By using the definitions of z_1 and z_2

given in Eq. (12), we have

$$y_1(x) = \frac{1}{2} z_1 + \frac{1}{2} z_2 = e^{-bx} \cos(\sqrt{c - b^2}x),$$

$$y_2(x) = -\frac{i}{2} z_1 + \frac{i}{2} z_2 = e^{-bx} \sin(\sqrt{c - b^2}x). \tag{13}$$

Both $y_1(x)$ and $y_2(x)$ are real-valued solutions.

Since $-b$ and $\pm\sqrt{c - b^2}$ are the real and imaginary parts of the characteristic roots, the real-valued solutions given in Eq. (13) can be obtained directly from the characteristic roots without first determining the complex exponential solutions.

Observe from Eqs. (12) and (13) that y_1 and y_2 are the real and imaginary parts of the complex-valued exponential solution $z_1 = e^{m_1 x}$, and y_1 and $-y_2$ are the real and imaginary parts of $z_2 = e^{m_2 x}$. Thus we can write the solutions given by Eq. (12) in the alternate form,

$$z_1 = e^{m_1 x} = y_1 + iy_2, \qquad z_2 = e^{m_2 x} = y_1 - iy_2.$$

EXAMPLE 2

Find real-valued solutions of

$$y'' - 2y' + 10y = 0.$$

solution. We look for exponential solutions. The characteristic roots are

$$m_1 = \frac{2 + \sqrt{4 - 40}}{2} = \frac{2 + i6}{2} = 1 + i3,$$

$$m_2 = \frac{2 - \sqrt{4 - 40}}{2} = 1 - i3.$$

Hence the DE has the complex-valued solutions

$$z_1 = e^{(1+i3)x} = e^x(\cos 3x + i \sin 3x),$$

$$z_2 = e^{(1-i3)x} = e^x(\cos 3x - i \sin 3x).$$

We see that $y_1 = e^x \cos 3x$ and $y_2 = e^x \sin 3x$ are real-valued solutions because they are the real and imaginary parts of z_1. Alternatively, y_1 and y_2 can be obtained by observing that the coefficient in the real exponential in Eq. (13), $-b = 1$, is the real part of the characteristic roots and the coefficient in the trigonometric functions $\sqrt{c - b^2} = 3$ is the imaginary part of the characteristic root, with a positive imaginary part.

Since any linear combination of two solutions of the DE is also a solution, we have obtained two possible forms of general solutions in Case

II. They are, from Eq. (12),

$$y = c_1 e^{m_1 x} + c_2 e^{m_2 x} \tag{14}$$

and, from Eq. (13),

$$y = c_1 e^{-bx} \cos(\sqrt{c - b^2}x) + c_2 e^{-bx} \sin(\sqrt{c - b^2}x). \tag{15}$$

In Eq. (14) the arbitrary constants c_1 and c_2 are, in general, complex numbers, since $e^{m_1 x}$ and $e^{m_2 x}$ are complex-valued solutions. We refer to Eq. (14) as a complex-valued form of a general solution. In Eq. (15) c_1 and c_2 are real numbers. We refer to Eq. (15) as a trigonometric form of a general solution. These are plausible definitions of general solutions because, as the reader should show, the solutions $e^{-bx} \cos(\sqrt{c - b^2}x)$ and $e^{-bx} \sin(\sqrt{c - b^2}x)$ are not proportional. In fact, in Section 4.5 we prove that Eqs. (14) and (15) are general solutions in Case II. In Example 3 we show how to use these general solutions to solve an IVP.

EXAMPLE 3

Solve the IVP

$$y'' - 2y' + 10y = 0, \qquad y(0) = 0,\ y'(0) = 1.$$

solution. In Example 2 we showed that $y_1 = e^x \cos 3x$ and $y_2 = e^x \sin 3x$ are solutions of this DE. Thus

$$y = c_1 y_1 + c_2 y_2 = c_1 e^x \cos 3x + c_2 e^x \sin 3x$$

is a solution of the DE for all values of the constants. We determine these constants so that this solution satisfies the initial conditions. We first compute y':

$$y' = c_1[-3e^x \sin 3x + e^x \cos 3x] + c_2[3e^x \cos 3x + e^x \sin 3x].$$

The initial conditions $y = 0$ and $y' = 1$ at $x = 0$ give

$$y(0) = c_1 = 0, \qquad y'(0) = c_1 + 3c_2 = 1.$$

By solving these algebraic equations we get $c_1 = 0$, $c_2 = 1/3$. We insert these values of c_1 and c_2 into the general solution to obtain

$$y = \frac{1}{3} e^x \sin 3x \tag{16}$$

as a solution of the IVP.

Exercises

1. Determine the real-exponential and complex-valued exponential solutions of:
   (a) $y'' + 2y' + (13/4)y = 0$ (b) $y'' + 9y = 0$

(c) $y'' + 2y' - 8y = 0$ (d) $y'' + 6y' + 13y = 0$
(e) $y'' + 2y' + 5y = 0$

2. Determine the trigonometric form of the general solution for each of the parts in Exercise 1 for which it is applicable.

3. Find a general solution of the DE $y'' + 4y' + ay = 0$ for all values of the constant a, except $a = 4$.

4. Solve the following IVPs using either the trigonometric or complex-valued form for the general solution. Find the limit as $x \to \infty$ of the solution.
(a) $y'' + 2y' + (13/4)y = 0, \quad y(0) = 1, \quad y'(0) = 2$
(b) $y'' + 6y' + 13y = 0, \quad y(1) = 0, \quad y'(1) = 3$
(c) $y'' + y = 0, \quad y(0) = 1, \quad y'(0) = 2$
(d) $y'' - 4y' + 29y = 0, \quad y(0) = 0, \quad y'(0) = 1/2$

5. Solve the IVP $y'' + cy = 0$, $y(0) = 1$, $y'(0) = 0$, where c is any constant. Discuss the limit as $x \to \infty$ of the solution.

6. Use the results of Appendix II on the derivatives of complex-valued functions $f(x)$ to show that

$$[c_1 f_1(x) + c_2 f_2(x)]' = c_1 f_1' + c_2 f_2',$$

where $f_1(x)$ and $f_2(x)$ are any two differentiable complex-valued functions, and c_1 and c_2 are any two complex numbers.

7. Use the results of Exercise 6 to show that if $z_1(x)$ and $z_2(x)$ are two complex-valued solutions of $y'' + 2by' + cy = 0$, then $z = c_1 z_1 + c_2 z_1$ is a complex-valued solution for all values of the complex constants c_1 and c_2.

8. Show that if $z(x) = u(x) + iv(x)$ is a complex-valued solution of $y'' + 2by' + cy = 0$, where b and c are *real* constants, then $u(x)$ and $v(x)$ are both real-valued solutions.

9. In many applications we are required to solve DEs with constant, but complex-valued, coefficients. Typical examples are $y'' + iy = 0$ and $y'' - iy = 0$. Give a simple argument to show that there are no real-valued solutions to these equations. Determine all the complex valued exponential solutions of these two DEs. (*Hint:* To determine $\pm\sqrt{i}$, use de Moivre's formula; see Exercise 6 of Appendix II.)

4.4 CASE III. REPEATED CHARACTERISTIC ROOTS

In Case III ($b^2 - c = 0$) the characteristic equation

$$m^2 + 2bm + c = 0$$

has only one distinct root,

$$m_1 = m_2 = -b. \tag{1}$$

Then

$$y = e^{mx} = e^{-bx} \tag{2}$$

is the only exponential solution of the DE

$$y'' + 2by' + cy = 0. \tag{3}$$

Consequently $y = c_1e^{-bx}$ is a solution for all values of the constant c_1. This solution is not a general solution because it contains only one arbitrary constant, and hence it is not possible to satisfy the initial conditions

$$y(x_0) = a_0 \qquad \text{and} \qquad y'(x_0) = a_1 \tag{4}$$

for all choices of a_0 and a_1. Specifically, we observe that

$$y = c_1e^{-bx} + c_2e^{-bx}$$

is *not* a general solution because we can rewrite it as

$$y = (c_1 + c_2)e^{-bx} = c_3e^{-bx}.$$

Therefore there is only one constant, c_3, available to satisfy both initial conditions given in Eq. (4).

We obtain a second solution of Eq. (3) by using the method of reduction of order. This method is useful for obtaining additional solutions when one solution is known. It will be discussed in detail in Section 11.3. The essence of the method is that when one solution

$$y = y_1 = e^{-bx}$$

of Eq. (3) is known, a second solution can be obtained in the form

$$y = u(x)y_1(x) = u(x)e^{-bx}, \tag{5}$$

where the function $u(x)$ is to be determined so that Eq. (5) satisfies DE (3). That is, we allow the constant c_1 in the solution $y = c_1e^{-bx}$ to vary with x. This is why the method also is called variation of constants or variation of parameters.

Differentiating Eq. (5) twice, we get

$$y' = -bue^{-bx} + u'e^{-bx} = (-bu + u')e^{-bx}$$

and

$$\begin{aligned} y'' &= -b(ue^{-bx})' + (u'e^{-bx})' \\ &= -b(-bue^{-bx} + u'e^{-bx}) - bu'e^{-bx} + u''e^{-bx} \\ &= (b^2u - 2bu' + u'')e^{-bx}. \end{aligned}$$

By substituting these expressions for y, y', and y'' in Eq. (3), we obtain, after some calculation,

$$e^{-bx}[u'' + (c - b^2)u] = 0.$$

Since $b^2 - c = 0$, this reduces to

$$u''e^{-bx} = 0. \tag{6}$$

We conclude from Eq. (6) that $u'' = 0$ because $e^{-bx} \neq 0$. Thus by seeking a solution in the form of Eq. (5), we observe that u satisfies the DE, $u'' = 0$ which is much simpler to solve than the original DE given in Eq. (3).

By integrating this equation for u twice we get

$$u = c_2x + c_1, \tag{7}$$

where c_1 and c_2 are arbitrary constants of integration. Hence if u is given by Eq. (7), then Eq. (5) will be a solution of the DE. We insert Eq. (7) into Eq. (5) and obtain the solution

$$y = u(x)e^{-bx} = (c_1 + c_2x)e^{-bx} = c_1e^{-bx} + c_2xe^{-bx} \tag{8}$$

This solution is a linear combination of the two functions

$$y_1 = e^{-bx} \qquad \text{and} \qquad y_2 = xe^{-bx}. \tag{9}$$

The function y_1 is the exponential solution already given in Eq. (2). Since Eq. (8) is a solution for all values of the arbitrary constants c_1 and c_2, we conclude, by taking $c_1 = 0$ and $c_2 = 1$ in Eq. (8), that y_2 is a solution of the DE. It is precisely x times the exponential solution.

The solution given in Eq. (8) is a candidate for a general solution because it is a solution for all values of the arbitrary constants c_1 and c_2, and the solutions y_1 and y_2 are not proportional, as the reader can verify. We shall use it as a general solution for solving IVPs. We shall prove in Section 4.5 that it is a general solution.

EXAMPLE

Solve the IVP

$$y'' - 6y' + 9y = 0, \qquad y(0) = 2, \; y'(0) = 1.$$

solution. The characteristic roots are

$$m_1 = m_2 = \frac{6 \pm \sqrt{36 - 36}}{2} = 3.$$

There is only one distinct root, and thus $e^{m_1x} = e^{3x}$ and xe^{3x} are solutions of the DE. The linear combination

$$y = c_1e^{3x} + c_2xe^{3x} \tag{10}$$

is a solution, where c_1 and c_2 are arbitrary constants. We determine these constants so that Eq. (10) satisfies the initial conditions. Since

$$y' = 3c_1e^{3x} + c_2(3x + 1)e^{3x},$$

we get

$$y(0) = c_1 = 2, \qquad y'(0) = 3c_1 + c_2 = 1.$$

Thus $c_1 = 2$ and $c_2 = 1 - 3c_1 = -5$, and the solution of the IVP is

$$y = 2e^{3x} - 5xe^{3x}.$$

Exercises

1. Obtain two solutions, which are not proportional, for each of the following DEs.
   (a) $y'' - 2y' + y = 0$ (b) $y'' + 8y' + 6y = 0$
   (c) $y'' + 7y = 0$ (d) $y'' + 2y' + y = 0$
   (e) $y'' - 4y' + 4y = 0$ (f) $6y'' + 3y' + 6y = 0$

2. Solve the following IVPs.
   (a) $y'' - 2y' + y = 0, \quad y(0) = 0, \quad y'(0) = 1$
   (b) $y'' + 2y' + y = 0, \quad y(0) = 1, \quad y'(0) = 0$
   (c) $y'' - 6y' + 9y = 0, \quad y(1) = 1, \quad y'(1) = 1$

3. Obtain a general solution of $y'' + 2by' + b^2y = 0$ for any constant b. Find the limit of this solution as $x \to \infty$.

4. Solve the IVP

$$y'' + 2by' + b^2y = 0, \qquad y(0) = a_0, \qquad y'(0) = 0,$$

   where a_0 and b are any prescribed constants.

5. (a) Show that all solutions of the DE $y'' + b_1y' + b_0y = 0$ approach zero as $x \to \infty$ if both b_0 and b_1 are positive constants.
   (b) Show that if b_0 and b_1 are not both positive, then all solutions of the DE do not behave as in (a).
   (c) What do the conditions $b_0 > 0$, $b_1 > 0$ imply about the real parts of the characteristic roots? If all solutions of a homogeneous linear DE approach zero as $x \to \infty$, then the DE is said to be asymptotically stable. This property is of great interest in the applications of DEs and has been the subject of extensive mathematical study.

*6. For repeated characteristic roots, obtain the second solution by using the following alternative argument. For $m_1 \neq m_2$, $y_1 = e^{m_1x}$, and $y_2 = e^{m_2x}$ are two solutions of the DE. Since any linear combination of two solutions is also a solution, the function

$$y_3 = \frac{e^{m_2x} - e^{m_1x}}{m_2 - m_1}$$

is also a solution if $m_1 \neq m_2$. In the limit as $m_2 \to m_1$, this function gives the "indeterminant form" (0/0). Set $m_2 = m_1 + \epsilon$ and show that in the limit as $m_2 \to m_1$, or as $\epsilon \to 0$, we get $y_3 = xe^{m_1x}$ by using l'Hospital's rule, with ϵ as the variable, to evaluate the indeterminate form.

4.5 GENERAL SOLUTIONS AND THE INITIAL VALUE PROBLEM

The IVP for the second-order homogeneous linear DE with constant coefficients is to determine solutions of

$$y'' + 2by' + cy = 0 \tag{1}$$

that satisfy the initial conditions,

$$y(x_0) = a_0 \qquad \text{and} \qquad y'(x_0) = a_1 \tag{2}$$

where x_0, a_0, and a_1 are prescribed real numbers. In the examples in the previous sections the IVPs were solved by finding two solutions y_1 and y_2 of the DE, which were not proportional. Then the linear combination

$$y = c_1y_1 + c_2y_2 \tag{3}$$

was used as a general solution. The constants c_1 and c_2 were determined by requiring the general solution to satisfy the initial conditions. This led to two algebraic equations for c_1 and c_2. In all the examples these algebraic equations had only one solution.

We now wish to study the IVP given in Eqs. (1) and (2) to determine when it can be solved for any given initial values. This will lead to a precise definition of a general solution.

Let y_1 and y_2 denote any two solutions of DE (1). We form the linear combination given in Eq. (3). It is a solution of the DE for all values of c_1 and c_2. In order for Eq. (3) to be a solution of the IVP, it must satisfy the initial conditions in Eq. (2). This gives

$$c_1y_1(x_0) + c_2y_2(x_0) = a_0 \tag{4}$$

and

$$c_1y_1'(x_0) + c_2y_2'(x_0) = a_1. \tag{5}$$

The coefficients in the algebraic equations (4) and (5) are y_1 and y_2, and their derivatives evaluated at $x = x_0$. They are known numbers. We solve Eqs. (4) and (5) by first eliminating c_2. Thus we multiply both sides of Eq. (4) by $y_2'(x_0)$ and both sides of Eq. (5) by $y_2(x_0)$. This gives

$$c_1[y_1(x_0)y_2'(x_0)] + c_2[y_2(x_0)y_2'(x_0)] = a_0y_2'(x_0)$$

and

$$c_1[y_1'(x_0)y_2(x_0)] + c_2[y_2'(x_0)y_2(x_0)] = a_1y_2(x_0).$$

We subtract these two equations and get

$$[y_1(x_0)y_2'(x_0) - y_1'(x_0)y_2(x_0)]c_1 = a_0y_2'(x_0) - a_1y_2(x_0). \tag{6}$$

We can solve Eq. (6) for c_1 if

$$W(x_0) = y_1(x_0)y_2'(x_0) - y_1'(x_0)y_2(x_0) \neq 0. \tag{7}$$

Then we can divide both sides of Eq. (6) by $W(x_0)$ to get

$$c_1 = \frac{a_0 y_2'(x_0) - a_1 y_2(x_0)}{W(x_0)}. \tag{8}$$

A similar expression is obtained for c_2 by eliminating c_1 from Eqs. (4) and (5). This gives

$$W(x_0)c_2 = -a_0 y_1'(x_0) + a_1 y_1(x_0). \tag{9}$$

If $W(x_0) \neq 0$, then

$$c_2 = \frac{-a_0 y_1'(x_0) + a_1 y_1(x_0)}{W(x_0)}. \tag{10}$$

Thus if $W(x_0) \neq 0$, there is one and only one solution of the algebraic equations (4) and (5), and it is given by Eqs. (8) and (10).

If $W(x_0) = 0$, then the left sides of Eqs. (6) and (9) are zero. Consequently the algebraic equations (4) and (5) have a solution only if the right sides of Eqs. (6) and (9) vanish. That is,

$$a_0 y_2'(x_0) - a_1 y_2(x_0) = 0 \tag{11}$$

and

$$-a_0 y_1'(x_0) + a_1 y_1(x_0) = 0. \tag{12}$$

Thus if $W(x_0) = 0$, the initial conditions will not be satisfied by Eq. (3) unless the initial data a_0 and a_1 are restricted to satisfy Eqs. (11) and (12). Consequently, if we select the solutions y_1 and y_2 of the DE so that $W(x_0) = 0$, then we can solve the IVP only for these special values of a_0 and a_1.

The preceding discussion shows that in order to solve the IVP for any a_0 and a_1, we should select the solutions y_1 and y_2 so that $W(x_0) \neq 0$. The function $W(x)$, which is defined by

$$W(x) = y_1(x)y_2'(x) - y_1'(x)y_2(x), \tag{13}$$

is called the Wronskian[1] of the two solutions y_1 and y_2. The number $W(x_0)$ is the value of the Wronskian at the initial point $x = x_0$. The Wronskian depends on the choice of the functions y_1 and y_2.

We now verify that y_1 and y_2, used in the previous sections to form candidates for general solutions, satisfy $W(x) \neq 0$ *for any* x. First, we consider Cases I and II, where the characteristic roots are unequal. We have $y_1 = e^{m_1 x}$ and $y_2 = e^{m_2 x}$, and hence

$$W(x) = m_2 e^{m_1 x} e^{m_2 x} - m_1 e^{m_1 x} e^{m_2 x} = (m_2 - m_1) e^{(m_1 + m_2)x}.$$

Since $m_2 \neq m_1$ and $e^{(m_1+m_2)x} \neq 0$, we have $W(x) \neq 0$ for any x. The same

[1] It is named after the Polish mathematician, H. Wronski (1778–1853).

result can be shown for the trigonometric form of the solution in Case II (see Exercise 2). Finally, for Case III, we recall that $y_1 = e^{-bx}$ and $y_2 = xe^{-bx}$. Then we have

$$W(x) = e^{-bx}[-bxe^{-bx} + e^{-bx}] + be^{-bx}xe^{-bx} = e^{-2bx} \neq 0,$$

for any x.

The demonstrations that $W(x) \neq 0$ for every x are special cases of a more general result that we prove in the following theorem.

theorem 1

If the Wronskian of two solutions y_1 and y_2 is zero at a single point, then it is zero for all values of x. This means that the Wronskian is either zero for all values of x or it is not zero for any value of x.

proof. Since y_1 and y_2 are solutions of DE (1), we have

$$y_1'' + 2by_1' + cy_1 = 0$$

and

$$y_2'' + 2by_2' + cy_2 = 0.$$

By multiplying the first equation by y_2 and the second by y_1 and subtracting the results, we get

$$(y_1y_2'' - y_2y_1'') + 2b(y_1y_2' - y_2y_1') = 0. \tag{14}$$

The term in the second bracket is precisely $W(x)$. We now show that the term in the first bracket is W'. Thus by differentiating $W(x)$, which is defined in Eq. (13), we get

$$W' = y_1y_2'' + y_1'y_2' - y_1'y_2' - y_1''y_2 = y_1y_2'' - y_1''y_2. \tag{15}$$

Consequently, Eq. (14) can be rewritten in terms of W as,

$$W' + 2bW = 0.$$

A general solution of this DE is

$$W(x) = ce^{-2bx}, \tag{16}$$

where c is an arbitrary constant. Since $e^{-2bx} \neq 0$ for any value of bx, we deduce from Eq. (16) that if $W(x) = 0$ for some value of x, then $c = 0$ and $W(x) = 0$ for every value of x. Conversely, if $W(x) \neq 0$ for some value of x, then $c \neq 0$ and $W(x) \neq 0$ for any value of x. This proves the theorem.

This theorem is useful when we wish to determine whether a candidate for a general solution is indeed a general solution. We can then evaluate the Wronksian at any convenient point.

We now have sufficient information to define a general solution of DE (1).

definition

A general solution of DE (1) is a linear combination

$$y = c_1y_1 + c_2y_2 \tag{17}$$

of two solutions y_1 and y_2 of the DE for which $W(x) \neq 0$ for at least one value of x.

Of course, Theorem 1 shows that if $W(x) \neq 0$ for one value of x, it is never zero. In Case I (the characteristic roots are real and distinct) a linear combination of the two exponential solutions is a general solution. In Case II (the characteristic roots are complex conjugate numbers) two general solutions are given by a linear combination of the two complex-valued exponential solutions, or by a linear combination of the two real-valued trigonometric solutions. Finally, in Case III (repeated characteristic root) a linear combination of the exponential solution $y_1 = e^{-bx}$ and $y_2 = xe^{-bx}$ is a general solution. Thus the IVP given by Eqs. (1) and (2) can always be solved using a general solution of the form given in Eq. (17). The constants c_1 and c_2 are uniquely determined by the initial conditions. In addition, we shall prove in Section 4.6 that this is the only solution of the IVP. This uniqueness result implies the next theorem.

theorem 2

Every solution of the DE

$$y'' + 2by' + cy = 0 \tag{18}$$

is contained in a general solution. The proof of this theorem is discussed in Exercise 6.

Exercises

1. Find a general solution for each of the following DEs. Show that your answer is a general solution.
   (a) $y'' - 5y = 0$
   (b) $y'' - y' + 3ay = 0$
   (c) $y'' - 2y' - \dfrac{5}{4}y = 0$
   (d) $y'' + 2y' + y = 0$
   (e) $y'' + 6y' = 0$
   (f) $y'' = 0$

2. We have shown in Section 4.3 that the functions

   $$y_1 = e^{-bx} \cos(\sqrt{c - b^2}x) \qquad \text{and} \qquad y_2 = e^{-bx} \sin(\sqrt{c - b^2}x)$$

   are real-valued solutions of the DE $y'' + 2by' + cy = 0$ when $b^2 - c < 0$ (Case II); see Eq. (13) in Section 4.3. By using the definition, show that the Wronskian of these solutions is not zero for any value of x.

3. (a) Show that the DE $y'' - y = 0$ has solutions $y_1 = e^x$, $y_2 = e^{-x}$, $y_3 = \sinh x$, and $y_4 = \cosh x$.

(b) Using the Wronskian show that all possible pairs of these solutions have nonzero Wronskians. (c) Give four different forms for a general solution.

4. For what values of a_0 and a_1 can the IVP $y'' + 2y' + 2y = 0$, $y(0) = a_0$, $y'(0) = a_1$ be solved if only the one parameter family of solutions $y = c_1e^{-x}\sin x$ is used?

5. For what values of a_0 and a_1 can the IVP

$$y'' + 3y' + 2y = 0, \qquad y(0) = a_0,\ y'(0) = a_1$$

be solved if we use only the linear combination $y = c_1y_1 + c_2y_2$, where $y_1 = e^{-x}$ and $y_2 = 4e^{-x}$? *Hint:* Use Eqs. (11) and (12).

*6. To prove Theorem 2, let $z(x)$ be any solution of the DE in Eq. (18). Observe that $z(x)$ satisfies the IVP consisting of Eq. (18) and the initial conditions $y = z(x_0)$, $y' = z'(x_0)$, where x_0 is any point. Show that the constants in the general solution $y = c_1y_1 + c_2y_2$ can be selected so that it also satisfies this IVP. Since there is at most one solution of the IVP, conclude that there is a choice, $c_1 = A$ and $c_2 = B$ such that $z = Ay_1 + By_2$.

4.5-1 The Wronskian and linear independence

In this section we demonstrate the relationship between the Wronskian and the concepts of linear independence and linear dependence of solutions.

definition

Two functions f_1 and f_2, which are defined on an interval $\alpha < x < \beta$, are linearly dependent if there are constants c_1 and c_2, not both zero, such that

$$c_1f_1(x) + c_2f_2(x) = 0, \qquad \text{for all } x \text{ in } (\alpha, \beta). \tag{1}$$

The functions f_1 and f_2 are linearly independent if they are not linearly dependent. That is, f_1 and f_2 are linearly independent if the only values of the constants for which Eq. (1) is valid are $c_1 = c_2 = 0$.

The condition of linear dependence of two functions is equivalent to the statement that the two functions are proportional. That is, there is a constant $A \neq 0$ such that $f_1(x) = Af_2(x)$, for all x in (α, β). The ideas of linear independence and linear dependence are general concepts that occur in many other branches of mathematics.

EXAMPLE 1

Show that the functions $f_1 = \sin 2x$ and $f_2 = \sin x \cos x$ are linearly dependent on the interval $(0, 2\pi)$.

solution. We must show that the linear combination

$$c_1f_1 + c_2f_2 = c_1 \sin 2x + c_2 \sin x \cos x = 0. \tag{2}$$

Since $\sin 2x = 2 \sin x \cos x$, Eq. (2) is reduced to

$$(2c_1 + c_2) \sin x \cos x = 0.$$

Therefore if we choose any constants c_1 and c_2 such that $2c_1 + c_2 = 0$, this equation holds for all values of x. Thus there are constants c_1 and c_2 not both zero such that Eq. (2) is satisfied for all x; that is, $\sin 2x$ and $\sin x \cos x$ are linearly dependent functions.

EXAMPLE 2

Show that the functions $f_1 = x$ and $f_2 = x^2$ are linearly independent on the interval $(0, 2)$.

solution. To show that these functions are linearly independent, we shall show that they are not linearly dependent. Thus we consider the linear combination,

$$c_1f_1 + c_2f_2 = c_1x + c_2x^2 = 0.$$

If these functions are linearly dependent, then there are constants c_1 and c_2, not both zero, such that this equation is true for all x in $(0, 2)$. However if we evaluate this equation at any two distinct points in $(0, 2)$, we can conclude that c_1 and c_2 must equal zero and hence x and x^2 are not linearly dependent. For example, using $x = \frac{1}{2}$ and $x = 1$, we get

$$\tfrac{1}{2}c_1 + \tfrac{1}{4}c_2 = 0 \qquad \text{and} \qquad c_1 + c_2 = 0.$$

The only solution for these algebraic equations is $c_1 = c_2 = 0$.

The relationship between linear independence of two solutions and the condition $W(x) \neq 0$ is given in the following theorem.

theorem

If y_1 and y_2 are two solutions of

$$y'' + 2by' + cy = 0 \tag{3}$$

for which $W(x) \neq 0$, then y_1 and y_2 are linearly independent functions. Conversely if y_1 and y_2 are two solutions that are linearly independent, then $W(x) \neq 0$.

The proof of this theorem is discussed in Exercise 4.

We can now restate our definition of a general solution: A general solution of the second-order DE (3) is a linear combination of two linearly independent solutions.

Exercises

Use the definition in the text to establish the results of Exercises 1 and 2.

1. Show that the following pairs of functions are linearly independent on any interval (a, b).
   (a) $e^{a_1 x}, e^{a_2 x}$, a_1 and a_2 constant, with $a_1 \neq a_2$
   (b) x^n, x^{n+1}, n a positive integer (c) $\sin 2x, \cos 2x$
   (d) $e^{\alpha x} \sin \beta x, e^{\alpha x} \cos \beta x$, $\beta \neq 0$

2. Are the following functions linearly independent or linearly dependent?
   (a) $y_1 = x, y_2 = a^{\log_a x}$, for $x > 0$
   (b) $y_1 = \ln x, y_2 = \ln x^2$, for $x > 0$
   (c) $\sin (x - \pi/2), \cos x$
   *(d) $P_1(x)e^{a_1 x}, P_2(x)e^{a_2 x}$, where $a_1 \neq a_2$ and P_1 and P_2 are arbitrary polynomials (not identically zero)
   (e) $\sin x, \sin 3x$.

3. Let y_1 and y_2 be two solutions of $y'' + 2by' + cy = 0$. (a) Prove that if y_1 and y_2 are both zero at the same point, they cannot be linearly independent. (b) Prove that if y_1 and y_2 have a minimum or a maximum value at the same point, then they are linearly dependent.

*4. Prove the theorem in the text by first considering the two solutions y_1 and y_2 for which $W \neq 0$. Then study the algebraic equation

$$c_1 y_1(x) + c_2 y_2(x) = 0$$

and its derivative

$$c_1 y_1'(x) + c_2 y_2'(x) = 0.$$

From $W(x) \neq 0$ for all x, conclude that $c_1 = c_2 = 0$. Prove the converse by contradiction. Thus assume that the Wronskian of the linearly independent solutions is zero at some point x_0. Show that this implies that the above algebraic equations have a nonzero solution $c_1 = A$ and $c_2 = B$ at x_0 in addition to the zero solution. Show that $f(x) = Ay_1 + By_2$ satisfies the IVP

$$f'' + 2bf' + cf = 0, \qquad f(x_0) = f'(x_0) = 0.$$

Verify that the uniqueness theorem of Section 4.6 implies the contradiction.

5. We can define the Wronskian for any two differentiable functions f_1 and f_2 (not necessarily for solutions of a DE) by Eq. (13) of Section 4.5; that is, $W(x) = f_1 f_2' - f_2 f_1'$. Find the Wronskian for the following functions.
   (a) $f_1 = x, f_2 = 1/x$
   (b) $f_1 = \sin (x - \pi/8), f_2 = \cos (x - \pi/8)$
   (c) $f_1 = 1/x, f_2 = e^{1/x}$
   (d) $y_1 = e^{-(x^2/2)}, e^{-(x^2/2)} \int_{x_0}^{x} e^{z^2/2}\, dz$

6. Prove that if $f_1(x)$ and $f_2(x)$ are linearly dependent and differentiable functions

(not necessarily solutions of a DE) on an interval (α, β), then the Wronskian of these functions is zero for every value of x in the interval.

*7. Use the functions $f_1 = x$, $f_2 = x^2$ to show that the statement "if f_1 and f_2 are two linearly independent functions on an interval (a, b) then their Wronskian $\neq 0$ for all x in (a, b)" is not true in general. This example does not contradict the theorem in the text because this theorem requires that the functions in question be solutions of the DE $y'' + 2by' + cy = 0$. We can conclude instead that x and x^2 cannot both be solutions of this DE in any interval containing $x = 0$.

*4.6 A UNIQUENESS THEOREM

In the previous sections we have shown how to solve IVPs by first constructing a general solution and then determining the constants in the general solution so that it satisfies the initial conditions. We shall now prove that this gives the only solution of the IVP. In the applications of DEs it is important to establish whether or not an IVP has a unique solution. If the physical problem for which the IVP is a mathematical model has a unique solution and this IVP does *not* have a unique solution, or vice versa, then the validity of the mathematical model must be seriously questioned.

theorem

The IVP

$$y'' + 2by' + cy = 0, \tag{1}$$

$$y(x_0) = a_0,\ y'(x_0) = a_1 \tag{2}$$

has at most one solution.

proof. We shall use the method of contradiction to prove the result. We assume that the IVP has two distinct solutions, $y = z_1(x)$ and $y = z_2(x)$. Thus z_1 and z_2 satisfy the IVPs

$$z_1'' + 2bz_1' + cz_1 = 0, \qquad z_1(x_0) = a_0,\ z_1'(x_0) = a_1, \tag{3}$$

$$z_2'' + 2bz_2' + cz_2 = 0, \qquad z_2(x_0) = a_0,\ z_2'(x_0) = a_1. \tag{4}$$

We subtract the DE in Eq. (3) from the DE in Eq. (4) and obtain

$$(z_2 - z_1)'' + 2b(z_2 - z_1)' + c(z_2 - z_1) = 0. \tag{5}$$

If we define the function $z(x)$ by

$$z \equiv z_2 - z_1, \tag{6}$$

then Eq. (5) is reduced to

$$z'' + 2bz' + cz = 0. \tag{7}$$

Similarly, by subtracting the initial conditions in Eqs. (3) and (4) and using Eq. (6), we get

$$z(x_0) = z_2(x_0) - z_1(x_0) = 0,$$
$$z'(x_0) = z_2'(x_0) - z_1'(x_0) = 0 \tag{8}$$

as the initial conditions for z. Thus $z(x)$ is a solution of the IVP given by Eqs. (7) and (8).

It is clear that $z = 0$ for all x is a solution. We will now show that it is the only solution. For simplicity of presentation we shall prove the theorem only for $b \geq 0$ and $c \geq 0$. The completion of the proof is outlined in Exercise 1. We multiply both sides of the DE in Eq. (7) by z'. This gives

$$z'z'' + 2b(z')^2 + cz'z = 0.$$

Since $\frac{1}{2}(z'^2)' = z'z''$ and $\frac{1}{2}(z^2)' = zz'$, this equation is equivalent to

$$\tfrac{1}{2}(z'^2 + cz^2)' + 2b(z')^2 = 0.$$

We integrate both sides of this equation from x_0 to x and obtain

$$\tfrac{1}{2}[z'^2(x) - z'^2(x_0)] + \frac{c}{2}[z^2(x) - z^2(x_0)] + 2b\int_{x_0}^{x} z'^2(s)\,ds = 0.$$

Since z satisfies the initial conditions given in Eq. (8), this equation simplifies to

$$z'^2 + cz^2 + 4b\int_{x_0}^{x} z'^2(s)\,ds = 0. \tag{9}$$

The left side of Eq. (9) is the sum of three nonnegative terms. Consequently, Eq. (9) implies that each of these terms must be zero. In particular we have $z' \equiv 0$, which implies that $z =$ constant. However the initial condition $z(x_0) = 0$ shows that this constant must be equal to zero and hence $z = 0$ for all x. The definition of z given in Eq. (6) implies that $z_1 = z_2$ for all values of x. This contradicts the assumption that z_1 and z_2 are distinct solutions and proves the theorem for $b \geq 0$ and $c \geq 0$.

Exercises

1. Complete the proof of the theorem by establishing the following lemma.

lemma

If $z(x)$ is a solution of the IVP

$$z'' + 2bz' + cz = 0, \qquad z(x_0) = z'(x_0) = 0,$$

then there is a constant k such that the function

$$v(x) = e^{-kx}z(x)$$

is a solution of the IVP

$$v'' + 2b_0v' + c_0v = 0, \qquad v(x_0) = v'(x_0) = 0,$$

where the constants b_0 and c_0 are positive.

2. Prove the theorem for the special case when $b = 0$ and $c \leq 0$. (*Hint:* First show that the difference z of two solutions of the IVP in Eqs. (1) and (2) satisfies

$$(z' - kz)(z' + kz) = 0, \qquad z(x_0) = 0,$$

where $c = -k^2$. Then deduce from this equation that $z = 0$ for all x.)

3. If $b \neq 0$, show by making the transformation of independent variables $z(x) = e^{-bx}Z(x)$ from z to Z that the DE in Eq. (7) can always be transformed to $Z'' + AZ = 0$, where A is a constant. Use this result, Exercise 2, and the result proved in the text to establish the theorem for any c and any $b \neq 0$.

4. Use the method of this section to prove that the IVP, $y' + q(x)y = 0$, $y(x_0) = a_0$, where $q(x)$ is a nonnegative continuous function, has no more than one solution.

chapter 5

APPLICATIONS OF THE SECOND-ORDER HOMOGENEOUS EQUATION

5.1 GENERAL PROPERTIES OF THE SOLUTIONS AND APPLICATIONS

In Chapter 4 we have observed that the nature of the solutions of

$$y'' + 2by' + cy = 0 \tag{1}$$

depends on the roots m_1 and m_2 of the characteristic equation $m^2 + 2bm + c = 0$. This is summarized in Table 5.1. We observe that in Case II we have already obtained two forms for a general solution. They are given by Eqs. (i) and (ii) in Table 5.1 To interpret the answers to many problems corresponding to Case II it is convenient to transform the general solution,

$$y = e^{-bx}[c_1 \cos \omega x + c_2 \sin \omega x], \qquad \omega \equiv (c - b^2)^{1/2}, \tag{2}$$

into either of the forms given by Eqs. (iii) or (iv) of Table 5.1. They are discussed in the following theorem.

TABLE 5.1 Solutions of the DE $y'' + 2by' + cy = 0$

Case Number	$b^2 - c$	*Roots*		*General Solution*
I	>0	$m_1 = -b + \sqrt{b^2 - c}$ $m_2 = -b - \sqrt{b^2 - c}$	real and distinct	$y = c_1 e^{m_1 x} + c_2 e^{m_2 x}$
II	<0	$m_1 = -b + i\sqrt{c - b^2}$ $m_2 = -b - i\sqrt{c - b^2}$	complex conjugate	(i) $y = c_1 e^{m_1 x} + c_2 e^{m_2 x}$ or (ii) $y = e^{-bx}[c_1 \cos(\sqrt{c - b^2}x) + c_2 \sin(\sqrt{c - b^2}x)]$ or (iii) $y = Ae^{-bx} \cos(\sqrt{c - b^2}x - \phi)$ or (iv) $y = Ae^{-bx} \sin(\sqrt{c - b^2}x - \psi)$
III	$=0$	$m_1 = m_2 = -b$,	real and equal	$y = c_1 e^{-bx} + c_2 x e^{-bx}$

theorem

Two other forms of general solutions of DE (1) in Case II are

$$y = Ae^{-bx} \cos(\omega x - \phi) \tag{3}$$

$$y = Ae^{-bx} \sin(\omega x - \psi). \tag{4}$$

The arbitrary constants A, ϕ, and ψ are defined by

$$A = (c_1^2 + c_2^2)^{1/2},$$

$$\phi = \tan^{-1} \frac{c_2}{c_1},$$

$$\psi = -\tan^{-1} \frac{c_1}{c_2} = \phi - \pi/2, \tag{5}$$

where the principal value (between $-\pi/2$ and $\pi/2$) of the arc tangent function is used.

Before proving the theorem we give a geometric interpretation of A and ϕ. It is apparent if we consider the c_1, c_2 plane, see Figure 5.1, that any point P in this plane represents possible values of the pair, c_1 and c_2. The abscissa and ordinate of P give c_1 and c_2, respectively. The length of the radius from the origin to P is

$$A = (c_1^2 + c_2^2)^{1/2}.$$

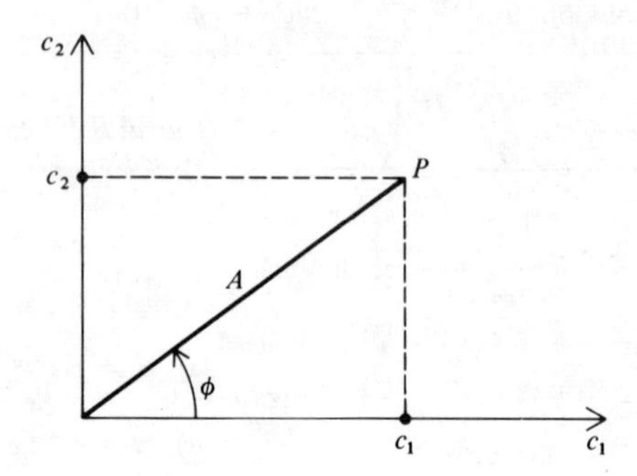

Figure 5.1

Furthermore, for the angle ϕ between the radius and the c_1 axis we have, $\tan \phi = c_2/c_1$. Since $\tan \phi$ is a periodic function, there are infinitely many values of ϕ that satisfy $\tan \phi = c_2/c_1$. Thus for a given value of c_2/c_1 there are an infinite number of values of $\phi = \tan^{-1} c_2/c_1$. We pick the value of ϕ which lies between $-\pi/2$ and $\pi/2$. This is called the principal value of the function $\tan^{-1} c_2/c_1$. It is clear from this discussion and the figure that the constants c_1 and c_2 are related to the constants A and ϕ by,

$$c_1 = A \cos \phi \qquad \text{and} \qquad c_2 = A \sin \phi. \tag{6}$$

proof. We insert the expressions given in Eq. (6) for c_1 and c_2 into the general solution given by Eq. (2). This gives

$$y = Ae^{-bx}[\cos \phi \cos \omega x + \sin \phi \sin \omega x]. \tag{7}$$

This expression can be further simplified by substituting $s = \omega x$ and $r = \phi$ in the elementary trigonometric identity

$$\cos (s - r) = \cos r \cos s + \sin r \sin s.$$

This gives $\cos (\omega x - \phi) = \cos \phi \cos \omega x + \sin \phi \sin \omega x$. Consequently, Eq. (7) is reduced to Eq. (3). A similar analysis can be used to establish the general solution given by Eq. (4).

The quantities A and ϕ in Eq. (3) are arbitrary constants if c_1 and c_2 are arbitrary constants. If c_1 and c_2 are specific numbers, then A and ϕ are evaluated by using Eq. (5). If either Eq. (3) or Eq. (4) is used as a general solution to solve an IVP, then the constants A and ϕ or A and ψ are determined from the initial conditions.

We shall now interpret the general solution given by Eq. (3). The quantity Ae^{-bx} in Eq. (3) is called the *amplitude* of the solution. If $b > 0$ (<0), then the amplitude exponentially decreases (increases) as x increases. If $b = 0$, then the amplitude is constant. In physical applications $b > 0$ usually corresponds to damping of the solution due to "friction" or, more generally, to some dissipation of energy. The "wave factor" $\cos(\omega x - \phi)$ is a periodic function of x; see Figure 5.2. The quantity ϕ is called the *phase angle* or *phase shift*. The fundamental period P of $\cos(\omega x - \phi)$ is

$$P = \frac{2\pi}{\omega}.$$

It is the distance between two consecutive maxima or two consecutive minima. Furthermore, $\omega = 2\pi/P$ is called the *circular frequency* of the periodic function. As we see in Figure 5.2, the distance of the nearest maximum to $x = 0$ is equal to ϕ/ω. This is an interpretation of the phase angle.

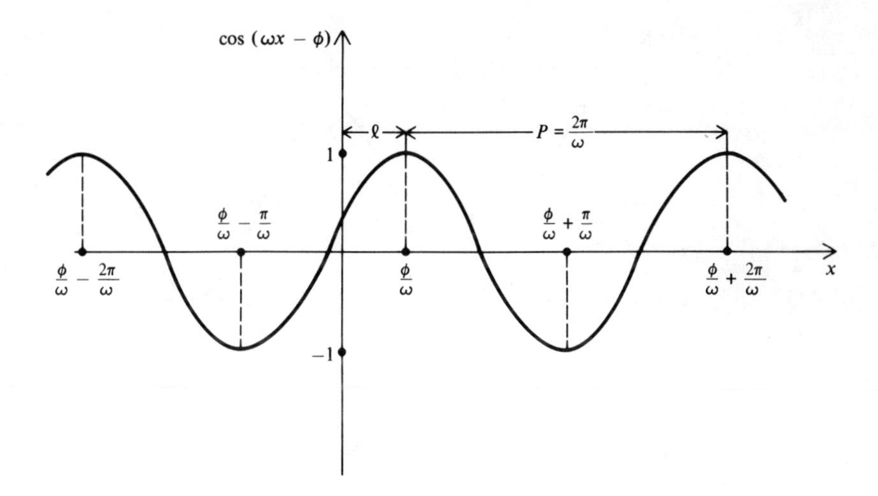

Figure 5.2 The wave factor $\cos(\omega x - \phi)$ is drawn for $\phi > 0$.

The general solution, Eq. (3), is the product of the exponentially varying amplitude and the periodic wave factor. We refer to it as an amplitude–phase angle form of a general solution. If $b > 0$, as in Figure 5.3a, then the solution is a damped oscillating function as $x \to +\infty$. Similarly, if $b < 0$, as in Figure 5.3c, the solution is an oscillating function whose amplitude becomes unbounded as $x \to \infty$. If $b = 0$ as in Figure 5.3b, then the solution is periodic, of period $2\pi/\omega$ and constant amplitude A. It is frequently called *simple harmonic motion*. If $b = 0$ for Case II, we observe that m_1 and m_2 are pure imaginary numbers.

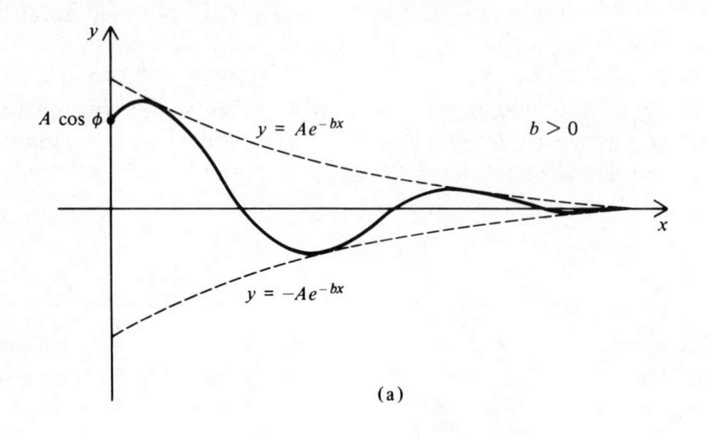

(a)

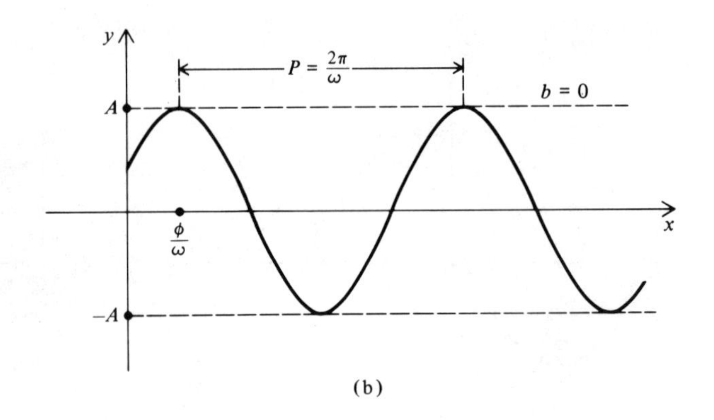

(b)

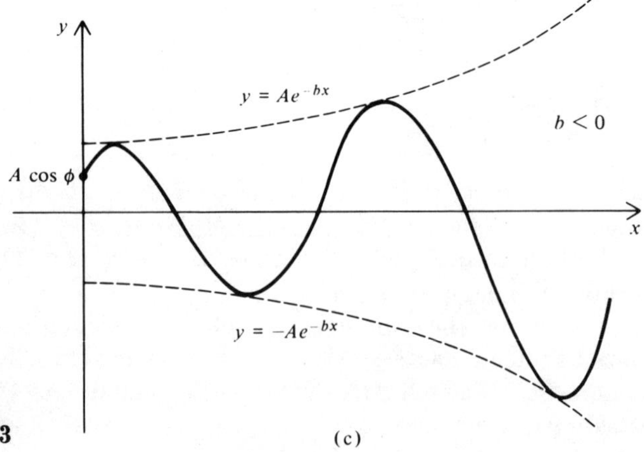

(c)

Figure 5.3

EXAMPLE

Determine an amplitude–phase angle form of a general solution of the DE

$$y'' + 4y = 0.$$

solution. The roots of the characteristic equation are $m_1 = i2$ and $m_2 = -i2$. Hence, e^{i2x} and e^{-i2x} are complex-valued exponential solutions. However, we prefer to use a real-valued form for the solution. Since $b = 0$ and $\omega = \sqrt{c - b^2} = 2$, we have the general solution

$$y = c_1 \cos 2x + c_2 \sin 2x. \tag{8}$$

We use the preceding theorem to transform the general solution given in Eq. (8), into the alternative form,

$$y = A \cos (2x - \phi). \tag{9}$$

This equation can also be obtained directly from Eq. (8) by using the trigonometric identity

$$\cos (2x - \phi) = \cos \phi \cos 2x + \sin \phi \sin 2x \tag{10}$$

and proceeding as in the proof of the theorem. To do this we rewrite Eq. (8) as

$$y = \sqrt{c_1^2 + c_2^2}\left[\frac{c_1}{\sqrt{c_1^2 + c_2^2}} \cos 2x + \frac{c_2}{\sqrt{c_1^2 + c_2^2}} \sin 2x\right] \tag{11}$$

Choosing

$$A = \sqrt{c_1^2 + c_2^2}, \qquad \cos \phi = \frac{c_1}{\sqrt{c_1^2 + c_2^2}}, \qquad \sin \phi = \frac{c_2}{\sqrt{c_1^2 + c_2^2}},$$

and using Eq. (10), we find that

$$y = A(\cos \phi \cos 2x + \sin \phi \sin 2x) = A \cos (2x - \phi).$$

Exercises

1. Solve the following IVPs and discuss the behavior of the solutions as $x \to \infty$.
   (a) $y'' - 4y' + 3y = 0, \quad y(0) = 0,\ y'(0) = 10$
   (b) $y'' + y' - 6y = 0, \quad y(0) = 1,\ y'(0) = -2$
   (c) $y'' + y' - 6y = 0, \quad y(0) = a_0,\ y'(0) = -2,$
   where a_0 is any constant
   (d) $y'' - 3y' = 0, \quad y(0) = 1,\ y'(0) = a_1,$ where a_1 is any constant
   (e) $y'' + 4y = 0, \quad y(0) = 0,\ y'(0) = 1$
   (f) $y'' + 6y' + 10y = 0, \quad y(0) = 1,\ y'(0) = 1$

2. Determine an amplitude–phase angle form of the following functions. Give the values of the amplitudes, the phase angles, the periods, and the circular frequencies.

(a) $y = e^{-x}(\cos x + 3 \sin x)$ (b) $y = \cos 2x + 6 \sin 2x$

(c) $y = e^{-3x}\left(\cos \frac{x}{2} + \sin \frac{x}{2}\right)$

(d) $y = Be^{-bx}(\cos \omega x + \sin \omega x)$, for all real-values of the constants B, b, and ω

3. Express the solutions of the following IVPs in amplitude–phase angle form. (*Hint:* First solve the IVPs using the trigonometric form of the general solution, as given in Eq. (2). Then use the theorem to transform the resulting solution to the required form.)
(a) $y'' + y = 0, \quad y(0) = 0, y'(0) = 3$
(b) $y'' - 4y' + 29y = 0, \quad y(0) = 1, y'(0) = 0$
(c) $y'' - 8y' + 25y = 0, \quad y(0) = 6, y'(0) = 48$

4. Solve the IVP $y'' + 2y' + 2y = 0, \quad y(0) = 1, \ y'(0) = 2$ by using the amplitude–phase angle form of the general solution directly.

5. Deduce the general solution, Eq. (4), from the trigonometric form, Eq. (2), of a general solution in Case II.

6. Discuss the behavior of the solutions of the IVP

$$y'' = 0, \qquad y(x_0) = a_0, \ y'(x_0) = a_1$$

as $x \to \infty$.

7. Use a general solution to show that if the real parts of the characteristic roots of DE (1) are both positive, then all solutions grow exponentially as $x \to \infty$.

8. Use a general solution to show that if the real parts of the characteristic roots of DE (1) are both negative, then all solutions decay exponentially as $x \to \infty$. In this case the DE is called asymptotically stable.

9. Determine the initial values a_0 and a_1 for which the solutions of the DE decay to zero as $x \to \infty$ if the roots of the characteristic equation satisfy the inequalities $m_1 > 0$, $m_2 < 0$. Determine a_0 and a_1 for which the solutions grow exponentially as $x \to \infty$.

10. If the coefficient $c = 0$ in DE (1), then $m_1 = 0$. Show that if $b > 0$ and $y'(0) \neq 0$, then all solutions of the DE decay to a constant value as $x \to \infty$. What happens if $y'(0) = 0$?

11. Show by making the transformation of the independent variable x, $t = x - x_0$, that the IVP

$$y'' + 2by' + cy = 0, \qquad y(x_0) = a_0, \ y'(x_0) = a_1$$

can always be reduced to the IVP consisting of the DE and the new initial condition $y(0) = b_0$, $y'(0) = b_1$ at the initial point $t = 0$. Determine the relations between a_0, a_1 and b_0, b_1.

5.2 POPULATION DYNAMICS: A PREDATOR-PREY PROBLEM

In Section 2.4-1 we studied a simple model for the growth and decay of a population of a single species. In many applications of population dynamics the primary interest is to study the interaction of two or more species. In fact, the principal concern of the science of ecology is the mutual interaction between organisms and their environment. A typical type of species interaction is the competition of different species for the same food sources. The interaction may be detrimental to some or all of the species or it may be beneficial to all of the species (symbiosis). A second common type of species interaction is the predator-prey relationship, in which one or more species are the prey for other species.

The mathematical study of predator-prey problems was initiated by the Italian mathematician V. Volterra. His son-in-law, the zoologist D'Ancona, was studying the records of the fish catches in the upper Adriatic Sea. He found that the relative numbers of sharks and soles that were caught varied with time. The food sources for soles are other aquatic organisms, which are available in essentially unlimited quantities. However, the soles are a food source for the sharks. Thus, the soles are the prey and the sharks are the predators. Another example of the predator-prey interaction is the problem of the lynx (predator) and the snowshoe rabbit (prey) of northern Canada. In this case, studies have been made using records of the number of pelts received by the Hudson Bay Company.

In order to derive a mathematical model for describing the predator-prey interaction, we denote the number of prey at time t by $y(t)$ and the number of predators by $z(t)$. The rates of change of these two species are y' and z', respectively. In the simple one-species population dynamics model that we studied in Section 2.4-1, we assumed that y' was proportional to y; that is, $y' = a_1 y$. The constant of proportionality, a_1, is equal to the difference between the birth and death rates of the prey. However, the presence of the predator will reduce the growth rate y' of the prey. Specifically, we assume that

$$y' = a_1 y - a_2 z. \tag{1}$$

The constant a_2 is positive because an increase in the number of predators should decrease the number of prey.

Similarly, we assume that z' is given by

$$z' = a_3 y - a_4 z. \tag{2}$$

In Eq. (2), $a_3 > 0$ is the rate of increase of predators because of the presence of the prey as a food source and $a_4 > 0$ is the difference between the natural death and birth rates of the predators when the prey are absent. We take

$a_4 > 0$ because in the absence of their food source, the predators would die out.

We observe that if $a_2 = a_3 = 0$, then DEs (1) and (2) are reduced to $y' = a_1 y$ and $z' = -a_4 z$, respectively. This means that y and z can be determined independently of each other; that is, there is no interaction between predator and prey. Each species grows or decays as if the other were absent. Thus a_2 and a_3 are interaction coefficients. If a_2 and a_3 are small (large), then there is weak (strong) interaction between the species.

We assume that at some initial time, which we take as $t = 0$, the numbers of predators and prey are known. Thus we have

$$y(0) = y_0 \qquad \text{and} \qquad z(0) = z_0 \tag{3}$$

Equations (1), (2), and (3) form an IVP for a system of two first-order DEs with constant coefficients. The variations of the predator and prey populations for $t > 0$ are determined by solving this problem, as we show in the next example.

EXAMPLE

In an isolated marine intertidal community, it was observed that a species of barnacles was a prey for a species of starfish. Suppose that from a series of observations, it was estimated that the rate coefficients are

$$a_1 = 2, a_2 = 5, a_3 = 4, a_4 = 2.$$

The units of the coefficients are numbers of members per unit time, per member. The initial value for the barnacles is $y_0 = 500$, and the initial value for the starfish is $z_0 = 100$. Determine the population size of the predators and prey at any $t > 0$.

solution. We must solve the IVP

$$y' = 2y - 5z, \tag{4}$$

$$z' = 4y - 2z, \tag{5}$$

$$y(0) = 500, z(0) = 100.$$

First we reduce this system of two differential equations to a single second-order DE for y by eliminating z from Eqs. (4) and (5). We solve Eq. (4) for z. This gives

$$z = \frac{2y}{5} - \frac{y'}{5}. \tag{6}$$

We substitute this expression for z and its derivative into DE (5) and get

$$y'' + 16y = 0.$$

A general solution of this DE is

$$y = c_1 \cos 4t + c_2 \sin 4t. \tag{7}$$

We determine z by inserting Eq. (7) into Eq. (6). This gives

$$z = \frac{2}{5}(c_1 - 2c_2)\cos 4t + \frac{2}{5}(c_2 + 2c_1)\sin 4t. \tag{8}$$

The constants c_1 and c_2 are determined by substituting $x = 0$ into Eqs. (7) and (8) for y and z, and then using the initial conditions. This gives

$$y(0) = c_1 = 500, \qquad z(0) = \frac{2}{5}(c_1 - 2c_2) = 100$$

The solution of these algebraic equations is $c_1 = 500$ and $c_2 = 500/4$. Substituting these values of c_1 and c_2 into Eqs. (7) and (8), we get the solution of the IVP as

$$y = 500\cos 4t + \frac{500}{4}\sin 4t$$

and

$$z = 100\cos 4t + \frac{900}{2}\sin 4t.$$

To interpret this result we write the solution in amplitude–phase angle form as

$$y = A_1\cos(4t - \phi_1) \quad \text{and} \quad z = A_2\cos(4t - \phi_2), \tag{9}$$

where the amplitudes A_1 and A_2 of the prey and predators are

$$A_1 = \left[(500)^2 + \frac{(500)^2}{16}\right]^{1/2} = 125\sqrt{17},$$

$$A_2 = \left[(100)^2 + \frac{(900)^2}{4}\right]^{1/2} = 50\sqrt{85},$$

and the phase angles are

$$\phi_1 = \tan^{-1}\frac{500/4}{500} = \tan^{-1}\frac{1}{4} \approx 14.5°,$$

$$\phi_2 = \tan^{-1}\frac{900/2}{100} = \tan^{-1}\frac{9}{2} \approx 77.5°.$$

Thus the numbers of predators and prey vary periodically with fundamental period $P = 2\pi/4 = \pi/2$. The phase angles differ considerably. A sketch of the solution is given in Figure 5.4. The number of prey reach their first maximum, $y = A_1$ at $t = \phi_1/4$, shortly after the initial instant. Then the increase in the predators diminish the number of prey until the prey population is so small that the food supply for the predators is seriously diminished. Thus the predators reach a maximum and then they must decrease. Consequently the predators reach their maximum after the prey have achieved a maximum. The difference in the phase angles $\phi_2 - \phi_1 \approx 63° \approx 1.1$ rad is proportional to the time lag of the predators' maximum with respect to the prey's maximum.

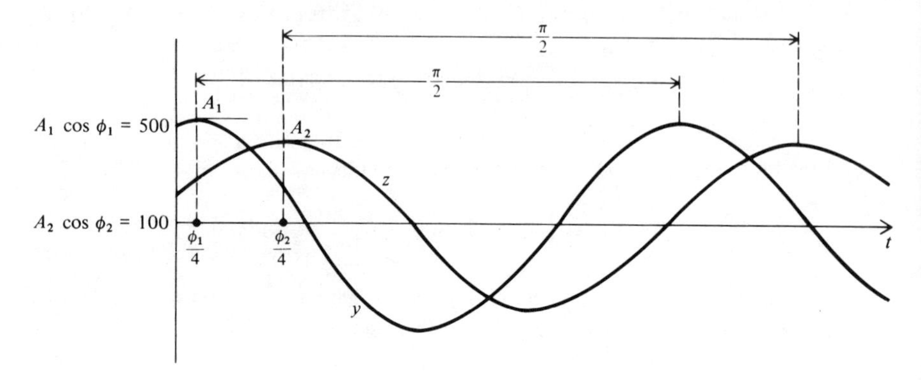

Figure 5.4

Equation (9) shows that $y < 0$ for all t for which $\cos(4t - \phi_1) < 0$. Since the number of prey y cannot be negative, this suggests that the mathematical model given by Eqs. (1), (2), and (3) is not sufficiently accurate to describe the interaction of the two populations except possibly for very small values of t. In a more sophisticated mathematical model of the predator-prey interaction, which is called the Volterra-Lotka theory, the differential equations are nonlinear and these difficulties do not occur. (See Exercise 3.)

Exercises

1. Suppose that for the marine community discussed in the example, the rate coefficients are

$$a_1 = 2,\ a_2 = 2,\ a_3 = 4,\ a_4 = 2,$$

and the initial numbers of prey and predators are $y_0 = 5000$ and $z_0 = 100$, respectively. Determine the variation in time of the prey and predator populations. If the variation is periodic, determine the circular frequency and the phase angle. What are the smallest values of t at which the number of prey and the number of predators is zero?

2. Suppose that for the marine community discussed in the example, the rate coefficients are

$$a_1 = 10,\ a_2 = 6,\ a_3 = 1316,\ a_4 = 4.$$

The initial numbers of prey and predators are $y(0) = y_0$ and $z(0) = z_0$, respectively. Determine the population size of the predators and prey at any $t > 0$. What are the population sizes as $t \to \infty$? Are the results physically meaningful? For what choices of y_0 and z_0 do the populations approach zero as $t \to \infty$? For what choices are they positive for all $t \geq 0$?

*3. The original theory describing the predator-prey interaction of two species was derived and studied independently by V. Volterra and the biologist A. Lotka.

The theory was derived from the following considerations. If each of the two species existed alone, we assume that they would obey the simple growth and decay model of Section 2.4-1. That is,

$$y' = k_1 y \qquad \text{and} \qquad z' = k_2 z,$$

where k_1 and k_2 are the differences in the birth and death rates of the prey and predator, respectively. However, if the species interact, that is, the predators eat the prey, then k_1 will diminish and k_2 will increase as a result of the interaction. The more numerous the predators the faster will k_1 diminish, and the more numerous the prey the faster will k_2 increase. A simple approximation of this fact is to assume that

$$k_1 = (\alpha_1 - \alpha_2 z) \qquad \text{and} \qquad k_2 = (-\beta_1 + \beta_2 y),$$

where the constants α_1, α_2, β_1, and β_2 are positive. Inserting this into the DEs, we get

$$y' = (\alpha_1 - \alpha_2 z)y \qquad \text{and} \qquad z' = (-\beta_1 + \beta_2 y)z. \tag{i}$$

These are the Volterra-Lotka equations. They are a coupled pair of nonlinear DEs because of the product term yz that appears in both equations.

A solution (Y_0, Z_0) of Eq. (i) is an equilibrium solution (rest point) if Y_0 and Z_0 are constants for all t; that is, $Y_0' = Z_0' = 0$. Thus all the equilibrium solutions are determined by setting the right sides of the DEs in (i) equal to zero. Show that there are exactly two equilibrium solutions: $Y_0 = Z_0 = 0$ and $Y_0 = \beta_1/\beta_2$, $Z_0 = \alpha_1/\alpha_2$.

A common practice in obtaining approximations to solutions of nonlinear equations is to seek solutions near an equilibrium solution by "linearizing" the nonlinear DEs. In particular, we shall consider the equilibrium solution $Y_0 = \beta_1/\beta_2$ and $Z_0 = \alpha_1/\alpha_2$. Thus we seek solutions in the form

$$y = \frac{\beta_1}{\beta_2} + \bar{y} \qquad \text{and} \qquad z = \frac{\alpha_1}{\alpha_2} + \bar{z}$$

where $\bar{y}$ and $\bar{z}$ are "small" quantities to be determined. Substituting this into the nonlinear DEs given by Eq. (i) we obtain

$$\bar{y}' = -\frac{\alpha_2\beta_1}{\beta_2}\bar{z} - \alpha_2\bar{z}\bar{y} \qquad \text{and} \qquad \bar{z}' = \frac{\beta_2\alpha_1}{\alpha_2}\bar{y} + \beta_2\bar{y}\bar{z}. \tag{ii}$$

If $\bar{y}$ and $\bar{z}$ are small, then $\bar{y}\bar{z}$ is small compared to either $\bar{y}$ or $\bar{z}$. Thus the last term on the right sides of both the DEs in Eq. (ii) are negligible, and we get the approximate linear DEs

$$\bar{y}' = -\frac{\alpha_2\beta_1}{\beta_2}\bar{z} \qquad \text{and} \qquad \bar{z}' = \frac{\beta_2\alpha_1}{\alpha_2}\bar{y}.$$

Solve these DEs by eliminating either the $\bar{y}$ or the $\bar{z}$ variable. Sketch this solution. Interpret the results in terms of the predator-prey interaction.

*4. Linearize the nonlinear Volterra-Lotka equations in the previous problem about the equilibrium solution $Y_0 = Z_0 = 0$. Solve the resulting linear DEs. Sketch the solution.

*5. We consider two species that live in the same environment and compete for the same food resources, but one is not a predator of the other. If $y(t)$ and $z(t)$ are the number of individuals in each of the two species, then Volterra proposed the following mathematical model for competition:

$$y' = [\alpha_1 - \alpha_2(y + z)]y$$
$$z' = [\beta_1 - \beta_2(y + z)]z$$

where the α's and β's are positive constants. Determine the equilibrium solutions for these equations if $A = \beta_2\alpha_1 - \alpha_2\beta_1 \neq 0$. Linearize the DEs about each of the equilibrium solutions. Solve the resulting linear DEs when $\alpha_1 = 2$, $\alpha_2 = \beta_1 = \beta_2 = 1$. Interpret the solutions as $t \to \infty$. An equilibrium solution is stable if all solutions of the linearized problem are bounded as $t \to \infty$. It is unstable if the linearized solution is unbounded. Which equilibrium solutions are unstable?

5.3 THE VIBRATING PENDULUM

The differential equation

$$\theta'' + \frac{g}{l}\theta = 0 \tag{1}$$

for the small-amplitude oscillations of the simple pendulum is derived in Appendix I.2. Here g is the gravitational constant, l is the hanging length of the pendulum, and $\theta(t)$ is the angle that the pendulum makes with the vertical direction, measured in radians; see Figure 5.5. The independent variable is the time t. Equation (1) is obtained by applying Newton's second law to the pendulum mass. We assume that at $t = 0$ the pendulum is given an initial angular displacement θ_0 and an initial angular velocity θ_1. This gives the initial conditions,

$$\theta(0) = \theta_0, \theta'(0) = \theta_1. \tag{2}$$

We determine the subsequent motion of the pendulum by solving the IVP given by Eqs. (1) and (2).

The roots of the characteristic equation of the DE are the complex numbers, $m_1 = i(g/l)^{1/2}$ and $m_2 = -i(g/l)^{1/2}$. A general solution is given by

$$\theta = c_1 \cos \omega t + c_2 \sin \omega t, \qquad \omega = \left(\frac{g}{l}\right)^{1/2}. \tag{3}$$

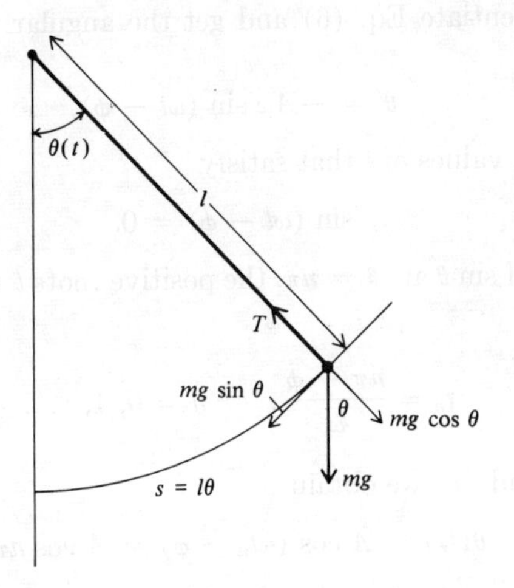

Figure 5.5

The constants c_1 and c_2 are determined in the usual way by requiring that this general solution satisfy the initial conditions. This yields

$$\theta(0) = c_1 = \theta_0 \quad \text{and} \quad \theta'(0) = \omega c_2 = \theta_1.$$

Solving these for c_1 and c_2 and substituting the results in Eq. (3), we obtain the solution of the IVP as

$$\theta = \theta_0 \cos \omega t + \frac{\theta_1}{\omega} \sin \omega t. \tag{4}$$

In order to study the qualitative features of the solution, it is convenient to transform it into the amplitude–phase angle form. From the definitions of A and ϕ in Eq. (9) of Section 5.1, we have

$$A = \left[\theta_0^2 + \left(\frac{\theta_1}{\omega}\right)^2\right]^{1/2} \quad \text{and} \quad \phi = \tan^{-1}\left(\frac{\theta_1}{\omega\theta_0}\right). \tag{5}$$

The solution given in Eq. (4) can be rewritten as

$$\theta(t) = A \cos(\omega t - \phi). \tag{6}$$

This is a simple harmonic motion. It has the period

$$P = \frac{2\pi}{\omega} = 2\pi\sqrt{\frac{l}{g}}.$$

The period increases as the square root of the length of the pendulum. This means, for example, that a longer pendulum requires more time to return to its initial position. Since $|\cos(\omega t - \phi)| \leq 1$, the values of θ always lie in the interval $-A \leq \theta(t) \leq A$.

We differentiate Eq. (6) and get the angular velocity of the pendulum as

$$\theta' = -A\omega \sin(\omega t - \phi).$$

It is zero for the values of t that satisfy

$$\sin(\omega t - \phi) = 0.$$

Since the zeros of $\sin\theta$ are $\theta = n\pi$, the positive roots $t = t_n$ of this equation are

$$t_n = \frac{n\pi + \phi}{\omega}, \qquad n = 0, 1, \ldots. \tag{7}$$

From Eq. (6) and (7) we obtain

$$\theta(t_n) = A\cos(\omega t_n - \phi) = A\cos n\pi.$$

Consequently we get $|\theta(t_n)| = A$. Thus the angular velocity of the pendulum is zero when the pendulum reaches its maximum amplitude. This is physically reasonable because the pendulum must change its direction of motion when it reaches its maximum amplitude A. Another interpretation of this observation is given in Exercise 5. The simple harmonic motion persists for all time because we have assumed that friction is not present. The effects of frictional forces on the motion are discussed in the next section.

EXAMPLE

A pendulum of length $l = 2$ ft is given no initial angular displacement and an initial angular velocity of 1 rad/sec. Neglecting all frictional effects, determine the subsequent motion.

solution. Since $g = 32$ ft/sec^2 and $l = 2$ ft, $g/l = 16$. The IVP describing the motion is

$$\theta'' + 16\theta = 0 \qquad \theta(0) = 0, \theta'(0) = 1.$$

The characteristic roots are $m_1 = i4$ and $m_2 = -i4$. A general solution is

$$\theta = c_1\cos 4t + c_2\sin 4t. \tag{8}$$

We require that Eq. (8) satisfy the initial conditions. This yields

$$\theta(0) = c_1 = 0 \qquad \text{and} \qquad \theta'(0) = 4c_2 = 1.$$

Therefore $c_1 = 0$ and $c_2 = 1/4$ and the solution of the IVP is

$$\theta = \tfrac{1}{4}\sin 4t.$$

The pendulum executes a simple harmonic motion with a frequency of $\omega = 4$ rad per sec and amplitude $A = \frac{1}{4}$ rad $\approx 14.3°$. The period is $P = 2\pi/\omega = \pi/2$ sec.

Exercises

1. A pendulum of length $l = 2$ ft is given an initial angular displacement of $\theta(0) = 1/10$ rad. It is released with zero initial angular velocity. Neglecting frictional effects, determine the subsequent motion. What are the amplitude, circular frequency, period, and phase angle?

2. (a) A 6-in.-long pendulum is given an initial angular displacement of $1/2$ rad. It is released with an angular velocity $\theta'(0) = 1$ rad/sec. Neglecting frictional effects, determine the subsequent motion. Find the amplitude, circular frequency, period, and phase angle of the motion.
   (b) Where is the pendulum and in which direction is it moving for $t = \pi/8$ sec? For $t = \pi$ sec?

3. Suppose that the pendulum of Exercise 2 has the same initial displacement but an initial angular velocity $\theta'(0) = -1$ rad/sec. (a) Determine the subsequent motion. (b) For what values of t is the mass at $\theta = \sqrt{17}/16$ rad and moving to the right? (c) For what values of t is the pendulum mass at $\theta = \sqrt{17}/16$ rad and moving to the left?

4. A pendulum of length l is started from its rest position of $\theta(0) = 0$ with an initial angular velocity of $\theta'(0) = \theta_1$, where $\theta_1 > 0$. Determine the value of θ_1 so that the maximum angular displacement of the resulting motion does not exceed α rad, where α is a given constant less than $\pi/2$ rad.

5. Multiply the DE $\theta'' + \omega^2\theta = 0$ of the frictionless, vibrating pendulum by θ' and use the identities $\theta'\theta'' = (\theta'^2/2)'$ and $\theta'\theta = (\theta^2/2)'$ to show that

$$\frac{1}{2}(\theta')^2 + \frac{\omega^2}{2}\theta^2 = E = \text{constant}. \tag{i}$$

   The first term on the left side of Eq. (i) is proportional to the kinetic energy of the pendulum because θ' is the angular velocity. The second term is proportional to the potential energy that is stored by the pendulum. Thus Eq. (i) states that the total energy, which is the sum of the kinetic and potential energies, is a constant; that is, it does not vary with time. This is a statement of the law of conservation of energy. Show by using Eq. (i) that the kinetic energy reaches its maximum value when the potential energy is zero, and conversely the potential energy reaches its maximum when the kinetic energy is zero. This is related to the fact that the pendulum reaches its maximum amplitude when the angular velocity is zero. Use the solution given in Eq. (6) to evaluate E.

6. For Exercise 1, determine the kinetic and potential energies and the total energy. Does the law of conservation of energy apply to this problem (see Exercise 5)?

7. A rigid body of mass m is resting on a frictionless plane, as shown in Figure 5.6. The mass is connected to a spring and set in horizontal motion. The spring is unextended (natural state) when the center of the mass is a distance d from

the rigid wall. Let $y(t)$ denote the additional displacement of the mass from its natural position. Assume that the force f exerted by the spring is given by Hooke's law, $f = ky$, where k is the spring constant. This relation is usually accurate only for small displacements of the mass. For large displacements, f and y are related nonlinearly. Apply Newton's second law of motion to the mass. Neglecting friction forces such as air resistance and using Hooke's law, show that the equation of motion is $my'' + ky = 0$.

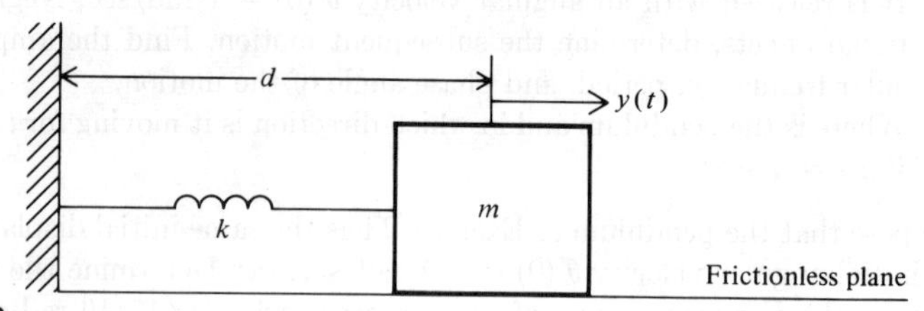

Figure 5.6

8. Assume that the mass in Exercise 7 weighs 3 lb and that the spring constant is $k = 10$ lb/in. The mass is given an initial horizontal displacement of 10 in. and released at zero velocity. Determine the resulting motion, the amplitude, the circular frequency, and the period. Determine the kinetic and potential energies. Does the law of conservation of energy hold?

9. In Exercise 8 the mass is started from its rest position $y(0) = 0$ with an initial velocity of $y'(0) = a_1$. Determine the maximum value of a_1 so that the amplitude of the subsequent motion never exceeds 2 in.

10. Derive the DE for the motion of a mass that hangs vertically from a spring. Assume that the spring obeys Hooke's law (see Exercise 7) and there is no air resistance.

5.4 THE VIBRATING PENDULUM WITH FRICTION

The oscillations of a real pendulum are restrained by frictional forces, such as the mechanical friction at the pivot and the resistance of the medium through which the pendulum is oscillating. These forces cause dissipation of energy and tend to "damp" the pendulum motion. To study these effects mathematically, we assume that the frictional forces are proportional to the velocity $l\theta'$ of the mass. Then the DE that describes the small amplitude motions of the pendulum is (see Appendix I.2),

$$\theta'' + 2s\theta' + \frac{g}{l}\,\theta = 0. \tag{1}$$

The specified constant s in Eq. (1) is equal to $(1/2)\mu/m$ where μ is the friction coefficient and m is the mass of the pendulum. It is positive because friction resists the motion. Typical initial conditions at $t = 0$ are that the pendulum is raised to the angle $\theta(0) = \theta_0$ and then it is released with zero initial angular velocity. Thus we have

$$\theta(0) = \theta_0, \qquad \theta'(0) = 0. \tag{2}$$

The motion of the pendulum is described by the IVP given by Eqs. (1) and (2).

To solve the IVP we first determine the characteristic roots corresponding to DE (1). They are

$$m_1 = -s + \sqrt{s^2 - \frac{g}{l}} \qquad \text{and} \qquad m_2 = -s - \sqrt{s^2 - \frac{g}{l}}. \tag{3}$$

The nature of the motion of the pendulum depends on the sign of $s^2 - g/l$. There are three possibilities, which we shall now proceed to analyze.

Case I

$$s^2 - \frac{g}{l} > 0. \tag{4}$$

Since $(s^2 - g/l)^{1/2} < s$, the roots are real and negative. Equation (4) implies that $s > \sqrt{g/l}$, which means that the damping is large. This would occur, for example, if the pendulum were immersed in a heavy oil or if the pivot of the pendulum were rusty. A general solution of the DE is

$$\theta = c_1 e^{m_1 t} + c_2 e^{m_2 t}.$$

The constants c_1 and c_2 are determined in the usual way by requiring the general solution to satisfy the initial conditions given by Eq. (2). This gives the solution of the IVP as

$$\theta = \frac{\theta_0}{2h}\left[(s + h)e^{-(s-h)t} - (s - h)e^{-(s+h)t}\right], \tag{5}$$

where the number h is defined by

$$h = \left(s^2 - \frac{g}{l}\right)^{1/2}.$$

Since the exponents in Eq. (5) are negative, the motion is decaying exponentially as t increases. The motion is said to be *overdamped*. As $t \to \infty$, the pendulum returns to its vertical equilibrium position, $\theta = 0$. If $\theta_0 > 0$, it is a monotonically decreasing function of t because $\theta'(t) < 0$ for $t > 0$, as we can show by differentiating Eq. (5). Thus the pendulum descends from its initial position θ_0 toward its vertical equilibrium position $\theta = 0$ as t increases, but it does not pass through $\theta = 0$ in a finite time. A sketch of $\theta(t)$ is given in Figure 5.7 for a typical value of θ_0.

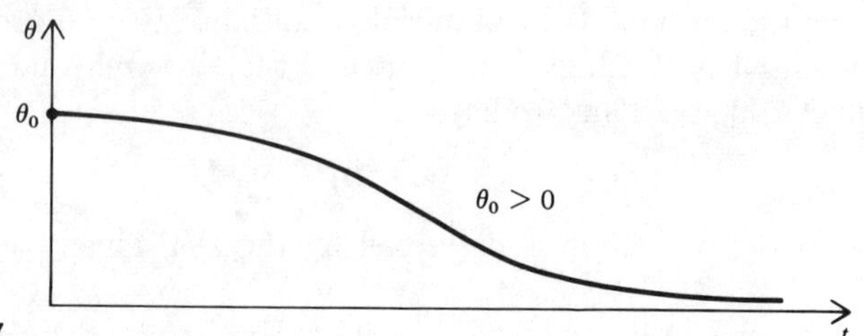

Figure 5.7

Case II

$$\frac{g}{l} - s^2 > 0\,. \tag{6}$$

This implies that $s < \sqrt{g/l}$, or that the friction coefficient is "small." The characteristic roots given in Eq. (3) are therefore the complex numbers

$$m_1 = -s + iq \qquad \text{and} \qquad m_2 = -s - iq,$$

where the real number q is defined by

$$q = \left(\frac{g}{l} - s^2\right)^{1/2}. \tag{7}$$

A general solution of the DE is

$$\theta = e^{-st}(c_1 \cos qt + c_2 \sin qt)\,.$$

The solution of the IVP in amplitude–phase angle form is

$$\theta = Ae^{-st} \cos\,(qt - \phi), \tag{8}$$

where the constants A and ϕ are given by

$$A = \frac{\theta_0}{q}\sqrt{s^2 + q^2} = \frac{\theta_0}{q}\left(\frac{g}{l}\right)^{1/2} = \theta_0\frac{\omega}{q}, \qquad \phi = \tan^{-1}\frac{s}{q}\,. \tag{9}$$

In Eq. (9) we have used $\omega = (g/l)^{1/2}$. We recall that ω is the frequency of oscillation of the undamped pendulum that was analyzed in Section 5.3; see Eqs. (3) and (6) in that section. Furthermore, we have from Eq. (7) that

$$q = \left(\frac{g}{l}\right)^{1/2}\left[1 - \frac{s^2}{g/l}\right]^{1/2} = \omega\left(1 - \frac{s^2}{\omega^2}\right)^{1/2}. \tag{10}$$

Thus as $s \to 0$, $q \to \omega$.

The pendulum is released from its initial position and it oscillates as described by Eq. (8) about $\theta = 0$. The amplitude of the motion decreases due to the exponential factor and

$$\lim_{t \to \infty} \theta(t) = 0.$$

The motion is a damped oscillatory motion, as we show in Figure 5.8. This is called the *underdamped* case. It is not a periodic motion. However, it can be shown by setting the derivative of the solution equal to zero (see Exercise 11) that the times between any two successive maxima or the times between any two successive minima are all equal and given by $2\pi/q$. We call $2\pi/q$ the quasi-period and hence q is the quasi-circular frequency. Since $\omega^2 = g/l$, Eq. (6) shows that $\omega^2 - s^2 > 0$ or, equivalently, that $s^2/\omega^2 < 1$. Then we conclude from Eq. (10) that q is always less than the frequency ω of the undamped oscillation.

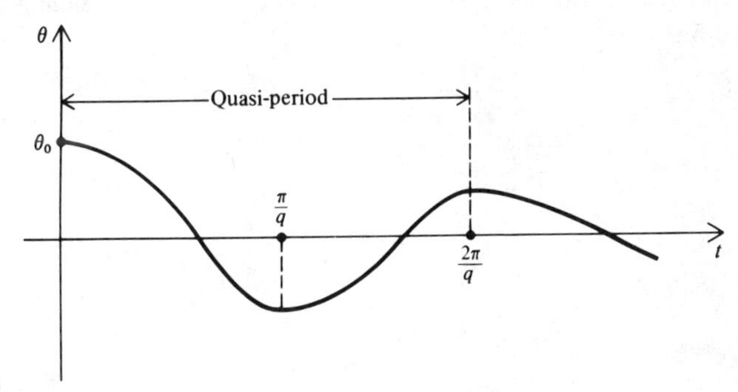

Figure 5.8

We can expand q given by Eq. (10) in a convergent Taylor series about $s = 0$ as

$$q = \omega - \frac{s^2}{2\omega} - \cdots$$

because $(s/\omega)^2 < 1$. Therefore, for small values of s, the quasi-frequency q differs from ω only by the *square of a small quantity*. That is why scientists frequently ignore the friction when they wish to approximate the frequency for problems with small friction.

Case III

$$\frac{g}{l} - s^2 = 0.$$

Since the roots of the characteristic equation are $m_1 = m_2 = -s$, the solution of the IVP is,

$$\theta = \theta_0(1 + st)e^{-st}. \tag{11}$$

This is called the *critically damped case*. The graph of Eq. (11) is similar to the overdamped case. However, the amplitude decreases slower than in the overdamped case because of the t factor in Eq. (11). Since $s > 0$ and $t > 0$, it follows from Eq. (11) that $\theta \neq 0$ for any finite value of $t > 0$. It approaches $\theta = 0$ as $t \to \infty$ because

$$\lim_{t\to\infty} te^{-st} = 0 \qquad \text{for } s > 0.$$

Furthermore, we have from Eq. (11) that $\theta' = -\theta_0 s^2 te^{-st}$. Thus θ decreases to zero monotonically as $t \to \infty$.

If the pendulum is given an initial velocity, that is, $\theta'(0) \neq 0$ in Eq. (2), then in the overdamped and critically damped cases the pendulum need not decrease monotonically to zero as $t \to \infty$. If the initial velocity is sufficiently large, then the pendulum "overshoots" the vertical position and approaches it through negative values of θ as $t \to \infty$. This is illustrated in Figure 5.9.

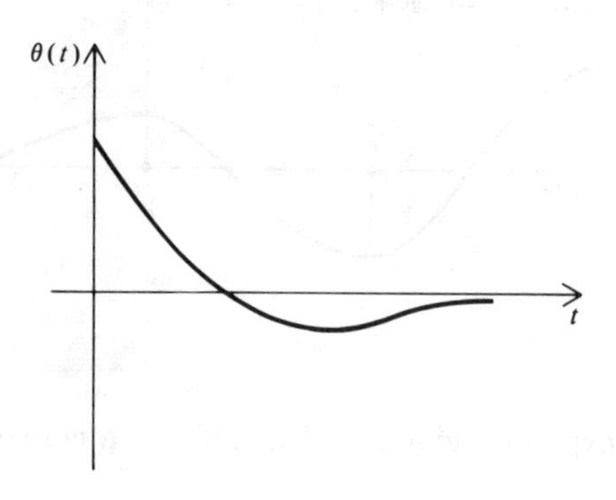

Figure 5.9

Exercises

1. A pendulum of length $l = 4$ ft oscillates through a viscous fluid. If $s = 3$ and the pendulum is given an initial angular displacement of $1/2$ rad and started from rest, $\theta'(0) = 0$, determine the subsequent motion. Is the motion overdamped, critically damped, or underdamped? Is the motion monotonic as t varies? Determine the maximum angular displacement.

2. A pendulum of length $l = 4$ ft oscillates through a viscous fluid whose friction constant is $s = 2$. The pendulum is given an initial angular displacement of $\theta(0) = 1/10$ rad and an initial angular velocity of $\theta' = -1/10$ rad/sec. Determine the subsequent motion. The motion is underdamped (Case II) because $s^2 - g/l = 4 - 32/4 = -4 < 0$. Determine the maximum amplitude, the quasi-period, and the quasi-circular frequency.

3. A pendulum of length $l = 4$ ft moves through a medium whose friction constant is $s = 2$. The pendulum is started from the vertical position with an initial angular velocity $\theta'(0) = \theta_1$. Determine the maximum value of θ_1 so that the angular displacement of the subsequent motion never exceeds 1/2 rad.

4. Show that if $s^2 - g/l \geq 0$ (the overdamped and critically damped cases), then the pendulum can pass through its rest position $\theta = 0$ for $t > 0$ no more than once for any initial values.

5. (a) Show that in the underdamped case the ratio of two successive maxima of the angular displacement is given by

$$\frac{\theta(t_n)}{\theta(t_{n+2})} = e^{2\pi s/q}$$

where q is defined in Eq. (7). The quantity $\delta = \ln [\theta(t_n)/\theta(t_{n+2})]$ is called the *logarithmic decrement of the motion*. Show that

$$s = \omega\delta(\delta^2 + 4\pi^2)^{-1/2}.$$

Since $\theta(t_n)$ and $\theta(t_{n+2})$ are usually easy to measure in a physical experiment, δ is easy to calculate. When δ is found, the friction coefficient s can be determined from this equation. (b) Compute δ for Exercise 2.

6. The total energy for the vibrating pendulum is equal to the sum of the kinetic and potential energies. It is proportional to

$$E = \frac{1}{2}\,\theta'^2 + \frac{1}{2}\,\omega^2\theta^2,$$

as we have shown in Exercise 5 of Section 5.3; there $\omega^2 = g/l$. By multiplying the DE $\theta'' + 2s\theta' + \omega^2\theta = 0$ of the vibrating pendulum with friction by θ' show that E satisfies

$$\frac{dE}{dt} = -2s(\theta')^2. \tag{i}$$

For $s > 0$ show that $E(t)$ is a monotonically decreasing function. Does the law of conservation of energy apply? Observe that the frictional resistance force is given by $R = 2s\theta'$. Interpret Eq. (i) as showing that the rate of decrease in energy of the pendulum is equal to the rate of doing work against the frictional force.

7. An experiment was performed on a vibrating pendulum having a length $l = 2$ ft and for which the frictional resistances were small (underdamped). It was found that two successive maxima of the angular displacement were $\theta_1 = \pi/15$ and $\theta_2 = \pi/18$ radians. Use the result of Exercise 5 to determine the friction coefficient of the pendulum.

8. The quasi-circular frequency q is given by Eq. (7). It is a function of the length of the pendulum and the friction coefficient s. For fixed s does it increase or decrease as l varies? Sketch a graph of q vs. l for fixed s.

9. (a) Show that the maxima and minima of θ given by Eq. (8) occur at those values $t = t_n$ that satisfy $\tan(qt - \phi) = -s/q$.
   (b) Use the result of (a) and Eq. (9) to show that $t_n = n\pi/q$, $n = 0, \pm 1, \pm 2, \ldots$.
   (c) Show that the maxima occur at $t_{2n} = 2n\pi/q$ and the minima occur at $t_{2n+1} = (2n+1)\pi/q$, $n = 0, 1, \ldots$.
   (d) Show that the quasi period is equal to $2\pi/q$.

5.5 THE DECAY OF A CHARGE IN A SERIES ELECTRIC CIRCUIT

Consider the series RCL electric circuit shown in Figure 5.10. At time $t = 0$ a charge y_0 is placed on the capacitor and the circuit is completed by closing the switch. We wish to determine the charge $y(t)$ for all $t > 0$.

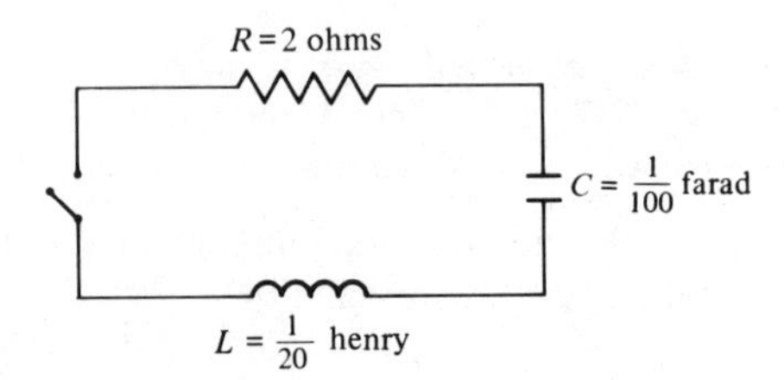

Figure 5.10

We have shown in Appendix I.3, by using Kirchhoff's law, that the DE for the charge in a RCL circuit is

$$y'' + \frac{R}{L}y' + \frac{1}{LC}y = 0. \tag{1}$$

We substitute the specific values of R, L, and C as given in Figure 5.10 into this equation. This gives

$$y'' + 40y' + 2000y = 0. \tag{2}$$

Since the initial charge is y_0 and there is no initial current in the circuit, we have the initial conditions

$$y(0) = y_0 \quad \text{and} \quad y'(0) = i(0) = 0. \tag{3}$$

The solution of the IVP given by Eqs. (2) and (3) is

$$y(t) = y_0 e^{-20t}(\cos 40t + \tfrac{1}{2}\sin 40t). \tag{4}$$

The reader should verify that it can be rewritten in the form

$$y(t) = Ae^{-20t}\cos(40t - \phi)$$

with $A = y_0\sqrt{5}/2$ and $\phi = \tan^{-1}\frac{1}{2} \approx 26.5°$. We see that the charge $y(t)$ and hence the current $i(t) = y'$ are damped oscillatory functions. The damping (or friction) term $(R/L)y' = 40y'$, in the DE, which arises because of the dissipation of energy in the resistor, causes the total current in the circuit to decay to zero as $t \to +\infty$. The oscillatory effect is caused by the discharging and charging of the capacitor. When the capacitor becomes fully charged, then the charge has a local maximum and the current is zero because $i = y'$.

Exercises

1. The capacitor of an RCL electric circuit with $R = 2$ ohms, $L = 1$ henry, and $C = 1/50$ farad, is given an initial charge $y(0) = 1$. There is no initial current in the circuit. (a) Determine the subsequent variation of the charge and the current in the circuit. Is the variation underdamped, critically damped, or overdamped? (b) Determine the maximum value of the charge for $t > 0$. (c) For what value of t is the charge first equal to zero?

2. An initial current $i(0) = 1/2$ amp is supplied to an RCL circuit with $R = 10$ ohms, $L = 1/4$ henry, and $C = 1/200$ farad. There is no initial charge on the capacitor. (a) Determine the subsequent variation of the charge. (b) Determine the maximum value of the charge for $t > 0$. (c) Is the charge ever equal to zero?

3. What conditions on the circuit parameters R, C, and L must be satisfied for the charge variation to be underdamped, critically damped, and overdamped?

4. What is the minimum value of the resistance in a discharging RCL circuit that will insure that the charge on the capacitor will never reach zero? Assume zero initial current.

5. Consider an LC circuit ($R = 0$) with zero initial current and initial charge y_0 on the capacitor. Find the current and the charge for all values of t.

chapter 6

BOUNDARY-VALUE AND EIGENVALUE PROBLEMS

6.1 SOLUTION OF BOUNDARY-VALUE PROBLEMS

In this section we shall show how to construct solutions to boundary-value problems (BVP). In a BVP we wish to determine solutions of the DE

$$y'' + 2by' + cy = 0$$

on an interval $x_0 < x < x_1$ that satisfy conditions (boundary conditions) at the *two* specified points, x_0 and x_1. As in the case of IVPs, we shall use general solutions and determine the constants so that the boundary conditions are satisfied. We illustrate the techniques by several examples.

EXAMPLE 1

Solve the BVP

$$y'' + 4y = 0, \qquad \text{for } 0 < x < \frac{\pi}{4} \tag{1}$$

$$y(0) = 0,\ y\left(\frac{\pi}{4}\right) = 7. \tag{2}$$

solution. A general solution is

$$y = c_1 \cos 2x + c_2 \sin 2x.$$

We require it to satisfy the boundary conditions given by Eq. (2). Thus by evaluating this general solution at $x = 0$ and $x = \pi/4$ and substituting the results in Eq. (2), we get the algebraic equations,

$$y(0) = c_1 = 0$$

and

$$y\left(\frac{\pi}{4}\right) = c_1 \cos \frac{\pi}{2} + c_2 \sin \frac{\pi}{2} = 7$$

to determine c_1 and c_2. Since $\sin(\pi/2) = 1$ and $\cos(\pi/2) = 0$, the solution is

$$c_1 = 0 \qquad \text{and} \qquad c_2 = 7.$$

A solution of the BVP is then obtained by substituting these values for c_1 and c_2 into the general solution. This gives,

$$y = 7 \sin 2x.$$

EXAMPLE 2

Solve the BVP

$$y'' + 4y = 0, \qquad \text{for } 0 < x < l \tag{3}$$

$$y(0) = 0,\ y(l) = 7. \tag{4}$$

Here l is any positive constant.

solution. The DE is the same as in Example 1 and the boundary conditions are the same when $l = \pi/4$. We now try to solve the BVP for any $l > 0$. A general solution of Eq. (3) is $y = c_1 \cos 2x + c_2 \sin 2x$ (see Example 1). We evaluate it at $x = 0$ and $x = l$ and then use the boundary conditions given by Eq. (4) to determine c_1 and c_2. This gives

$$y(0) = c_1 = 0 \tag{5}$$

$$y(l) = c_2 \sin 2l = 7. \tag{6}$$

For any l such that $\sin 2l \neq 0$ we can solve Eq. (6) to obtain

$$c_2 = \frac{7}{\sin 2l}.$$

Thus by substituting $c_1 = 0$ and this value of c_2 into the general solution, we get the solution to the BVP as

$$y = \frac{7}{\sin 2l} (\sin 2x), \qquad \text{when } \sin 2l \neq 0.$$

When $l = n\pi/2$, for $n = 1, 2, \ldots$, then $\sin 2l = 0$ and there is no value of c_2 that will satisfy the boundary condition given in Eq. (6). Thus the BVP has no solution when $l = n\pi/2$. This shows that the solution of a BVP depends crucially on the interval on which it is defined. The solution may cease to exist when the interval is changed.

In Section 4.5 we have shown that the IVP for the second-order linear DE with constant coefficients always has a solution. Example 2 shows that even a simple BVP need not have a solution. In contrast to this, we now consider another BVP, which possesses several solutions.

EXAMPLE 3

Solve the BVP

$$y'' + 4y = 0, \qquad \text{for } 0 < x < \pi$$

$$y(0) = 4,\ y(\pi) = 4.$$

solution. The DE is the same as in the two preceding examples. We require the general solution

$$y = c_1 \cos 2x + c_2 \sin 2x$$

to satisfy the boundary conditions. This gives the algebraic equations

$$y(0) = c_1 = 4$$

and

$$y(\pi) = c_1 \cos 2\pi + c_2 \sin 2\pi = 4.$$

Since $\cos 2\pi = 1$ and $\sin 2\pi = 0$, we conclude that

$$c_1 = 4 \qquad \text{and} \qquad c_2 \text{ is arbitrary.}$$

By c_2 arbitrary, we mean that both boundary conditions are satisfied for any value of c_2. Inserting these values of c_1 and c_2 into the general solution, we find that

$$y = 4 \cos 2x + c_2 \sin 2x$$

is a solution of the BVP for all values of c_2. Thus there are infinitely many solutions of the BVP—one for each value of c_2.

For the simple BVPs considered in this section, it is possible to develop a general theory that can predict the number of solutions without actually solving the BVP. However, the theory is difficult and beyond the scope of this book.

Exercises

1. Solve the following BVPs.
   (a) $y'' - 3y' + 2y = 0, \qquad \text{for } 0 < x < 1$
   $y(0) = 1,\ y(1) = 0$

(b) $y'' - 6y' + 25y = 0, \quad \text{for } 0 < x < 1$
$y(0) = y(1) = 0$

(c) $y'' - 6y' + 25y = 0, \quad \text{for } 0 < x < \pi/4$
$y'(0) = 1,\ y(\pi/4) = 0$

(d) $y'' - y = 0, \quad \text{for } 0 < x < 1$
$y(0) = A,\ y'(1) = 0$, where A is a prescribed number

(e) $y'' - K^2y = 0, \quad \text{for } 0 < x < 1$
$y(0) = 0,\ y'(1) = 1/K$, where $K \neq 0$ is a constant

2. For which values of the constants A and B (if any) does the BVP

$$y'' + 4y = 0 \qquad \text{for } 0 < x < \pi$$

$$y(0) = A,\ y(\pi) = B$$

have (a) no solution, (b) exactly one solution, and (c) more than one solution?

3. Solve the BVP $y'' + 4y = 0$ for $0 < x < l$, $y(0) = 4$, $y(l) = 4$, where l is any constant. Find values of l, if any, for which the BVP has (a) exactly one solution, (b) many solutions, and (c) no solution.

4. Boundary conditions that involve both the unknown function and its derivative at the same point occur often in applications. A typical example is

$$y'' - y = 0, \qquad 0 < x < 1$$

$$y'(0) + 3y(0) = 0,\ y'(1) + y(1) = 1.$$

Solve this BVP.

5. Solve the BVP

$$y'' + y = 0, \qquad 0 < x < 2$$

$$y'(0) + y(0) = 10,\ y'(2) + 6y(2) = 1.$$

6. Show (without solving) that any solution of the BVP

$$y'' + y = 0, \qquad 0 < x < \pi, \qquad y(0) = A,\ y(\pi) = B$$

can be written as $y = AY_1 + BY_2$, where Y_1 and Y_2 satisfy the given DE and the boundary conditions $y(0) = 1$, $y(\pi) = 0$, and $y(0) = 0$, $y(\pi) = 1$, respectively.

7. Determine the values of the constants α, A, and B such that the general BVP

$$y'' + 2by' + cy = 0, \qquad \text{for } 0 < x < \alpha \tag{i}$$

$$y(0) = A,\ y(\alpha) = B \tag{ii}$$

with $q = b^2 - c < 0$ has only one solution.

*8. An alternative but equivalent procedure for studying BVPs, such as in Eqs. (i) and (ii) of Exercise 7, is first to define a function $z(x)$ as a solution of the IVP

$$z'' + 2bz' + cz = 0 \qquad \text{and} \qquad z(0) = A,\ z'(0) = k \tag{iii}$$

where b, c, and A are the same given constants as in the BVP given by Eqs.

(i) and (ii). For each prescribed value of the constant k, the IVP given in Eq. (iii) has only one solution. We denote this solution by $z(x; k)$ to emphasize the dependence of the solution on the "free" parameter k. Show that if a value of $k = k_0$ can be determined such that at $x = \alpha$

$$z(\alpha; k_0) = B, \tag{iv}$$

then $y(x) = z(x; k_0)$ is a solution of the BVP. Show, assuming for simplicity that $b^2 > c$, that the number of solutions of the BVP is equal to the number of k that satisfies Eq. (iv).

This technique of solving BVPs by considering an "equivalent" IVP and adjusting the "free" initial values at one boundary point so that the boundary condition at the other boundary point is satisfied is called the shooting method. When implemented on a computer the shooting method is one of the most powerful techniques for the numerical solution of boundary-value problems.

*9. Consider, $y'' + y = 0$, $0 < x < 1$, $y(0) = 0$, $y'(1) = 1$. Apply the procedure outlined in Exercise 8 to solve this BVP.

6.1-1 Diffusion of pollutants

While attending a lecture a scientist, who is a nonsmoker, is annoyed by the high concentration of smoke in the auditorium. The auditorium is connected to the outdoors by a long narrow hall (see Figure 6.1). The hall has a ventilation system to remove smoke. It is possible to hear the lecture from the hall. Since the concentration of smoke is high in the auditorium and zero outdoors, it varies along the hall. The scientist wishes to determine how far down the hall he must stand to reduce the smoke concentration below a preassigned safe value. It is known from experiments that the harmful level of smoke concentration is 0.002 cubic feet of smoke per cubic foot of air.

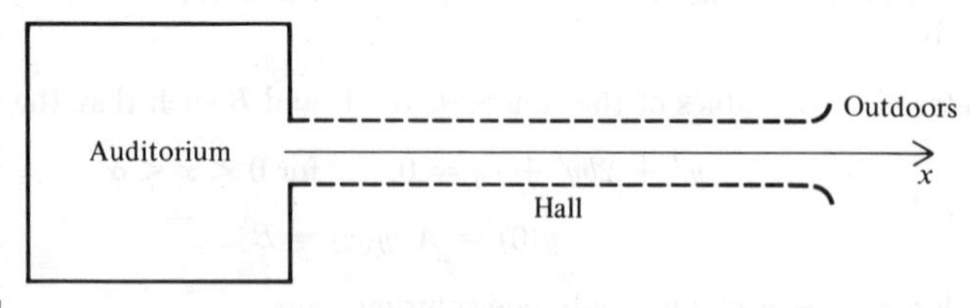

Figure 6.1

Sufficient time has elapsed so that the concentration of smoke in the hall does not vary with time. The hall is long and narrow. Thus we assume that the concentration of smoke y is constant over the cross section of the hall and consequently varies only with the distance x along the hall.

Furthermore, we assume that the smoke moves through the hall by diffusion. On the microscopic level, diffusion is a result of the thermal motion of the individual gas molecules. The basic law of diffusion states that the amount, $F(x)$, of smoke that passes through a unit area per unit time is proportional to the spatial rate of change of the concentration at x; that is,

$$F(x) = -Dy'(x). \tag{1}$$

Here the proportionality constant $D > 0$ is called the diffusion coefficient. The minus sign in Eq. (1) is essential because the smoke diffuses from regions of high concentration to regions of low concentration. Thus if $y' < 0$ (y is a decreasing function), the smoke will diffuse in the positive x direction. We assume that the ventilation system in the hall removes smoke from the hall at the position x at a rate proportional to its concentration at x.

To determine $y(x)$ we consider a section of the hall located between x and $x + \Delta x$; see Figure 6.2. The net flow of smoke into this element must be zero, since y is independent of time. Specifically, we have the following balance equation:

$$\left[\begin{array}{l}\text{The amount of smoke}\\ \text{diffusing into the}\\ \text{element at } x \text{ per}\\ \text{unit time}\end{array}\right] - \left[\begin{array}{l}\text{The amount of smoke}\\ \text{diffusing out of the}\\ \text{element at } x + \Delta x\\ \text{per unit time}\end{array}\right]$$

$$- \left[\begin{array}{l}\text{The amount of smoke}\\ \text{removed by the venti-}\\ \text{lation system per}\\ \text{unit time}\end{array}\right] = 0. \tag{2}$$

Let $V(x)$ be the amount of smoke removed by the ventilation system at station x. It is proportional to $y(x)$, and hence the total amount of smoke removed between x and $x + \Delta x$ is

$$\int_x^{x+\Delta x} V(z)\, dz = k \int_x^{x+\Delta x} y(z)\, dz.$$

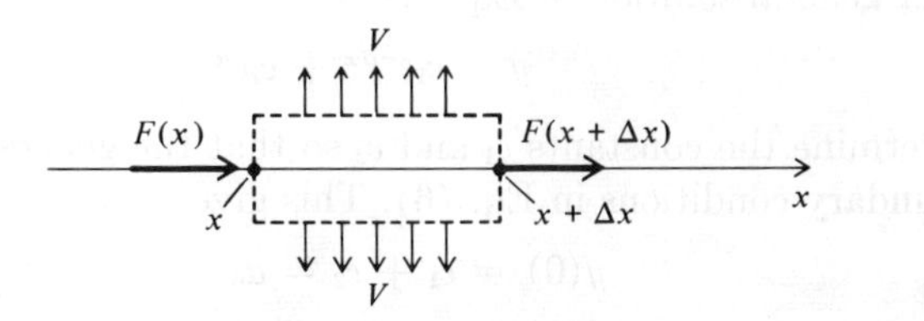

Figure 6.2

Here the given proportionality constant $k > 0$ is called the absorption coefficient of the ventilation system. Since Δx is small, we can approximate the integral by the value of the integrand at the midpoint $x + (\Delta x/2)$ of the section times the length Δx of the section. Thus

$$k \int_x^{x+\Delta x} y(z)\, dz \approx ky\left(x + \frac{\Delta x}{2}\right) \Delta x. \tag{3}$$

If A is the cross-sectional area of the hall, then using Eq. (1) and (3) in the balance equation, Eq. (2), we get the approximate equation

$$-Dy'(x)A - [-Dy'(x + \Delta x)A] - ky\left(x + \frac{\Delta x}{2}\right) \Delta x = 0.$$

We divide both sides of this equation by Δx and find that

$$AD\left[\frac{y'(x + \Delta x) - y'(x)}{\Delta x}\right] - ky\left(x + \frac{\Delta x}{2}\right) = 0. \tag{4}$$

We now take the limit of both sides of Eq. (4) as $\Delta x \to 0$. The bracketed term is the difference quotient for y'. Thus in the limit it is y''. Furthermore,

$$\lim_{\Delta x \to 0} y\left(x + \frac{\Delta x}{2}\right) = y(x).$$

This gives the DE

$$y'' - \alpha^2 y = 0, \tag{5}$$

where the constant α is defined by

$$\alpha = \left(\frac{k}{AD}\right)^{1/2}.$$

We must solve DE (5) subject to the conditions that the concentration at $x = 0$ is equal to the concentration a_0 of the smoke in the auditorium and that the concentration at $x = L$ equals the outdoor concentration, which we approximate as zero. Thus we obtain the boundary conditions

$$y(0) = a_0 \quad \text{and} \quad y(L) = 0. \tag{6}$$

The BVP given by Eqs. (5) and (6) determines the variation of the smoke concentration in the hall.

A general solution of Eq. (5) is

$$y = c_1 e^{-\alpha x} + c_2 e^{\alpha x}.$$

We determine the constants c_1 and c_2 so that the general solution satisfies the boundary conditions in Eq. (6). This gives

$$y(0) = c_1 + c_2 = a_0$$

$$y(L) = c_1 e^{-\alpha L} + c_2 e^{\alpha L} = 0.$$

We solve these algebraic equations and get

$$c_1 = \frac{-e^{2\alpha L}a_0}{1 - e^{2\alpha L}} \quad \text{and} \quad c_2 = \frac{a_0}{1 - e^{2\alpha L}}.$$

The denominator $1 - e^{2\alpha L}$ does not equal zero, because $e^{2\alpha L} > 1$ for $\alpha L > 0$. Then the solution of the BVP is

$$y(x) = \frac{-a_0 e^{2\alpha L}}{1 - e^{2\alpha L}} e^{-\alpha x} + \frac{a_0}{1 - e^{2\alpha L}} e^{\alpha x}.$$

Using the definitions

$$\sinh \alpha L = \frac{e^{\alpha L} - e^{-\alpha L}}{2} \quad \text{and} \quad \sinh \alpha(L - x) = \frac{e^{\alpha(L-x)} - e^{-\alpha(L-x)}}{2}$$

of the hyperbolic functions, we can rewrite the solution as

$$y(x) = a_0 \frac{\sinh \alpha(L - x)}{\sinh \alpha L}.$$

The solution is sketched in Figure 6.3 for several values of α. We see that it decays from a_0 at $x = 0$ to 0 at $x = L$. From Eq. (5) we observe that α is proportional to $k^{1/2}$. We recall that k is the rate at which the ventilation system removes smoke. An efficient ventilation system has a large value of k. As we expect, the solution decays more rapidly for larger values of k.

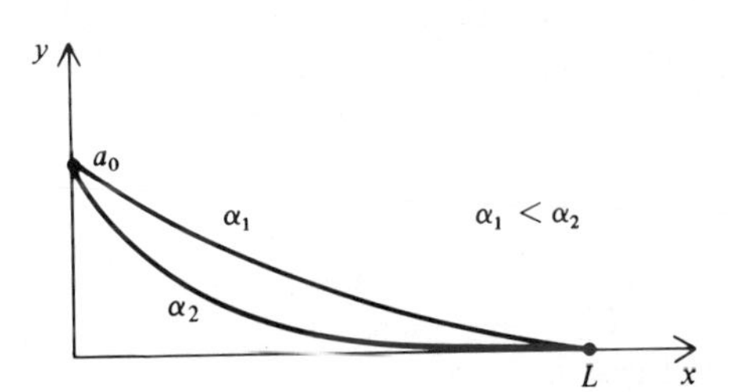

Figure 6.3

Exercises

1. Suppose that in the foregoing example $\alpha = 0.0001$, $L = 100$ ft, and $a_0 = .004$ ft^3. Find the concentration in the hall and the value of x at which the concentration equals the harmful level of 0.002.

2. Suppose that the flux of smoke $F(0) = -Dy'(0)$ at the auditorium end of the hall is known, instead of the actual concentration. Assuming that all other quantities are the same as in Exercise (1), determine the concentration of smoke in the hall and the position in the hall where it equals the harmful level. Express your answers in terms of $F(0)$ and D.

3. A long river flows through a populated region with uniform velocity U. Sewage is continuously injected at a constant rate at the beginning of the river, $x = 0$. The sewage is convected down the river by the flow and it is simultaneously decomposed by bacteria and other biological activity. Assume that the river is sufficiently narrow so that the concentration y of sewage is uniform over the cross section and that the polluting has been going on for a long time, so that y does not depend on time. Thus y is a function only of the distance x downstream from the sewage plant. If the rate of decomposition at position x is proportional, with proportionality constant k, to the concentration $y(x)$ at x, then show that y must satisfy

$$y'' - uy' - \alpha^2 y = 0$$

where $u = U/D$ and $\alpha = (k/AD)^{1/2}$. If the concentration $y(0) = y_0$ is known and the concentration at $x = L$ is measured as $y(L) = y_1 < y_0$, then determine the concentration in the stream for $0 \leq x \leq L$.

4. Solve Exercise 3 when the flux from the sewage plant $F(0) = -Dy'(0)$ is given instead of the concentration $y(0)$.

5. The equation of equilibrium of a tightly stretched and initially straight elastic string imbedded in an elastic foundation of modulus $k > 0$ is given by

$$y'' - \frac{k}{T} y = 0$$

where $y(x)$ is the vertical (transverse) deflection of the string. Here the weight of the string is neglected. In addition, the deflections are assumed to be small and the tension T is considered as a constant. The end $x = 0$ of the string is fixed [$y(0) = 0$] and the opposite end $x = L$ is given a vertical displacement $y(L) = a_0$. Solve the BVP to determine the deflection of the string. Determine the maximum deflection.

6. A gas diffuses into a liquid in a narrow pipe. The gas is absorbed by the liquid at a rate proportional to $y'(x)$, where $y(x)$ is the concentration of the gas at the station x in the pipe. Furthermore, the gas reacts chemically with the liquid. The amount of gas that disappears as a result of this reaction is proportional to $y(x)$. Show that the balance equation for y is $y'' - (k/D)y = 0$ where k is the reaction rate and D is the diffusion coefficient. The concentration is given at $x = 0$ as $y(0) = a_0$. At $x = L$ the gas is entirely absorbed by the liquid, and thus $y(L) = 0$. Solve the resulting BVP. Determine the value of x where the flux $F = -Dy'$ of gas is largest.

7. Solve Exercise 6, with the condition $y'(L) = a_1$ replacing $y(L) = a_0$.

6.2 EIGENVALUE PROBLEMS

In the problems that we have studied for second-order DEs, the coefficients b and c in the DEs were considered as specific numbers. However, there is an important special class of BVPs in which the coefficients are unspecified constants. A typical example is: Determine the values of the constant coefficient K for which the BVP

$$y'' + Ky = 0, \qquad \text{for } 0 < x < 1$$

$$y(0) = y(1) = 0$$

has solutions that are not zero everywhere in the interval $0 \leq x \leq 1$. These values of K are called *eigenvalues,* and the corresponding solutions that are not identically zero are the *eigenfunctions.* The BVP is then called an eigenvalue problem (EVP). Eigenvalue problems are characterized by homogeneous DEs and zero boundary conditions. We observe that $y \equiv 0$ is a solution of the EVP for all values of K. This solution is called the trivial solution. It is not an eigenfunction because it is zero for all x in $[0, 1]$. Solutions that are not identically zero in $[0, 1]$ are called nontrivial solutions.

Eigenvalue problems occur frequently in a variety of applications and they are important in many mathematical studies of DEs. We shall illustrate some of their properties by studying several examples and applications. In the first example we solve the problem originally posed.

EXAMPLE 1

Determine the eigenvalues and eigenfunctions of

$$y'' + Ky = 0, \qquad \text{for } 0 < x < 1, \tag{1}$$

$$y(0) = y(1) = 0. \tag{2}$$

solution. Clearly, $y \equiv 0$ is a solution of the EVP given by Eqs. (1) and (2) for all values of K. We observe that if $y(x)$ is an eigenfunction, then $z(x) = Ay(x)$ is also an eigenfunction for all values of the constant A. To verify this statement, we insert z into DE (1). This gives

$$z'' + Kz = Ay'' + KAy = A[y'' + Ky].$$

The bracketed term on the right side of this equation is zero because y is a solution of the DE. Thus z is a solution of the DE for all values of A. Furthermore z satisfies the boundary conditions given by Eq. (2) because y satisfies them. Thus the eigenfunctions are determined only within an arbitrary multiplicative constant.

Since K is a constant (though unknown), Eq. (1) is a linear DE with constant coefficients. The general solution of Eq. (1) will have different forms for $K = 0$, $K > 0$, and $K < 0$. We consider each possibility separately.

First we consider $K = 0$. Then the DE is reduced to $y'' = 0$. The solution of this equation is

$$y = c_1x + c_2, \tag{3}$$

where c_1 and c_2 are arbitrary constants. We require (3) to satisfy the boundary conditions in Eq. (2). This gives $c_1 = c_2 = 0$. These are the only values of c_1 and c_2 for which Eq. (3) can satisfy Eq. (2). Therefore $y \equiv 0$ is the only solution of Eqs. (1) and (2) corresponding to $K = 0$. Since this is the trivial solution, it is not an eigenfunction. Consequently, $K = 0$ is not an eigenvalue.

We consider $K > 0$. The characteristic equation corresponding to Eq. (1) is $m^2 + K = 0$. A general solution is

$$y = c_1 \cos \sqrt{K}x + c_2 \sin \sqrt{K}x. \tag{4}$$

We require Eq. (4) to satisfy the boundary conditions given by Eq. (2). This yields

$$y(0) = c_1 = 0 \qquad \text{and} \qquad y(1) = c_1 \cos \sqrt{K} + c_2 \sin \sqrt{K} = 0.$$

Thus we conclude that

$$c_1 = 0 \tag{5}$$

and

$$c_2 \sin \sqrt{K} = 0. \tag{6}$$

Equation (6) implies that either $c_2 = 0$ or

$$\sin \sqrt{K} = 0 \qquad \text{and} \qquad c_2 \text{ is arbitrary.} \tag{7}$$

We consider both possibilities. If $c_2 = 0$, then both c_1 and c_2 are zero and Eq. (4) implies that $y \equiv 0$. This is the trivial solution and thus it is not an eigenfunction. If $\sqrt{K}$ satisfies Eq. (7) and $K > 0$, then $\sqrt{K} = \pm n\pi$, $n = 1, 2, \ldots$, and hence

$$K = (n\pi)^2, \qquad n = 1, 2, \ldots. \tag{8}$$

The solution, Eq. (4), is then given by

$$y = y_n = a_n \sin n\pi x, \qquad n = 1, 2, \ldots, \tag{9}$$

where a_n is an arbitrary constant.[1] Thus $K = (n\pi)^2$ are the eigenvalues and $a_n \sin n\pi x$ are the corresponding eigenfunctions. There are an infinite number of eigenvalues and eigenfunctions, one pair for each positive integer n.

If $K < 0$, then $-K > 0$, and we have the real characteristic roots $m_1 = \sqrt{-K}$ and $m_2 = -\sqrt{-K}$. A general solution of the DE is

$$y = c_1 e^{rx} + c_2 e^{-rx} \tag{10}$$

where $r = \sqrt{-K}$. Applying the boundary conditions, we find that c_1 and c_2 must satisfy the algebraic equations,

$$c_1 + c_2 = 0 \tag{11}$$

$$c_1 e^r + c_2 e^{-r} = 0. \tag{12}$$

Equation (11) shows that $c_1 = -c_2$. By inserting this result in Eq. (12), we find that

$$c_2(-e^r + e^{-r}) = c_2 e^{-r}(1 - e^{2r}) = 0. \tag{13}$$

[1] We have used a_n to denote the arbitrary constants instead of c_2.

Since $r \neq 0$, then $e^{-r}(1 - e^{2r}) \neq 0$, and Eq. (13) implies that $c_2 = 0$. From Eq. (11) we conclude that $c_1 = 0$. Thus the only solution of the EVP with $K < 0$ is $y \equiv 0$ and hence there are no eigenvalues $K < 0$.

We shall now show that the eigenfunctions given in Eq. (9) possess an important property, which is shared by the eigenfunctions of a wide class of other EVPs. We consider the following integrals:

$$I_{mn} = \int_0^1 y_m(x) y_n(x)\, dx.$$

We insert the eigenfunctions from Eq. (9) into this integral. This gives

$$I_{mn} = a_n a_m \int_0^1 \sin n\pi x \sin m\pi x\, dx. \tag{14}$$

Since the integral equals zero if $m \neq n$ and it equals $\frac{1}{2}$ if $m = n$ (see Exercise 2), we obtain from Eq. (14) the following basic property of the eigenfunctions

$$I_{mn} = \begin{cases} \dfrac{a_m{}^2}{2}, & \text{if } m = n \\ 0, & \text{if } m \neq n. \end{cases} \tag{15}$$

In general, a set of functions $z_1(x)$, $z_2(x)$, . . ., which are defined on an interval $[x_0, x_1]$ and satisfy

$$I_{mn} \equiv \int_{x_0}^{x_1} z_m(x) z_n(x)\, dx = \begin{cases} k_m, & \text{if } m = n \\ 0, & \text{if } m \neq n, \end{cases} \tag{16}$$

where $k_m \neq 0$ are constants, is called an orthogonal set of functions. Therefore it follows from Eqs. (15) and (16) that y_1, y_2, . . . form an orthogonal set of functions.

EXAMPLE 2

Determine the eigenvalues and eigenfunctions of

$$y'' + Ky = 0, \qquad \text{for } 0 < x < 1 \tag{17}$$

$$y'(0) = y(1) = 0. \tag{18}$$

solution. Here, again, $y \equiv 0$ is a solution of the EVP for all values of K. A general solution of the DE for $K > 0$ is, see Eq. (4),

$$y = c_1 \cos \sqrt{K}x + c_2 \sin \sqrt{K}x. \tag{19}$$

We require this solution to satisfy the boundary conditions. This gives

$$y'(0) = c_2\sqrt{K} = 0 \tag{20}$$

and

$$y(1) = c_1 \cos\sqrt{K} + c_2 \sin\sqrt{K} = 0. \tag{21}$$

Since $K \neq 0$, Eq. (20) shows that $c_2 = 0$, and hence Eq. (21) is reduced to

$$c_1 \cos\sqrt{K} = 0.$$

The solution $c_1 = 0$ of this equation implies that $y \equiv 0$, since $c_2 = 0$. Therefore, in order to have eigenfunctions, K must satisfy

$$\cos\sqrt{K} = 0. \tag{22}$$

The roots of Eq. (22) are $\sqrt{K} = \pm(2n+1)(\pi/2)$, $n = 0, 1, \ldots$. Consequently, the eigenvalues are

$$K = (2n+1)^2\left(\frac{\pi}{2}\right)^2, \qquad n = 0, 1, \ldots. \tag{23}$$

The corresponding eigenfunctions are obtained from Eq. (19) by using $c_2 = 0$ and Eq. (23). We obtain

$$y = y_n = a_n \cos(2n+1)\frac{\pi x}{2}, \qquad n = 0, 1, \ldots,$$

where a_n are arbitrary constants.

The reader should verify that for $K \leq 0$ there are no eigenvalues.

Exercises

1. Determine the eigenvalues and eigenfunctions for the following EVPs. They consist of the DE

$$y'' + Ky = 0, \qquad \text{for } 0 < x < 1$$

and the boundary conditions

(a) $y(0) = y'(1) = 0$

(b) $y'(0) = y'(1) = 0$

(c) $y(0) = y'(1) + y(1) = 0$

(d) $y'(0) + y(0) = y'(1) = 0$

(e) $y(0) = y(1)$ and $y'(0) = y'(1)$

2. Use the trigonometric substitution

$$\sin m\pi x \sin n\pi x = \tfrac{1}{2}[\cos(m-n)\pi x - \cos(m+n)\pi x]$$

to show that

$$\int_0^1 \sin n\pi x \sin m\pi x \, dx = \begin{cases} 0, & \text{if } m \neq n \\ \tfrac{1}{2}, & \text{if } m = n \end{cases}$$

3. Prove that if $y = y_1(x)$ is an eigenfunction of the EVP

$$y'' + Ky = 0, \qquad 0 < x < 1$$
$$y'(0) + b_1 y(0) = 0,\ y'(1) + b_2 y(1) = 0$$

then so is Ay_1 for all constants A.

4. Find the eigenvalues and eigenfunctions for the EVP

$$y'' + y' + Ky = 0, \qquad 0 < x < 1, \qquad y(0) = y(1) = 0.$$

5. Find the eigenvalues and eigenfunctions for the EVP

$$y'' + Ky' + y = 0, \qquad 0 < x < 1, \qquad y(0) = y(1) = 0.$$

In the following exercise we show that it is possible to prove that the eigenfunctions (if any) of a given eigenvalue problem are orthogonal without actually solving for these eigenfunctions.

6. Consider the EVP

$$y'' + Ky = 0 \qquad \text{for } 0 < x < 1$$
$$y'(0) + b_1 y(0) = 0,\ y'(1) + b_2 y(1) = 0,$$

where b_1 and b_2 are specified constants. Denote the eigenvalues and eigenfunctions by K_n and $y_n(x)$, respectively. Show that if $K_n \neq K_m$, then

$$\int_0^1 y_n(x) y_m(x)\, dx = 0$$

(*Hint:* (K_n, y_n) and (K_m, y_m) satisfy the DEs

$$y_n'' + K_n y_n = 0 \qquad \text{and} \qquad y_m'' + K_m y_m = 0.$$

Multiply the first equation by y_m and the second by y_n, integrate the results from $x = 0$ to $x = 1$ using integration by parts, and then apply the boundary conditions. Subtract the resulting two equations.)

7. Determine the eigenvalues and eigenfunctions of the EVPs consisting of the DE $y'' + Ky = 0$ on the infinite interval $0 < x < \infty$ and the boundary conditions
   (a) $y(0) = 0,\ \lim_{x\to\infty} y(x) = 0$
   (b) $y(0) = 0,\ y(\infty)$ is bounded (not necessarily equal to zero)
   (c) $y'(0) = 0,\ \lim_{x\to\infty} y'(x) = 0$

6.2-1 Buckling of elastic columns

Eigenvalue problems were discovered by L. Euler in the eighteenth century. He considered the "buckling" of a long and slender elastic rod that is deformed by compressive forces P acting axially on the ends of the rod; see Figure 6.4. Slender rods are conventionally used as columns in civil, aircraft, and marine structures. For example, in a steel-framed building the vertical columns support the weight and loads of the structure above them. This weight induces the force P on the ends of the column. If P is small, then the column deflects only slightly from its natural straight state. For a sufficiently large value of P the column suddenly bows out of the straight state, with deflections of large amplitude. This is called buckling. Once the column buckles it is unable to support the load. Clearly, the buckling of a single column in a building can have a disastrous effect on the entire structure.

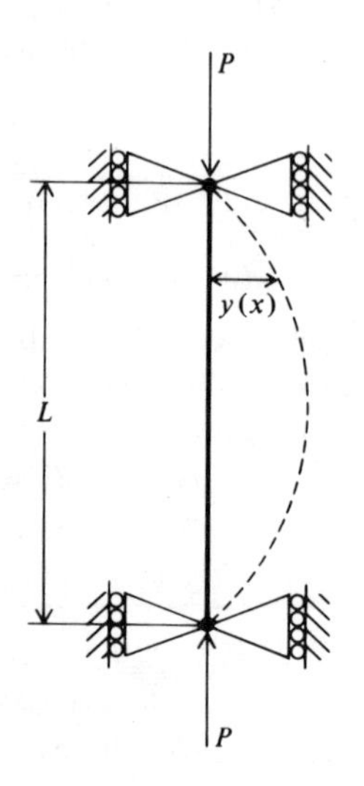

Figure 6.4

Columns were used extensively in Greek and Roman structures. They were designed according to empirical formulas developed by the architects of these ancient civilizations. It was not until Euler's mathematical studies of columns that engineers developed rational and efficient procedures for their design.

If the ends of the column are free to rotate,[2] then the EVP derived by Euler to describe the buckling of the rod is

$$y'' + Ky = 0, \qquad y(0) = y(1) = 0. \tag{1}$$

Here $y(x)$ is the lateral displacement of the column (see Figure 6.4), and

[2] This condition is called simply supported or pinned ends by engineers.

the constant K is defined by

$$K = \frac{P}{EI}. \tag{2}$$

We have taken the length of the column $L = 1$ for simplicity. $E > 0$ is the elastic modulus (Young's modulus) of the rod and $I > 0$ is a moment of inertia of the rod's cross section. The elastic modulus is a measure of the stiffness of the material used in the construction of the column. Both E and I are specified constants. In Eq. (2) the applied force P and, consequently, K are positive for compression and negative for tension. Compression means that the column is "squeezed" by the force. Tension "stretches" the column. We wish to determine the values of P and, consequently, of K for which Eq. (1) has nonzero solutions. These nonzero solutions will be the bowed shapes of the column. That is, we wish to determine the eigenvalues and eigenfunctions of the EVP given by Eq. (1). This problem has already been solved in Example 1 of Section 6.2. We shall now interpret the results of this example in terms of the column.

The trivial solution, $y \equiv 0$, means that the column is straight. It is usually called the unbuckled equilibrium state. Thus the unbuckled state is a possible equilibrium state for all values of K. For $K < 0$ it is the only solution of Eq. (1) as we have shown in Example 1 of Section 6.2. Since $P < 0$ if $K < 0$, the interpretation of this result is that a stretched column will not buckle.

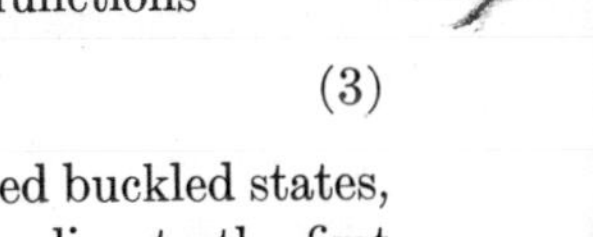

For $K \neq (n\pi)^2$, the trivial solution $y \equiv 0$ is the only solution of the EVP. For $K = (n\pi)^2$, other solutions given by the eigenfunctions

$$y = y_n = a_n \sin n\pi x \tag{3}$$

are possible states of equilibrium of the rod. They are called buckled states, since the rod is not straight. The eigenfunctions corresponding to the first three eigenvalues are sketched in Figure 6.5.

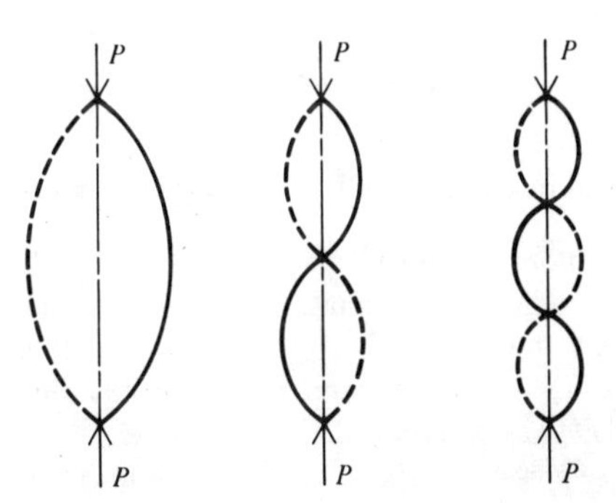

Figure 6.5

For $K < \pi^2$ the unbuckled state is the only possible equilibrium state of the rod. As P is increased from zero, the column will remain straight for $K < \pi^2$. When K reaches the lowest eigenvalue π^2, nontrivial solutions occur and the column will buckle. For this reason, $P = \pi^2 EI$ is called the buckling load of the rod. The amplitudes a_n of the buckled states are not determined, since eigenfunctions are known only within arbitrary multiplicative constants. In engineering practice, rods usually deform with large amplitudes when the buckling load is reached. If the amplitude is sufficiently large, the material of the column weakens and the column breaks. Consequently, in the actual design and analysis of structures involving columns, engineers must design the columns so that $K = P/EI < \pi^2$. For a fixed P this can be achieved by making I sufficiently large. The moment of inertia is large for cross sections whose area is "far" from the centroid. This accounts for the widespread use of columns with cross sections resembling the letters I and H.

Exercises

1. Determine the buckling load for a column of length L.

2. Consider a pinned end column, with constant EI, which is acted on by a *fixed* axial load P. Show that if the column is to remain straight, then its length L must satisfy $L^2 < EI\pi^2/P$. Thus, as P is increased, the column must be made shorter in order to support the larger load and still remain straight.

3. Consider the eigenvalue problem $y'' + Ky = 0$, and $y(0) = y(1) = 0$. Multiply both sides of the DE by y and integrate the result from 0 to 1. Then use integration by parts and the boundary conditions to show that

$$-\int_0^1 (y')^2\,dx + K\int_0^1 y^2\,dx = 0.$$

Thus if y is an eigenfunction and K is an eigenvalue, then

$$K = \int_0^1 (y')^2\,dx \Big/ \int_0^1 y^2\,dx \tag{i}$$

This ratio of two integrals is called the Rayleigh quotient.

*4. It is shown in more advanced textbooks that if a function $y = \phi(x)$, which satisfies the boundary conditions, makes the Rayleigh quotient in Exercise 3 smaller than for all other functions satisfying the boundary conditions, then $\phi(x)$ is the eigenfunction corresponding to the smallest eigenvalue of the EVP in Exercise 3. Furthermore, the quotient in Eq. (i), with $y = \phi$, is precisely equal to the lowest eigenvalue. This observation by Rayleigh is the basis of the widely used Rayleigh method for the approximate determination of eigenvalues and eigenfunctions. In this method a set of functions $\phi_1(x)$, $\phi_2(x), \ldots, \phi_n(x)$ is selected. Each function satisfies the boundary conditions. Then the linear combination $y = a_1\phi_1(x) + a_2\phi_2(x) + \cdots + a_n\phi_n(x)$, where the coefficients $a_1, \ldots, a_n$ are arbitrary constants, is formed. It is a guess for the eigenfunction. The linear combination satisfies the boundary conditions because each $\phi_i(x)$ satisfies them. (Show this.) We determine the a_i by first substituting the linear combination for y in the Rayleigh quotient. Then by performing all the indicated integrations, the quotient is an algebraic function $Q(a_1, \ldots, a_n)$. The a_i are evaluated from the condition that $Q(a_1, \ldots, a_n)$ is a minimum. This minimum is obtained by setting $\partial Q/\partial a_i = 0$ for $i = 1, \ldots, n$. This gives n linear algebraic equations for the n coefficients $a_1, \ldots, a_n$. If $a_i = a_i^0$, for $i = 1, \ldots, n$, is the solution of these equations, then $K = Q(a_1^0, \ldots, a_n^0)$ is an approximation of the lowest eigenvalue. (a) For the EVP in Exercise 3, $\phi_1(x) = x(1 - x)$ satisfies the boundary conditions. Use $y = a_1\phi_1$ as a guess for the eigenfunction. Use the Rayleigh method to obtain an approximation for the lowest eigenvalue. Compare with the exact value obtained in Example 1. (b) Give a linear combination of two functions that is a guess for the eigenfunction for the Rayleigh method. Set up the Rayleigh quotient but do not evaluate the integrals. Explain how to determine the approximation for the eigenvalues.

5. It can be shown that the frequencies and modes of the free vibrations of a tightly stretched elastic string of length $l = 1$ are obtained from the eigenvalues and eigenfunctions of the following EVP: $y'' + Ky = 0$, and $y(0) = y(1) = 0$. The eigenvalues K_n are proportional to the square root of the frequency of vibration. The eigenfunctions are the modes of vibration. The boundary conditions imply that the ends of the string are fixed. Determine the eigenvalues and eigenfunctions.

6. Assume that the end $x = 1$ of the string in Exercise 5 is resting on a spring, with spring constant $c > 0$. Then the boundary condition $y(1) = 0$ in Exercise 5 is replaced by $y'(1) - cy(1) = 0$. Determine the eigenvalues and eigenfunctions. Discuss the variation of the eigenvalues with the spring constant c.

chapter 7

THE SECOND-ORDER INHOMOGENEOUS EQUATION WITH CONSTANT COEFFICIENTS

7.1 INTRODUCTION

We now develop methods to solve the inhomogeneous DE

$$y'' + 2by' + cy = r(x). \tag{1}$$

The prescribed function $r(x)$ is called the inhomogeneous term or the forcing function. For simplicity of presentation, we assume that $r(x)$ is a continuous function on an interval (α, β).

In Section 3.2 we showed that every solution of the first-order inhomogeneous DE

$$y' + q(x)y = r(x) \tag{2}$$

is given by $y = y_c + y_p$, where y_c is a general solution of the homogeneous DE $y' + qy = 0$ and y_p is a solution of the inhomogeneous DE given by Eq. (2). We establish an analogous result for the second-order DE in the next two theorems.

theorem 1

Let y_1 and y_2 be two linearly independent solutions of the homogeneous DE $y'' + 2by' + cy = 0$. Furthermore let y_p be a solution of inhomogeneous DE (1). Then, any solution y of the inhomogeneous DE is given by

$$y = c_1y_1 + c_2y_2 + y_p$$

for suitable constants c_1 and c_2.

proof. Let y be a solution of DE (1). We define the function Y by $Y = y - y_p$. Thus

$$y = Y + y_p. \tag{3}$$

We now show that the function Y is a solution of the homogeneous DE by substituting Eq. (3) into DE (1). This gives

$$(Y + y_p)'' + 2b(Y + y_p)' + c(Y + y_p) = r.$$

Then by performing the indicated differentiations and regrouping terms we get

$$(Y'' + 2bY' + cY) + y_p'' + 2by_p' + cy_p = r(x). \tag{4}$$

Since y_p is a solution of the inhomogeneous DE, then $y_p'' + 2by_p' + cy_p = r(x)$, and hence Eq. (4) is reduced to

$$Y'' + 2bY' + cY = 0.$$

This shows that Y is a solution of the homogeneous DE. Thus there are constants c_1 and c_2 such that $Y = c_1y_1 + c_2y_2$. The proof is completed by inserting this expression for Y into Eq. (3).

theorem 2

Let y_c be a general solution of the homogeneous DE corresponding to DE (1) and y_p be any solution of (1). Then

$$y = y_c + y_p \tag{5}$$

is a solution of DE (1).

proof. We substitute Eq. (5) into the left side of Eq. (1). This gives

$$y'' + 2by' + cy = [y_c'' + 2by_c' + cy_c] + [y_p'' + 2by_p' + cy_p].$$

Since y_c is a solution of the homogeneous DE, the first square-bracketed term in the right side of Eq. (6) is zero. The second square-bracketed term equals $r(x)$ because y_p is a solution of DE (1). Thus we have $y'' + 2by' + cy = r(x)$. Hence Eq. (5) is a solution and the proof is completed.

These results suggest the following definition.

definition

A general solution of

$$y'' + 2by' + cy = r$$

is given by $y = y_c + y_p$, where y_c is a general solution of the homogeneous DE and y_p is any solution of inhomogeneous DE.

The functions y_c and y_p are called *complementary* and *particular* solutions of DE (1). We wish to emphasize that the particular solution is not a uniquely determined function. Because if y_p is any particular solution and $z(x)$ is any solution of the homogeneous DE, then it is clear from Theorem 2 that $y = y_p + z$ is also a particular solution.

This suggests the technique for finding general solutions of the inhomogeneous DE. First find a general solution of the homogeneous DE. Then find a particular solution by one of the methods we shall describe in this chapter. The sum is a general solution of the inhomogeneous DE.

To solve IVPs and BVPs, the two arbitrary constants in the general solution are determined by making the general solution satisfy the initial or boundary conditions. In the usual way this leads to two algebraic equations to determine the two constants. The solution thus obtained for the IVP is the only solution (see Exercise 9).

We conclude this section by establishing a superposition property of the solutions of DE (1) that is useful in constructing particular solutions. The property is valid because the DE is linear.

theorem 3

If the functions $y_{p,1}$ and $y_{p,2}$ are particular solutions of the DEs

$$y'' + 2by' + cy = r_1(x) \tag{6}$$

and

$$y'' + 2by' + cy = r_2(x), \tag{7}$$

respectively, then

$$y_p = y_{p,1} + y_{p,2} \tag{8}$$

is a particular solution of

$$y'' + 2by' + cy = r_1(x) + r_2(x). \tag{9}$$

proof. To establish this result we substitute Eq. (8) into the DE (9). This gives

$$y_p'' + 2by_p' + cy_p = (y_{p,1}'' + 2by_{p,1}' + cy_{p,1}) + (y_{p,2}'' + 2by_{p,2}' + cy_{p,2}).$$

Since $y_{p,1}$ and $y_{p,2}$ are solutions of DEs (6) and (7), we conclude that y_p is a particular solution corresponding to $r(x) = r_1(x) + r_2(x)$.

For the first-order DE (2) we obtained a particular solution in terms of an integral of $r(x)$. A corresponding result is obtained for the second-order DE (1) in Section 7.3 by the method of variation of parameters. That is, we express y_p in terms of integrals of $r(x)$. This gives a particular solution for any choice of $r(x)$. However, for a special class of functions $r(x)$, which occur frequently in applications, it is possible to obtain particular solutions more readily by the method of undetermined coefficients. In this method we develop a systematic procedure for determining an appropriate y_p if $r(x)$ belongs to this special class. This is discussed in the next section.

Exercises

1. In each of the following cases verify that the given function y_p is a particular solution of the DE. Then give a general solution of the DE.
(a) $y'' + y = x, \quad y_p = x$ (b) $y'' + y = e^x, \quad y_p = e^x/2$

(c) $y'' - 3y' - 4y = 4x^2, \quad y_p = -x^2 + \dfrac{3}{2}x - \dfrac{13}{8}$

(d) $y'' + 4y' - 2y = 8 \sin 2x, \quad y_p = -\dfrac{4}{25}(3 \sin 2x + 4 \cos 2x)$

2. In the DE $y'' + 3y' = 2$, the term containing y is absent; that is, $c = 0$. Thus the left side of the DE can be written as $y'' + 3y' = (y' + 3y)'$. Determine a general solution of this DE by first integrating it once. Then solve the resulting first-order DE. What are the complementary and particular solutions?

3. Repeat Exercise 2 for the DE $y'' + y' = e^{-x}$.

4. Obtain a general solution of $y'' + by' = r(x)$, which has the y term missing. Here $r(x)$ is any continuous function.

5. Prove that if $y = y_p(x)$ is a particular solution of
$$y'' + p(x)y' + q(x)y = r(x),$$
where $p(x)$, $q(x)$, and $r(x)$ are prescribed functions, and $y = u(x)$ is any solution of the corresponding homogeneous equation, then $y = u + y_p$ is a solution of the inhomogeneous DE.

6. Use Theorem 3 and Exercises 1a and b to show, without any calculation, that $y_p = x + (e^x/2)$ is a particular solution of $y'' + y = x + e^x$.

7. Suppose that, for a given inhomogeneous term $r_0(x)$, DE (1) has a particular solution y_{p0}. Show that if $r_0(x)$ is multiplied by any constant A, then a particular solution is given by Ay_{p0}.

8. Prove that if $y_{p,i}$, for $i = 1, 2, \ldots, n$, are particular solutions of
$$y'' + 2by' + cy = r_i(x)$$
for $i = 1, 2, \ldots, n$, respectively, then $y_p = y_{p,1} + y_{p,2} + \cdots + y_{p,n}$ is a par-

ticular solution of

$$y'' + 2by' + cy = r_1(x) + r_2(x) + \cdots + r_n(x).$$

*9. Given that $y = 0$ for all x is the only solution of the IVP

$$y'' + 2by' + cy = 0, \qquad y(x_0) = y'(x_0) = 0$$

(see Section 4.6), prove that the IVP given by Eqs. (1) and (2), with $r(x)$ a given function and b, c, and x_0 given numbers, has no more than one solution for each choice of initial values a_0 and a_1. (*Hint:* Let z_1 and z_2 be two solutions of the IVP and consider $z = z_1 - z_2$).

7.2 THE METHOD OF UNDETERMINED COEFFICIENTS

The special inhomogeneous terms

$$r(x) = (k_0 + k_1 x + \cdots + k_n x^n) e^{\alpha x} \sin \beta x$$

and

$$r(x) = (k_0 + k_1 x + \cdots + k_n x^n) e^{\alpha x} \cos \beta x,$$

where $k_0, k_1, \ldots, k_n$, α, and β are given constants, occur frequently in applications. Important special cases are $e^{\alpha x} \sin \beta x$, $e^{\alpha x} \cos \beta x$, and x^n. We shall now develop a systematic method for determining a particular solution of

$$y'' + 2by' + cy = r(x),$$

where $r(x)$ has one of these two special forms. We shall first consider simple cases to show how this procedure works. Although the method is not deductive and is limited essentially to these inhomogeneous terms, it is widely used because of its relative simplicity and the common occurrence of these forcing functions.

The idea is to look for a particular solution in the form

$$y_p = B_1 f_1(x) + B_2 f_2(x) + \cdots + B_n f_n(x),$$

where the $f_i(x)$ are specific functions and the constant coefficients B_i are to be determined. That is why this procedure is called the method of undetermined coefficients. We shall show how to choose the $f_i(x)$ for a given $r(x)$. The coefficients are then evaluated from the condition that y_p must satisfy the DE.

7.2-1 Exponential functions

We first study

$$y'' + 2by' + cy = ke^{\alpha x}, \tag{1}$$

where k and a are given real constants. Since $(e^{ax})' = ae^{ax}$ and $(e^{ax})'' = a^2e^{ax}$, we know that if $y = Be^{ax}$ is inserted into the left side of Eq. (1), then e^{ax} will be a common factor on both sides of the resulting equation. Thus we seek a particular solution in the form

$$y_p = Be^{ax}, \tag{2}$$

where B is a coefficient to be determined. We insert this guess into DE (1). This gives

$$(a^2 + 2ba + c)Be^{ax} = ke^{ax}.$$

We divide both sides of this equation by e^{ax} and get

$$(a^2 + 2ba + c)B = k. \tag{3}$$

If

$$a^2 + 2ba + c \neq 0, \tag{4}$$

we solve Eq. (3) for B and get $B = k/(a^2 + 2ba + c)$. The particular solution given by Eq. (2) is then

$$y_p = \frac{k}{a^2 + 2ba + c}\, e^{ax}. \tag{5}$$

If $a^2 + 2ba + c = 0$ that is, if a is a characteristic root, then there is no value of B that will satisfy Eq. (3). Hence, in this case, there is no particular solution in the form given by Eq. (2). We wish to emphasize that the condition given by Eq. (4) is equivalent to the statement that y_p given in Eq. (2) is not a solution of the homogeneous DE corresponding to Eq. (1).

If a satisfies Eq. (4), then Eq. (5) is a formula for a particular solution of Eq. (1). We recommend that the reader *not* memorize this formula. A particular solution should be obtained in any example in the same way as we derived Eq. (5). The procedure is illustrated in the next example.

EXAMPLE 1

Find a particular solution of

$$y'' - 3y' + 2y = 7e^{4x}.$$

solution. A general solution of the homogeneous DE is

$$y_c = c_1e^x + c_2e^{2x}.$$

The choice

$$y_p = Be^{4x} \tag{6}$$

is appropriate for a particular solution because it is not a solution of the homogeneous DE. We substitute Eq. (6) into the DE and get

$$(16 - 12 + 2)Be^{4x} = 7e^{4x}.$$

We divide both sides of this equation by e^{4x} and conclude that $B = 7/6$ and thus $y_p = (7/6)e^{4x}$.

We shall now consider DE (1) when a is a characteristic root. Then the choice given in Eq. (2) for a particular solution fails because it is a solution of the homogeneous DE. It is clear that an appropriate y_p in this case must be selected so that when it is inserted in the left side of Eq. (1), e^{ax} is a factor in the resulting expression. Thus we assume that

$$y_p = f(x)e^{ax}. \tag{7}$$

We determine f by requiring that Eq. (7) satisfy DE (1). Any f for which this is true will provide a particular solution. The derivatives of y_p are $y_p' = (f' + af)e^{ax}$ and $y_p'' = (f'' + 2af' + a^2f)e^{ax}$. We substitute these expressions for y_p and its derivatives into DE (1). This gives

$$[f'' + 2(a + b)f' + (a^2 + 2ba + c)f]e^{ax} = ke^{ax}.$$

By canceling the factor e^{ax} and observing that $a^2 + 2ba + c = 0$, because a is now a characteristic root, this equation is reduced to

$$f'' + 2(a + b)f' = k, \qquad \text{if } a + b \neq 0, \tag{8}$$

$$f'' = k, \qquad \text{if } a + b = 0. \tag{9}$$

If a is a repeated root of the characteristic equation, then it follows from the solution formula for the quadratic equation that $a = -b$ or equivalently that $a + b = 0$. Thus Eq. (9) holds if a is a repeated characteristic root and Eq. (8) holds if it is a simple characteristic root.

We first consider Eq. (8). The term proportional to f is absent from the left side. Therefore Eq. (8) is easily solved by first integrating the equation and then solving the resulting first-order DE. We set the integration constants equal to zero because any solution of Eq. (8) is sufficient. In this way we get

$$y_p = \frac{k}{2(a + b)} xe^{ax} \tag{10}$$

as a particular solution when e^{ax} is a solution of the homogeneous DE. We observe that in this case y_p is proportional to x times e^{ax}.

When a is a repeated characteristic root, then we know that e^{ax} and xe^{ax} are solutions of the homogeneous DE. In this case $a + b = 0$ and the expression in Eq. (10) is not defined. To find y_p we integrate both sides of Eq. (9) twice. Since we are seeking any solution of Eq. (9), we set the integration constants equal to zero. This gives $f = (k/2)x^2$, and hence from Eq. (7) we get

$$y_p = \frac{k}{2} x^2 e^{ax}. \tag{11}$$

Thus when x and xe^{ax} are solutions of the homogeneous DE, then a particular solution is proportional to x times xe^{ax}.

Equations (10) and (11) should not be memorized. In any problem, particular solutions should be obtained, as we illustrate in the next example.

EXAMPLE 2

Determine a particular solution of

$$y'' - 3y' + 2y = 3e^x.$$

solution. A complementary solution is

$$y_c = c_1e^{2x} + c_2e^x.$$

Since Be^x is a solution of the homogeneous DE, it is an inappropriate choice for y_p. Thus we try

$$y_p = Bxe^x. \tag{12}$$

It is a suitable choice because xe^x is not a solution of the homogeneous DE. We determine the constant B by substituting Eq. (12) into the DE. We get

$$[(2B + Bx) - 3(B + Bx) + 2Bx]e^x = 3e^x.$$

Hence $B = -3$ and $y_p = -3xe^x$.

7.2-2 Trigonometric functions

In the following examples we illustrate the techniques of choosing the form of particular solutions when $r(x) = k \sin \beta x$ or $r(x) = k \cos \beta x$.

EXAMPLE 1

Find a particular solution of

$$y'' - 4y' + 5y = 2 \cos x.$$

solution. In analogy with the results of Section 7.2-1, we guess that $y_p = B \cos x$. Substituting this into the DE gives

$$4B \cos x + 4B \sin x = 2 \cos x. \tag{1}$$

There is no value of B that will satisfy this equation because of the $\sin x$ term. We conclude that $y_p = B \cos x$ is not a particular solution. However, Eq. (1) suggests that we should add a term proportional to $\sin x$ to our guess to balance the term

$4B \sin x$. Therefore we try

$$y_p = B_1 \cos x + B_2 \sin x.$$

Substituting this into the DE and combining like terms, we get

$$-(4B_2 - 4B_1) \cos x + (4B_1 + 4B_2) \sin x = 0.$$

Since the functions $\cos x$ and $\sin x$ are linearly independent (see Section 4.5-1), the coefficients of $\cos x$ and $\sin x$ must be equal to zero. Thus we obtain the two algebraic equations

$$4(B_1 - B_2) = 2 \qquad \text{and} \qquad 4(B_1 + B_2) = 0.$$

Their solution is $B_1 = \frac{1}{4}$ and $B_2 = -\frac{1}{4}$, and hence

$$y_p = \tfrac{1}{4} \cos x - \tfrac{1}{4} \sin x.$$

EXAMPLE 2

Find a particular solution of

$$y'' + y = 6 \sin x.$$

solution. The choice $y_p = B_1 \cos x + B_2 \sin x$ is inappropriate because it is a solution of the homogeneous DE.

As in the exponential case, we shall seek a new choice for y_p by multiplying the old one by x. Thus we try $y_p = x(B_1 \cos x + B_2 \sin x)$. We substitute this into the DE and get, after some simplification,

$$2B_2 \cos x - 2B_1 \sin x = 6 \sin x.$$

Therefore by equating coefficients of $\cos x$ and $\sin x$ on both sides of this equation, we get $B_1 = -3$ and $B_2 = 0$, and the particular solution is $y_p = -3x \cos x$.

A generalization of the results of Examples 1 and 2 is given in line 2 of Table 7.1 in Section 7.2-4. They could also be obtained directly from the results of the previous section by taking a in e^{ax} as the imaginary number $a = \pm i\beta$ (see Exercise 4).

7.2-3 Polynomials

We now wish to obtain a particular solution of

$$y'' + 2by' + cy = r(x) = k_0 + k_1 x + \cdots + k_n x^n, \tag{1}$$

where $k_0, \ldots, k_n$ are given constants and n is a nonnegative integer. Thus $r(x)$ is an nth-degree polynomial. It includes the special case when $r(x)$ is a single power; that is, $r(x) = k_p x^p$, where p is a nonnegative integer.

If $c \neq 0$, it should be clear that an nth-degree polynomial is a reasonable choice for y_p. That is,

$$y_p = B_0 + B_1 x + \cdots + B_n x^n, \tag{2}$$

where $B_0, \ldots, B_n$ are constants to be determined. This polynomial should have the same degree as $r(x)$ because when (2) is inserted in (1), cB_nx^n will be the only term on the left side to balance k_nx^n on the right side. We illustrate the technique in Example 1.

EXAMPLE 1

Determine a particular solution of

$$y'' + 2y' + 5y = x^2 + 3. \tag{3}$$

solution. The inhomogeneous term in Eq. (3) is a second-degree polynomial. Since $c \neq 0$, we seek a particular solution of the form

$$y_p = B_0 + B_1x + B_2x^2. \tag{4}$$

We determine the constants B_0, B_1, and B_2 by substituting Eq. (4) into DE (3). After some calculations we get,

$$5B_2x^2 + [4B_2 + 5B_1]x + [2B_2 + 2B_1 + 5B_0] = x^2 + 3.$$

This equation will be satisfied for all values of x if the coefficients of the same powers of x on both sides are equal. This gives

$$5B_2 = 1, \qquad 4B_2 + 5B_1 = 0, \qquad 2B_2 + 2B_1 + 5B_0 = 3.$$

By solving these three algebraic equations for B_0, B_1, and B_2 and substituting the result in Eq. 4, we obtain the particular solution,

$$y_p = \frac{73}{125} - \frac{4}{25}x + \frac{x^2}{5}.$$

If $c = 0$, then DE (1) is given by

$$y'' + 2by' = k_0 + k_1x + \cdots + k_nx^n. \tag{5}$$

Equation (2) is an inappropriate form for a particular solution of DE (5) because y_p' and y_p'' are then polynomials of degree less than n, and the term cy_p is absent in Eq. (5). Thus there is no term proportional to x^n on the left side of Eq. (5) to balance k_nx^n on the right side. The appropriate choice if $b \neq 0$ is the $(n+1)$th-degree polynomial

$$y_p = B_1x + \cdots + B_nx^n + B_{n+1}x^{n+1}. \tag{6}$$

The constant term B_0 is omitted from Eq. (6) because it is a solution of the homogeneous equation. If $b = 0$, then a polynomial of degree $n + 2$,

$$y_p = B_2x^2 + \cdots + B_nx^n + B_{n+1}x^{n+1} + B_{n+2}x^{n+2}, \tag{7}$$

is the proper choice. The terms B_0 and B_1x are omitted from Eq. (7) because they are both solutions of the homogeneous equation. We note that if $c = b = 0$, then Eq. (5) reduces to

$$y'' = k_0 + \cdots + k_n x^n,$$

which can be integrated directly.

The general rule is: If any term in the polynomial given in Eq. (2) is a solution of the homogeneous DE (specifically B_0), then the appropriate form for y_p is found by multiplying Eq. (2) by x. If any term in the resulting polynomial is a solution of the homogeneous DE, then this choice must in turn be multiplied by x.

7.2-4 Products and summary

We shall study inhomogeneous terms that are products of exponentials, trigonometric functions, and polynomials. The results of the previous subsections for exponentials, trigonometric functions, and polynomials are summarized in Table 7.1.

We first consider

$$y'' + 2by' + cy = G(x)e^{ax} \tag{1}$$

where $G(x)$ is a specified function. A particular solution of Eq. (1) must give an exponential factor when it is substituted in the left side of the DE. Therefore we seek a particular solution in the form

$$y_p = f(x)e^{ax}. \tag{2}$$

We substitute Eq. (2) into Eq. (1) and get

$$f'' + 2(a + b)f' + (a^2 + 2ba + c)f = G(x). \tag{3}$$

Therefore, if a particular solution f of Eq. (3) is known, then Eq. (2) is a particular solution of Eq. (1). Specifically, if $G(x)$ is $\sin \beta x$, $\cos \beta x$, or a polynomial, then we can use the results of the preceding subsections to obtain a particular solution of Eq. (3).

Finally, if $r(x)$ is the triple product of a polynomial, an exponential, and either $\sin \beta x$ or $\cos \beta x$, then by analogous reasoning we obtain the particular solution listed in line 5 of Table 7.1 (see Exercise 9). We emphasize that, as in our previous work, if any of the terms in a possible choice for y_p is a solution of the homogeneous DE, then the complete choice must be multiplied by x.

TABLE 7.1 Forms for Particular Solutions of $y'' + 2by' + cy = r(x)$

$r(x)$	y_p
(1) ke^{ax}	Be^{ax}
(2) $k \sin \beta x$ or $k \cos \beta x$	$B_1 \cos \beta x + B_2 \sin \beta x$
(3) kx^n or $k_0 + k_1x + \cdots + k_nx^n$	$B_0 + B_1x + \cdots + B_nx^n$
(4) $(k_0 + k_1x + \cdots + k_nx^n)e^{ax}$	$(B_0 + B_1x + \cdots + B_nx^n)e^{ax}$
(5) $(k_0 + k_1x + \cdots + k_nx^n)e^{ax} \sin \beta x$ or $(k_0 + k_1x + \cdots + k_nx^n)e^{ax} \cos \beta x$	$(B_0 + B_1x + \cdots + B_nx^n)e^{ax} \cos \beta x + (D_0 + D_1x + \cdots + D_nx^n)e^{ax} \sin \beta x$

If any term in the sums given in this table for y_p is a solution of the homogeneous DE, then the appropriate choice should be x times the tabulated y_p. If any term in xy_p is a solution of the homogeneous DE, then x^2y_p is the appropriate form.

EXAMPLE

Determine the form for y_p for the DE

$$y'' + 2y' + 5y = e^{-x} \sin 2x.$$

solution. A general solution of the homogeneous DE is

$$y_c = c_1e^{-x} \cos 2x + c_2e^{-x} \sin 2x. \tag{4}$$

Since $r = e^{-x} \sin 2x$, we get from Table 7.1 that $y_p = B_1e^{-x} \cos 2x + B_2e^{-x} \sin 2x$. However, $e^{-x} \cos 2x$ and $e^{-x} \sin 2x$ are solutions of the homogeneous DE. This choice is inappropriate, and we must multiply it by x. Therefore we have

$$y_p = B_1xe^{-x} \cos 2x + B_2xe^{-x} \sin 2x. \tag{5}$$

This is a suitable choice because neither term in Eq. (5) is a solution of the homogeneous DE, as we see from Eq. (4).

Exercises

1. Use the method of undetermined coefficients to obtain a particular solution for each of the following inhomogeneous DEs.

(a) $y'' + 4y = 4e^{2x}$
(b) $y'' + 3y' = e^x$
(c) $y'' - y = e^x$
(d) $y'' - 3y' + 2y = 4e^{4x} + 3e^x$
(e) $y'' + y' - y = 3e^x$
(f) $y'' - y = \sinh x$
(g) $y'' + y = \sin 3x$
(h) $y'' + 2y' + y = \cos 2x$
(i) $y'' + y = \sin x$
(j) $y'' - 4y' + 5y = 2e^{3x} + \cos x$
(k) $y'' + p^2y = \sin \alpha t, \quad p \neq \alpha$
(l) $y'' + p^2y = \sin \alpha t, \quad p = \alpha$
(m) $y'' - 3y' = 3x^3$
(n) $y'' - 3y' + 2y = 3x^3 + x$
(o) $y'' + y' + y = x(1 + x)e^x$
(p) $y'' + 4y' + 5y = 10e^{-2x} \cos x$

2. For each of the following equations use the method of undetermined coefficients to determine an appropriate form of y_p. Do not evaluate the constants. Find a general solution.
   (a) $y'' + 3y' + y = e^x(x^2 + \sin x) + x^3e^{2x} \sin 3x$
   (b) $y'' + y = \sin^2 x$ (c) $y'' - y = (x + 3x^2)e^{2x} \cos x$
   (d) $y'' + y' + 2y = (1 + 2x)e^{-x} \sin x$

3. Solve the following IVPs.
   (a) $y'' - 5y' + 6y = (12x - 7)e^{-x}, \quad y(0) = y'(0) = 0$
   (b) $y'' - 4y' + 5y = 2x^2e^x, \quad y(0) = 2,\ y'(0) = 3$
   (c) $y'' + 4y = 4(\sin 2x + \cos 2x), \quad y(\pi) = y'(\pi) = 2\pi$

4. Obtain particular solutions of $y'' + 2by' + cy = k \sin \beta x$ by first using Euler's formulas to express $\sin \beta x$ as a sum of exponential functions e^{ax} where a is a pure imaginary number. Then use the results of Section 7.2-1.

5. Show for the DE $y'' + 2by' + cy = K_0 + K_1x + \cdots + K_nx^n$, where $c \neq 0$, that by choosing $y_p = B_0 + B_1x + \cdots + B_px^p$, where $p > n$, we obtain $B_{n+1} = B_{n+2} = \cdots = B_p = 0$. Thus if we include extra powers in our choice for y_p, we shall still obtain the correct result.

6. Prove that $A_0 + A_1x + \cdots + A_nx^n = 0$ for all x in the interval (α, β), where $A_0, A_1, \ldots, A_n$ are constants, implies that $A_0 = A_1 = \cdots = A_n = 0$.

7. For the DE $y'' + 2by' + cy = r(x)$ where $r(x) = K \sin \beta x$ or $r(x) = K \cos \beta x$, prove the following statements.
   (a) If $i\beta$ is not a characteristic root or, equivalently, if $r(x)$ is not a solution of the homogeneous DE, then $y_p = B_1 \cos \beta x + B_2 \sin \beta x$.
   (b) If $i\beta$ is a characteristic root, then $r(x)$ is a solution of the homogeneous DE, and $y_p = x(B_1 \cos \beta x + B_2 \sin \beta x)$.

8. For the DE $y'' + 2by' + cy = G(x)e^{ax}$, where $G(x)$ is a polynomial of degree $n \geq 0$, prove that $y_p = f(x)e^{ax}$, where $f(x)$ is a polynomial. It is of degree n if a is not a characteristic root; of degree $n + 1$ if a is a simple characteristic root; and of degree $n + 2$ if a is a repeated characteristic root.

*9. Derive the particular solution listed in line 5 of Table 7.1 by first obtaining y_p for the DE $y'' + 2by' + cy = e^{\alpha x}$, where $\alpha = a \pm i\beta$, and using the techniques for exponential functions given in Section 7.2-1. Then obtain a particular solution of the DE $y'' + 2by' + cy = G(x)e^{\alpha x}$, in the form $f(x)e^{\alpha x}$.

7.3 VARIATION OF PARAMETERS

Whenever possible, the method of undetermined coefficients should be used to find a particular solution of

$$y'' + 2by' + cy = r(x).$$

However, we have seen in the preceding section that it is readily applicable only for a limited class of inhomogeneous terms.

We now present another method, which is called *variation of parameters,* to find a particular solution. It is related to the method of reduction of order which was used in Section 4.4. We shall derive a formula for the particular solution which is valid for any continuous inhomogeneous term. We first illustrate the method by considering a simple example.

EXAMPLE

Find a particular solution of

$$y'' + y = \tan x, \qquad \text{for } -\frac{\pi}{2} < x < \frac{\pi}{2} \tag{1}$$

by the method of variation of parameters. We stipulate the interval $(-\pi/2, \pi/2)$ because $\tan x$ is discontinuous at $x = \pm\pi/2$.

solution. To apply the method we use the general solution

$$y = c_1 \cos x + c_2 \sin x, \tag{2}$$

of the homogeneous equation corresponding to Eq. (1). In the method of variation of parameters we seek a particular solution in the form

$$y_p = u(x) \cos x + v(x) \sin x, \tag{3}$$

where we replace the constants in Eq. (2) by functions; that is, we let the parameters c_1 and c_2 vary. To determine the two unknown functions u and v we need two equations. Only one equation is obtained by requiring that Eq. (3) satisfy DE (1). We choose the second equation so that the resulting two equations for u and v are easier to solve than DE (1). Specifically, we shall try to make u and v satisfy first-order DEs because we know how to solve them. Of course, the two equations should not be inconsistent.

To substitute Eq. (3) into DE (1) we first compute y_p' and then y_p''. This gives

$$y_p' = -u \sin x + v \cos x + [u' \cos x + v' \sin x]. \tag{4}$$

The first derivatives of u and v appear in the bracketed term on the right side of Eq. (4). When we compute y_p'' by differentiating Eq. (4), it will contain u'' and v''. Thus when we substitute Eq. (3) into Eq. (1), the resulting equation will contain u'' and v''. To avoid these second derivatives, we set the bracketed term in Eq. (4) equal to zero. This gives the first of the required two equations as

$$u' \cos x + v' \sin x = 0. \tag{5}$$

Then y_p' is given by

$$y_p' = -u \sin x + v \cos x. \tag{6}$$

Differentiation of Eq. (6) yields

$$y_p'' = -u \cos x - u' \sin x - v \sin x + v' \cos x. \tag{7}$$

We substitute Eqs. (3) and (7) into DE (1) and obtain

$$-u \cos x - u' \sin x - v \sin x + v' \cos x + (u \cos x + v \sin x) = \tan x.$$

By combining terms in this equation we get the second of the required two equations,

$$-u' \sin x + v' \cos x = \tan x. \tag{8}$$

We consider the two first-order DEs Eqs. (5) and (8), as two algebraic equations for u' and v'. To solve them for u' and v', we first multiply Eq. (5) by $\sin x$ and Eq. (8) by $\cos x$ and add the resulting equations. This gives,

$$v' = \sin x. \tag{9}$$

Similarly, we obtain

$$u' = -\frac{\sin^2 x}{\cos x} = \cos x - \sec x. \tag{10}$$

The functions u and v can now be found by direct integration of Eqs. (9) and (10). Thus[1]

$$u = \sin x - \ln(\tan x + \sec x) + b_1,$$

$$v = -\cos x + b_2, \tag{11}$$

where b_1 and b_2 are arbitrary constants of integration. Finally we substitute Eq. (11) into Eq. (3) and get

$$y_p = b_1 \cos x + b_2 \sin x - \cos x \ln(\tan x + \sec x), \tag{12}$$

as a particular solution for all values of the constants b_1 and b_2. Since $b_1 \cos x + b_2 \sin x$ is a general solution of the homogeneous DE, it is customary in the application of variation of parameters to take $b_1 = b_2 = 0$. Then we get a particular solution

$$y_p = -\cos x \ln(\tan x + \sec x).$$

We shall now apply variation of parameters to determine a formula for a particular solution of the inhomogeneous equation

$$y'' + 2by' + cy = r(x), \tag{13}$$

for any given continuous $r(x)$ and any given coefficients b and c. We assume that a general solution

$$y_c = c_1 y_1(x) + c_2 y_2(x) \tag{14}$$

[1] We use the integration formula,

$$\int^x \sec z \, dz = \ln|\tan x + \sec x|.$$

The absolute value signs are omitted in Eq. (11) because $\tan x + \sec x$ is positive in the interval $-\pi/2 < x < \pi/2$.

of the corresponding homogeneous equation has been obtained. According to the method of variation of parameters, we seek a particular solution of Eq. (13) in the form

$$y_p = u(x)y_1(x) + v(x)y_2(x). \tag{15}$$

We must determine the unknown functions $u(x)$ and $v(x)$ so that Eq. (15) is a solution.

Differentiation of Eq. (15) yields

$$y_p' = u'y_1 + uy_1' + v'y_2 + vy_2' = uy_1' + vy_2' + [u'y_1 + v'y_2].$$

We require that the bracketed term in this equation (which contains u' and v') be zero. This gives

$$u'y_1 + v'y_2 = 0. \tag{16}$$

Then y_p' is given by

$$y_p' = uy_1' + vy_2'.$$

Differentiation of this expression gives y_p''.

We substitute these expressions for y_p, y_p', and y_p'' into DE (13) and get

$$(y_1'' + 2by_1' + cy_1)u + (y_2'' + 2by_2' + cy_2)v + u'y_1' + v'y_2' = r. \tag{17}$$

Since y_1 and y_2 are solutions of the homogeneous DE, the bracketed terms in Eq. (17) are zero and we find that

$$u'y_1' + v'y_2' = r. \tag{18}$$

Thus Eqs. (16) and (18) are the two algebraic equations for u' and v'. As the reader can verify, in order to solve these equations for u' and v' we must have $y_1y_2' - y_1'y_2 \neq 0$. However, by the definition of a general solution (see Section 4.5), the functions y_1 and y_2 must satisfy the conditions

$$W = y_1y_2' - y_1'y_2 \neq 0,$$

where W is the Wronskian. Thus we can uniquely solve algebraic equations (16) and (18) for u' and v'. We get

$$u' = -\frac{y_2r}{W} \qquad \text{and} \qquad v' = \frac{y_1r}{W}. \tag{19}$$

We integrate Eq. (19) to obtain

$$u = -\int^x \frac{y_2r}{W}\,dz + b_1 \qquad \text{and} \qquad v = \int^x \frac{y_1r}{W}\,dz + b_2, \tag{20}$$

where b_1 and b_2 are arbitrary constants. Finally, by substituting Eq. (20) into Eq. (15), we obtain

$$y_p = b_1y_1 + b_2y_2 - y_1\int^x \frac{y_2r}{W}\,dz + y_2\int^x \frac{y_1r}{W}\,dz. \tag{21}$$

Since $b_1y_1 + b_2y_2$ is a general solution of the homogeneous DE, we set $b_1 = b_2 = 0$ in Eq. (21) and get the particular solution,

$$y_p = -y_1 \int^x \frac{y_2 r}{W}\, dz + y_2 \int^x \frac{y_1 r}{W}\, dz. \tag{22}$$

In many instances, the integrals in Eq. (22) cannot be evaluated explicitly. Then it is necessary to leave y_p in the integral form given in (22).

Exercises

1. Use the method of variation of parameters to obtain a general solution of the following equations.

   (a) $y'' - y = e^x$

   (b) $y'' + 3y' + 2y = x$

   (c) $y'' + y = \sec x$, for $-\dfrac{\pi}{2} < x < \dfrac{\pi}{2}$

   (d) $y'' + y = \cot x$, for $-\dfrac{\pi}{2} < x < \dfrac{\pi}{2}$

   (e) $y'' + y' - 2y = \ln x, \quad x > 0$

   (f) $y'' + y = 1/(1 + \sin x)$

   (g) $y'' + 3y' + 2y = \dfrac{1}{1 + e^x}$

   (h) $y'' - 3y' + 2y = \dfrac{e^{2x}}{1 + e^{2x}}$

   (i) $y'' - y = \dfrac{1}{e^x - e^{-x}}$

2. Try to solve Exercise 1(c) using the method of undetermined coefficients. What are the difficulties in the application of this method?

3. Find y_p for $y'' + y = r(x)$, where $r(x)$ is any continuous function.

*4. Use the result of Exercise 3 to show that the solution of the IVP $y'' + y = r(x)$, $y(0) = y'(0) = 0$ can be written as

$$y(x) = \int_0^x r(z) \sin (x - z)\, dz.$$

*5 Use the result of Exercise 4 to show that the solution of the IVP $y'' + y = r(x)$, $y(0) = a_0$, $y'(0) = a_1$ can be written as

$$y(x) = \int_0^x r(z) \sin (x - z)\, dz + a_0 \cos x + a_1 \sin x.$$

chapter 8

APPLICATIONS OF THE SECOND-ORDER INHOMOGENEOUS EQUATION

8.1 PROPERTIES OF THE SOLUTIONS AND APPLICATIONS

We shall study properties of the solution of the IVP for

$$y'' + 2by' + cy = r(x) \tag{1}$$

by considering specific applications to a variety of problems. First we study some simple examples.

EXAMPLE 1

This example involves the response to a constant forcing function. Solve the IVP,

$$y'' + 2by' + y = k, \qquad y(0) = 1, y'(0) = 0,$$

where b and k are specified constants, and $|b| < 1$.

solution. Since $|b| < 1$, a complementary solution (a general solution of the homogeneous equation) is

$$y_c = e^{-bx}(c_1 \cos qx + c_2 \sin qx) \qquad \text{and} \qquad q = \sqrt{1 - b^2}.$$

Since $r(x) = k$ is a constant, we employ the method of undetermined coefficients and seek a particular solution in the form

$$y_p = B.$$

We substitute this guess into the DE and get $B = k$. Therefore

$$y = y_c + y_p = e^{-bx}(c_1 \cos qx + c_2 \sin qx) + k$$

is a general solution. We determine the arbitrary constants c_1 and c_2 by requiring the general solution to satisfy the initial conditions. This gives

$$y(0) = c_1 + k = 1$$

and

$$y'(0) = -bc_1 + qc_2 = 0.$$

We solve these algebraic equations and substitute the result into the general solution. This gives

$$y = e^{-bx}\left[(1 - k) \cos qx + \frac{b}{q}(1 - k) \sin qx\right] + k.$$

We rewrite this solution in the amplitude–phase angle form as

$$y = Ae^{-bx} \cos (qx - \phi) + k \tag{2}$$

where the amplitude A and the phase angle ϕ are defined by

$$A = (1 - k)\left(1 + \left(\frac{b}{q}\right)^2\right)^{1/2} \quad \text{and} \quad \phi = \tan^{-1}\frac{b}{q}.$$

Here ϕ is the principal value of $\tan^{-1}(b/q)$ (see Section 5.1). The solution is sketched in Figure 8.1 for a representative value of k.

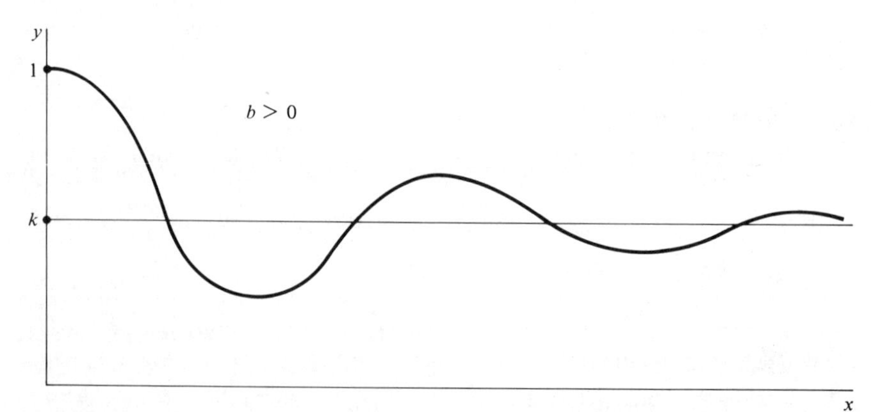

Figure 8.1

We now discuss the qualitative behavior of the solution. If $b > 0$, then the solution represents a damped oscillation about $y = k$ as $x \to \infty$. The term in Eq. (2) that is multiplied by A approaches zero as $x \to \infty$. It is the transient response. Thus, for sufficiently large x,

$$y \approx y_s(x) = k,$$

where y_s is the steady-state response (see Sections 3.2 and 3.3 for a discussion of transient and steady-state responses). Thus when b is in the interval $0 < b < 1$, the steady-state response is equal to the forcing function. It can be shown that when $b \geq 1$, the steady-state response is also equal to the forcing function; see Exercise 1(d).

If $b = 0$ (no damping), the solution is

$$y = y_c + y_p = (1 - k) \cos x + k.$$

It is a periodic function that represents an oscillation about the constant $y_p = k$. Since no term in the solution approaches zero as $x \to \infty$, there is no transient decay. Thus the full solution is a steady-state response. If $b < 0$ (negative damping), $Ae^{-bx} \cos (qx - \phi)$ grows exponentially as x increases. It dominates the particular solution $y_p = k$ as x increases. There is no transient response in this case.

EXAMPLE 2

This example involves the response to a linear function. Solve the IVP

$$y'' + 2by' + y = x \qquad y(0) = 1, y'(0) = 0,$$

where $| b | < 1$.

solution. A complementary solution is given in Example 1. We use the method of undetermined coefficients to find a particular solution. We choose

$$y_p = B_1x + B_2$$

and substitute this into the DE to determine the constants B_1 and B_2. Thus a general solution of the DE is

$$\begin{aligned} y &= y_c + y_p \\ &= e^{-bx}(c_1 \cos qx + c_2 \sin qx) + x - 2b, \qquad q = \sqrt{1 - b^2}. \end{aligned} \tag{3}$$

The constants c_1 and c_2 are evaluated by requiring this general solution to satisfy the initial conditions. We solve the resulting algebraic equations and get,

$$c_1 = 1 + 2b \qquad \text{and} \qquad c_2 = \frac{b + 2b^2 - 1}{q}.$$

The solution of the initial-value problem is obtained by substituting these expressions for c_1 and c_2 into the general solution, Eq. (3).

If $b > 0$, then the complementary solution is an exponentially damped oscillation. It is a transient response. The full solution approaches $y_p = x - 2b$ as $x \to \infty$, as we see in Figure 8.2. Thus y_p is the steady-state response. If $b = 0$, then the y_c persists for all x. The full solution represents a periodic oscillation about y_p. There is no transient term.

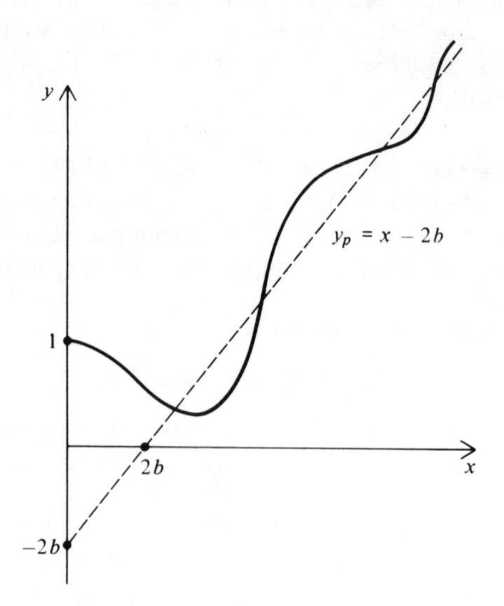

Figure 8.2

Exercises

1. Solve each of the following IVPs. Determine the transient and steady responses. Is the solution a periodic function? If it is a periodic function, determine its period. Determine the behavior of the solution as $x \to \infty$.
(a) $y'' + y = \sin 2x, \qquad y(0) = y'(0) = 0$
(b) $y'' + y = \sin x, \qquad y(0) = 0,\ y'(0) = 2$
(c) $y'' + 2by' + y = e^{-x} \sin 3x, \qquad y(0) = y'(0) = 0,$
for $0 \le b < 1$
(d) $y'' + 2by' + y = k, \qquad y(0) = 1,\ y'(0) = 0,$
for $b \ge 1$ and k any constant

2. Solve the following BVPs.
(a) $y'' - y = 1, \qquad y(0) = y(1) = 0$
(b) $y'' - y = \sin x, \qquad y(0) = y(1) = 0$
(c) $y'' + 2y' + y = x^2 + 3, \qquad y(0) = 1,\ y(1) = 0$

3. Show that the solution of the general IVP

$$y'' + 2by' + cy = r(x), \qquad y(x_0) = a_0,\ y'(x_0) = a_1$$

is given by $y = H(x) + F(x)$, where H and F are solutions of the following IVPs:

$$H'' + 2bH' + cH = 0, \qquad H(x_0) = a_0,\ H'(x_0) = a_1$$

$$F'' + 2bF' + cF = r(x), \qquad F(x_0) = F'(x_0) = 0.$$

The function $H(x)$ is the response to the initial conditions, and the function $F(x)$ is the response to the forcing function $r(x)$. Thus we can study the effects of the initial conditions and the forcing function separately and add the results.

4. Find $H(x)$ and $F(x)$ for the following IVPs.
   (a) $y'' + y = k, \quad y(0) = 1, y'(0) = 2,$
   where k is a constant
   (b) $y'' + 2y' + y = \sin x, \quad y(0) = 0, y'(0) = 1$
   (c) $my'' + \mu y' = mg, \quad y(0) = 0, y'(0) = a_1$
   This DE describes the motion of a particle of mass m falling through a resisting medium with friction coefficient μ. Here a_1 is the initial velocity and g is the gravity constant.

5. In many applications a forcing function only acts for a fixed length of time. For example, consider the IVP $y'' + y = r(x), \quad y(0) = 1, y'(0) = 0$, where

$$r(x) = \begin{cases} 1 - x, & \text{for } 0 < x < 1 \\ 0, & \text{for } \quad x \geq 1. \end{cases}$$

   (a) Find the solution of this IVP by first solving the DE in the two intervals $0 < x < 1$ and $x > 1$. Then require that the two solutions are equal and have equal derivatives at $x = 1$.
   (b) Sketch the solution.

6. Solve the following IVPs and sketch the solutions.
   (a) $y'' + y = r(x), \quad y(0) = y'(0) = 0$, with

$$r(x) = \begin{cases} \sin 2x, & \text{for } 0 \leq x < \pi/2 \\ 0, & \text{for } \quad x \geq \pi/2 \end{cases}$$

   (b) $y'' + y = r(x), \quad y(0) = y'(0) = 0$, with

$$r(x) = \begin{cases} 1, & \text{for } 0 \leq x < 1 \\ 0, & \text{for } \quad x \geq 1 \end{cases}$$

   In this case $r(x)$ is a discontinuous function. Since y is continuous, this implies that y'' is discontinuous.

8.2 A RESOURCE MANAGEMENT PROBLEM

A large artificial lake was created during the completion of a hydroelectric dam project. The government wishes to make the lake and the adjoining property a public recreation area with particular emphasis on sportfishing. The state conservation department will stock the lake periodically with fish. The insects and vegetation in the lake provide an unlimited food supply for the fish. The government wishes to make the lake attractive to fishermen, but not overly attractive because of the adverse effects on the local population and environment. Thus the problem is to determine the amount and frequency of stocking the lake in order to keep the number of fishermen within predetermined limits.

This problem is related to the predator-prey problem discussed in Section 5.2. Thus we denote the number of fish (prey) at time t by $y(t)$ and the number of fishermen (predators) by $z(t)$. We denote the rate of stocking by $S(t) \geq 0$.

To derive a simple mathematical model for this problem, we assume that $y'(t)$, the rate of change of the number of fish in the lake at time t, is proportional to the number of fish present at t. Furthermore, we assume y' decreases as the number of fishermen increases. Thus we have

$$y' = Ay - Bz + S(t), \tag{1}$$

where A is the difference between the natural birth and death rates of the fish and $B > 0$ is the "catch" rate of the fishermen. Similarly, we assume that

$$z' = Dy - Ez. \tag{2}$$

In Eq. (2) we have assumed that z', the rate of change of fishermen, is proportional ($D > 0$) to the number of fish present in the lake and decreases proportionately to the number of fishermen present ($E > 0$). We assume that at the initial time, which we take as $t = 0$, there are a_0 fish and b_0 fishermen present in the lake. Thus

$$y(0) = a_0 \qquad \text{and} \qquad z(0) = b_0, \tag{3}$$

where a_0 and b_0 are given numbers. The IVP that we shall study consists of the two DEs, Eqs. (1) and (2), and the initial conditions given in Eq. (3).

Since A depends on the birth and death rates of the particular species of fish, fairly accurate estimates of A may be available from laboratory and field studies. The other constants probably are known less accurately. We must seek positive solutions of the IVP because the dependent variables are the numbers of fish and fishermen.

EXAMPLE

From field studies and observations at other lake sites, suppose that the following estimates are made of the rates A, B, D, and E:

$$A = 1, \qquad B = 4, \qquad D = 1, \qquad E = 3. \tag{4}$$

Initially the lake is stocked with 10,000 fish, and there are no fishermen. Thus we have the initial conditions

$$y(0) = 10{,}000 \qquad \text{and} \qquad z(0) = 0. \tag{5}$$

The stocking of the lake is approximated by the following periodic function

$$S(t) = S_0 \sin^2 \frac{t}{2}, \tag{6}$$

where S_0 is the amplitude of the stocking. A sketch of $S(t)$ is given in Figure 8.3. The state conservation department and local environmental groups have agreed that no more than 200 fishermen should be at the lake at any one time. The problem is to determine the amplitude S_0 of the stocking to meet this restriction.

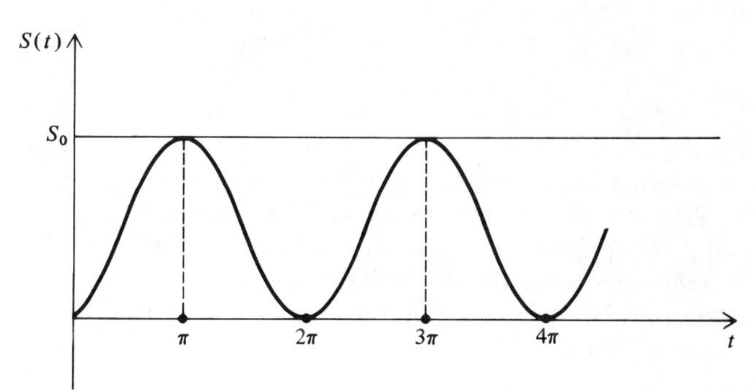

Figure 8.3

solution. By inserting the coefficients given in Eq. (4) and the function given by Eq. (6) into Eqs. (1), (2), and (3), we find that the IVP is

$$y' = y - 4z + S_0 \sin^2 \frac{t}{2} \tag{7}$$

$$z' = y - 3z \tag{8}$$

$$y(0) = 10{,}000, \qquad z(0) = 0. \tag{9}$$

We reduce this system of two first-order DEs to a single second-order DE by eliminating y from these equations. By solving Eq. (8) for y,

$$y = z' + 3z, \tag{10}$$

and substituting Eq. (10) and its derivative into Eq. (7) we get

$$z'' + 2z' + z = S_0 \sin^2 \frac{t}{2}. \tag{11}$$

From Eqs. (9) and (10) we get the initial conditions

$$z(0) = 0 \qquad \text{and} \qquad z'(0) = y(0) - 3z(0) = 10{,}000. \tag{12}$$

We now solve the IVP given by Eqs. (11) and (12). A complementary solution is

$$z_c = (c_1 + c_2 t)e^{-t} \tag{13}$$

We shall use the method of undetermined coefficients to find a particular solution of Eq. (11). We observe that $r(t) = S_0 \sin^2 (t/2)$ is not included in Table 7.1. However, by using the identity $\sin^2 A = \frac{1}{2}(1 - \cos 2A)$ we get

$$r(t) = \frac{S_0}{2}(1 - \cos t) = \frac{S_0}{2} - \frac{S_0}{2}\cos t.$$

Thus the form for the particular solution is $z_p = A + B \cos t + C \sin t$. The constants A, B, and C are evaluated by the method of undetermined coefficients. This leads to the general solution,

$$z = z_c + z_p = (c_1 + c_2 t)e^{-t} + \frac{S_0}{2} - \frac{S_0}{4}\sin t. \tag{14}$$

The constants c_1 and c_2 are determined by requiring that the general solution satisfies the initial condition given by Eq. (12). This leads to two algebraic equations whose solution is $c_1 = -S_0/2$, and $c_2 = 10{,}000 - S_0/4$.

We observe that z_c is a transient response because it approaches zero as $t \to \infty$, and z_p is the steady-state response. For sufficiently large t the solution is accurately approximated by the steady-state response.

To determine S_0 from the condition that $z(t) \leq 200$ we shall use z_p as an approximation to z. We observe that

$$\max z_p(t) = \max\left(\frac{S_0}{2} - \frac{S_0}{4}\sin t\right) = \frac{S_0}{2}\max\left(1 - \frac{\sin t}{2}\right)$$

$$= \frac{S_0}{2}\left(1 + \frac{1}{2}\right) = \frac{3S_0}{4}$$

because the maximum of $-\sin t = 1$. Thus the condition that $z_p \leq 200$ gives $\frac{3}{4}S_0 \leq 200$ or $S_0 \leq 800/3$ as the required estimate of the stocking amplitude.

For $S_0 = 800/3$, $c_1 = -400/3$ and $c_2 = 10{,}000 - 200/3 = 29{,}800/3$ and the transient response is

$$z_c = \tfrac{1}{3}(-400 + 29{,}800t)e^{-t}.$$

The maximum value of z_c occurs at $t \approx 1$, as we can show by using calculus. Then the maximum value of z_c is approximately 3700. We have already shown that the maximum value of z_p is 200. These results demonstrate that for "small" values of t, the transient may dominate the steady-state response. This transient effect may be an important environmental consideration. The size of the transient response is related to the initial number of fish present (see Exercise 7).

Exercises

1. For the example in the text: (a) Determine the number of fish $y(t)$ from Eqs. (10) and (14). (b) Determine the transient and steady-state responses for the fish population. (c) If the steady-state response is periodic, what are its period and maximum amplitude? (d) Are the fish and fishermen populations nonnega-

tive for all $t \geq 0$? (e) Determine the average number of fishermen on the lake over one period of the steady-state response (neglect the transient response).

2. Determine the steady-state response for the problem in the example if the rate coefficients are $A = 1$, $B = 16/7$, $D = 1$, and $E = 9/7$.

3. For the solutions to Exercise 2: (a) If the steady-state response is periodic, determine its maximum value, period, and phase angle. (b) Is the steady-state response ever negative?

4. Repeat Exercises 2 and 3 with the rate coefficients $A = 1$, $B = 5$, $D = 1$, and $E = 1$.

5. Discuss the qualitative features of the solution of the IVP consisting of Eqs. (1), (2), and (3) if $E < A$ and $BD - EA > 0$. What is the interpretation of these results for the fish-and-fisherman problem?

6. Repeat Exercise 5 if $E = A$ and $BD - EA < 0$.

7. For the example in the text, assume that the initial number of fish is a_0 instead of 10,000. Find the maximum value of the transient response z_c. Show in particular that if a_0 is small, the transient response will be small.

8.3 FORCED OSCILLATIONS OF A PENDULUM

We consider the vibrations of the simple pendulum whose mass is subjected to an applied tangential force

$$r(t) = r_0 \cos at$$

for $t > 0$. Here the constant r_0 is the prescribed amplitude of the force. The period of the force is $P = 2\pi/a$ and the circular frequency is a. Typical initial conditions are: At $t = 0$ the pendulum is given an initial angular displacement θ_0, and it is then released with zero initial angular velocity. We wish to determine the subsequent motion. In Section 5.3 we studied the oscillations of the pendulum when $r(t) \equiv 0$. This motion is called the free vibrations of the pendulum.

The equation of motion for the small-amplitude forced oscillations of the pendulum, neglecting friction is presented in Appendix I.2. It is

$$\theta'' + \omega^2\theta = \frac{r}{ml} = k \cos at, \tag{1}$$

where

$$k = \frac{r_0}{ml} \quad \text{and} \quad \omega = \left(\frac{g}{l}\right)^{1/2}. \tag{2}$$

The quantity ω, which is the circular frequency of the free vibrations, is called the *natural frequency* of the pendulum. The initial conditions are

$$\theta(0) = \theta_0 \qquad \text{and} \qquad \theta'(0) = 0. \tag{3}$$

A complementary solution of DE(1) is

$$\theta_c = c_1 \cos \omega t + c_2 \sin \omega t. \tag{4}$$

We obtain a particular solution by the method of undetermined coefficients. If $a = \omega$, then, as we see from Eq. (4), the inhomogeneous term $k \cos at = k \cos \omega t$ is a solution of the homogeneous DE. Thus the form of a particular solution is different for $a \neq \omega$ and $a = \omega$. We consider the two cases separately.

We first consider $a \neq \omega$. Then an appropriate guess for θ_p is

$$\theta_p = B_1 \sin at + B_2 \cos at. \tag{5}$$

The constants B_1 and B_2 are determined in the usual way by substituting Eq. (5) into Eq. (1). Thus we obtain θ_p and the general solution

$$\theta = \theta_c + \theta_p = c_1 \cos \omega t + c_2 \sin \omega t + \frac{k}{\omega^2 - a^2} \cos at.$$

We determine c_1 and c_2 by requiring this solution to satisfy the initial conditions given in Eq. (3). This yields

$$\theta = \theta_0 \cos \omega t + \frac{k}{\omega^2 - a^2} (\cos at - \cos \omega t) \tag{6}$$

as the solution of the IVP.

EXAMPLE

Suppose that a pendulum initially at rest in the vertical position is subjected to a force $r(t) = (15/2) \cos 3t$. If the weight of the mass is 2 lb and the length of the pendulum is 8 ft, find the response of the pendulum.

solution. Since $l = 8$, $g = 32$, $r_0 = 15/2$, and $ml = \text{weight} \times l/g$, the IVP can be written as

$$\theta'' + 4\theta = 15 \cos 3t, \qquad \theta(0) = \theta'(0) = 0.$$

A complementary solution is

$$\theta_c = c_1 \cos 2t + c_2 \sin 2t.$$

A particular solution is $\theta_p = -3 \cos 3t$ and hence a general solution is

$$\theta_1 = \theta_c + \theta_p = c_1 \cos 2t + c_2 \sin 2t - 3 \cos 3t.$$

Applying the initial conditions, we find that $c_1 = 3$ and $c_2 = 0$. Thus the oscillations of the pendulum are given by

$$\theta(t) = 3(\cos 2t - \cos 3t).$$

The function $\cos 2t$ is periodic, with periods $n\pi$, $n = 1, 2, \ldots$. The function $\cos 3t$ is also periodic, with periods $m(2\pi/3)$, $m = 1, 2, \ldots$. The smallest common period occurs when $m = 3$ and $n = 2$. Thus $\theta(t)$ is a periodic motion, with period 2π; see Figure 8.4.

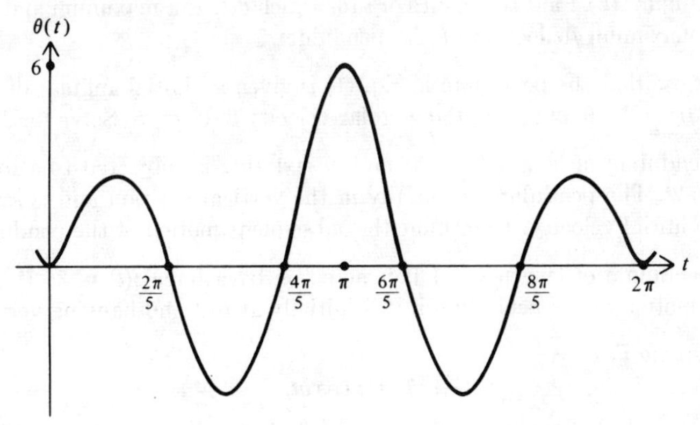

Figure 8.4

We now solve the IVP given by Eqs. (1) and (3) when $a = \omega$, that is, when the circular frequencies of the forcing function and the free vibrations coincide. The IVP is

$$\theta'' + \omega^2\theta = k \cos \omega t, \qquad \theta(0) = \theta_0,\ \theta'(0) = 0.$$

Since the inhomogeneous term $k \cos \omega t$ is a solution of the homogeneous DE, we take for a particular solution

$$\theta_p = B_1 t \cos \omega t + B_2 t \sin \omega t.$$

We determine the constants B_1 and B_2 by substituting θ_p into the DE. Thus we find that a general solution when $a = \omega$ is

$$\theta = \theta_c + \theta_p = c_1 \cos \omega t + c_2 \sin \omega t + \frac{k}{2\omega} t \sin \omega t.$$

We determine the constants c_1 and c_2 from the initial conditions. Then the solution of the IVP is

$$\theta = \theta_0 \cos \omega t + \frac{k}{2\omega} t \sin \omega t. \tag{7}$$

A discussion of this solution is given in Section 8.3-1.

Exercises

1. A pendulum of length $l = 2$ ft and $m = 1$ slug is subjected to a force $r(t) = (24/5) \cos t$. It is given an initial angular displacement $\theta(0) = 1/5$ rad and released with zero angular velocity. (a) Determine the subsequent motion of the pendulum. (b) Find the values of t for which $\theta(t)$ is a maximum and determine the maximum deflection of the pendulum.
2. Suppose that the pendulum in Eq. (1) is given an initial angular displacement of $\theta(0) = 1/10$ and an initial angular velocity $\theta'(0) = a_1$. Solve for $\theta(t)$.
3. A pendulum of length $l = 8$ ft and $m = 1$ slug is subjected to a force $r(t) = 4 \sin 2t$. The pendulum is initially in the vertical position and is started with zero initial velocity. Determine the subsequent motion of the pendulum.
4. A pendulum of length $l = 1$ ft is acted on by a force $r(t) = 3e^{-2t}$. Determine the motion of the pendulum if it is initially at rest and hanging vertically.
5. Solve the IVP
$$\theta'' + \omega^2\theta = k \cos at, \qquad a \neq \omega,$$
$$\theta(0) = 0, \theta'(0) = \theta_1,$$
where ω, k, a, and θ_1 are prescribed constants.
6. Solve the IVP in Exercise 5 when $a = \omega$.
7. A pendulum of length $l = 2$ ft is subjected to a force $r(t) = 3 \cos t + \cos 2t$. The pendulum is hanging vertically and has zero angular velocity. Determine its subsequent motion. Determine the fundamental period of the motion.

8.3-1 Beats and resonance

We shall now study the solutions for the forced oscillations of the pendulum given in Section 8.3. We first consider the solution

$$\theta = \theta_0 \cos \omega t + \frac{k}{\omega^2 - a^2} (\cos at - \cos \omega t) \tag{1}$$

for $\omega \neq a$, given by Eq. (6) in Section 8.3. If $k = 0$, there is no forcing term and the solution is then given by

$$\theta = H(t) = \theta_0 \cos \omega t,$$

which describes the free vibrations of the pendulum.

If $\theta_0 = 0$ in Eq. (1), then it is clear that

$$\theta = F(t) = \frac{k}{\omega^2 - a^2} (\cos at - \cos \omega t) \tag{2}$$

is a solution of the IVP

$$\theta'' + \omega^2\theta = k \cos at, \qquad \theta(0) = \theta'(0) = 0. \tag{3}$$

$F(t)$ is called the forced response of the pendulum because it is the response of a pendulum that is initially at rest. Thus the solution given in Eq. (1) is the sum of the free response $H(t)$ and the forced response $F(t)$.

We shall now analyze the forced response given by Eq. (2). It is the sum of periodic functions with different circular frequencies, ω and a. We recall that $\cos \omega t$ is a periodic function, with fundamental period $2\pi/\omega$. But $\cos \omega t$ is also a periodic function, with periods $2n\pi/\omega$ for $n = 1, 2, 3, \ldots$; these are all the integer multiples of the fundamental period. Similarly, $\cos at$ is a periodic function with periods $2m\pi/a$, for $m = 1, 2, 3, \ldots$. If a period $2n_0\pi/\omega$ of $\cos \omega t$ is equal to a period $2m_0\pi/a$ of $\cos at$, then the forcing frequency is related to the natural frequency by

$$a = \left(\frac{m_0}{n_0}\right)\omega. \tag{4}$$

Then $F(t)$ is a periodic function with period $2n_0\pi/\omega = 2m_0\pi/a$. Circular frequencies ω and a that satisfy Eq. (4) are said to be commensurate. When a and ω are commensurate, the fundamental period of the forcing function is $2\pi/a$ and the fundamental period of $F(t)$ is $2m_0\pi/a$ for some integer m_0. A typical forced response with $m_0 = 2$ and $n_0 = 1\,(a = 2\omega)$ is sketched in Figure 8.5.

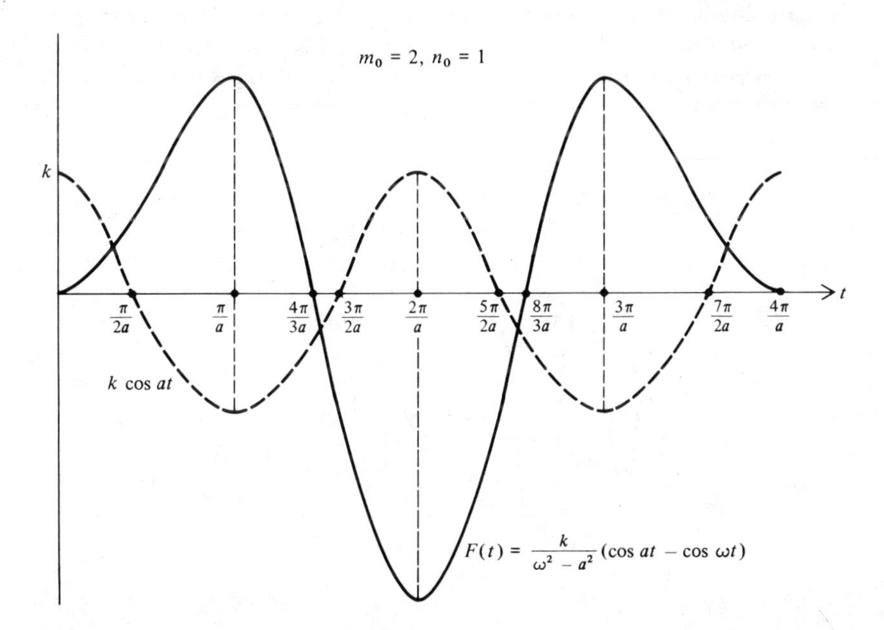

Figure 8.5

If a and ω do not satisfy Eq. (4), then the circular frequencies are said to be incommensurate, and $F(t)$ is the sum of periodic functions with incommensurable circular frequencies. It is not a periodic function. It is called an almost periodic function or a quasi-periodic function. The forced response is then a complicated motion that is difficult to describe. In the special case that a and ω are nearly equal, then it is possible to derive the qualitative features of $F(t)$. This leads to a description of the well-known phenomenon of beats, which is easily detected when two tuning forks of nearly equal frequencies are simultaneously struck.

To study beats we rewrite Eq. (2) by using the trigonometric identity

$$\cos A - \cos B = -2 \sin \frac{(A - B)}{2} \sin \frac{(A + B)}{2} .$$

Thus by setting $A = at$ and $B = \omega t$, we get

$$F(t) = \frac{-2k}{\omega^2 - a^2} \sin \frac{(a - \omega)t}{2} \sin \frac{(a + \omega)t}{2} \tag{5}$$

The function given by Eq. (5) is sketched as the solid curve in Figure 8.6. It is proportional to the product of the periodic functions $g_- = \sin [(a - \omega)t/2]$ and $g_+ = \sin [(a + \omega)t/2]$. If a and ω are nearly equal, then $| a - \omega |$ is small. Since the fundamental periods of g_- and g_+ are $4\pi/| a - \omega |$ and $4\pi/| a + \omega |$ and $| a - \omega |$ is small, the period of g_- is much longer than the period of g_+. Thus g_+ oscillates much more rapidly than g_-. This discussion suggests that we consider $F(t)$ as the product of

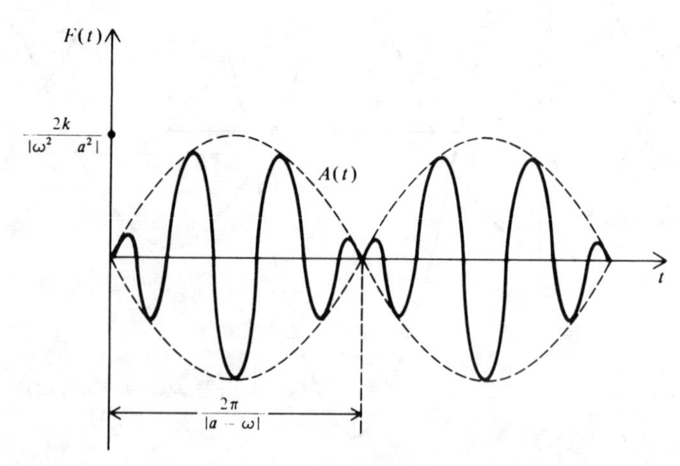

Figure 8.6

the slowly oscillating amplitude

$$A(t) = \frac{-2k}{\omega^2 - a^2} \sin \frac{(a - \omega)t}{2}$$

and the rapidly oscillating function $\sin [(a + \omega)t/2]$.

$F(t)$ is zero at the zeros of $A(t)$ as well as at the zeros of $\sin [(\omega + a)t/2]$. The amplitude $A(t)$ is sketched as the dashed curve in Figure 8.6.

An oscillating motion whose amplitude is periodic and slowly varying is a mathematical description of the phenomenon of beats. In the case of two tuning forks the variation in the amplitude of the sound is easily heard. In applications to electric circuits the slowly varying amplitude is usually called an amplitude modulation (AM).

We shall now analyze the solution

$$\theta = \theta_0 \cos \omega t + \frac{k}{2\omega} t \sin \omega t \tag{6}$$

given in Eq. (7) of Section 8.3 for the forced oscillations of the pendulum when $a = \omega$.[1] The first term on the right side of Eq. (6), $H = \theta_0 \cos \omega t$, is the free vibration. The last term is the forced response $F(t) = (k/2\omega)t\cdot \sin \omega t$ because it satisfies the IVP given by Eq. (3) with $a = \omega$. A sketch of the forced response is given in Figure 8.7. Thus when $a = \omega$, the forced response is not a periodic function. It is a growing oscillatory function and hence the solution is unbounded as $t \to \infty$. We refer to this condition as *resonance*, and the value $a = \omega$ is called the resonant circular frequency of the pendulum. For sufficiently large t, the forced vibration dominates and the free response makes a negligible contribution. Thus resonance occurs when the pendulum is driven by a forcing function whose circular frequency is precisely equal to the circular frequency of the free vibration of the pendulum. We observe also that beats occur near resonance.

There are several factors that inherently limit the possibility of resonance in physical problems. In the derivation of the pendulum DE in Appendix I.2, we approximated $\sin \theta$ by θ. This is an accurate approximation only for "small" values of θ. Thus as θ becomes large, which would occur at or near resonance, the present theory is inaccurate and the original nonlinear equation

$$\theta'' + \left(\frac{g}{l}\right) \sin \theta = r(t)/ml$$

[1] The forced response for $a = \omega$ can also be obtained from the forced response given by Eq. (5) for $a \neq \omega$ by taking the limit of Eq. (5) as $a \to \omega$.

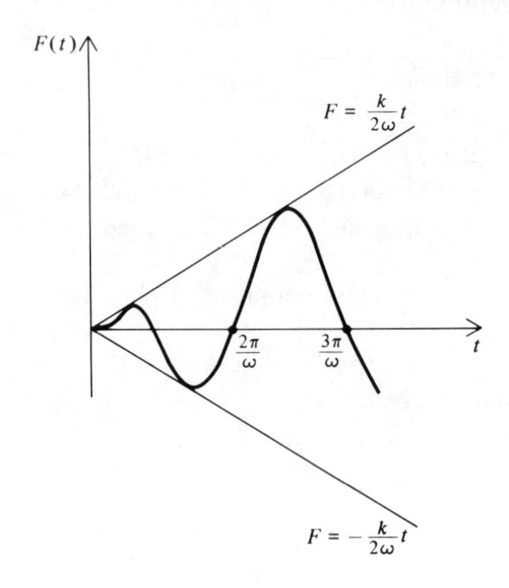

Figure 8.7

should be used. Furthermore, a small amount of energy must be dissipated during the motion of the pendulum due to friction in the pivot and air resistance. The effect of damping is to remove the resonance (see Exercise 7).

Exercises

1. A pendulum of length $l = 2$ ft is subjected to a force $r(t) = 3 \cos t$. It is given an initial angular displacement $\theta(0) = 1/10$ rad and released with zero angular velocity. (a) Determine the subsequent motion of the pendulum. (b) Determine the forced response. Is the forced response a periodic motion? If it is, what is the fundamental period? (c) Sketch the forced response.

2. Repeat Exercise 1 for a pendulum of length $l = 2$ ft that is initially at rest and is subjected to a force $r(t) = 4 \sin (4/3)t$.

3. Repeat Exercise 1 with a forcing function $r(t) = \sin 4t$. What is the behavior of the solution for large values of t?

4. A pendulum of length $l = 6$ inches, which is initially at rest and hanging vertically, is acted upon by a force $r(t) = Ae^{-t}$.
   (a) Determine the motion of the pendulum.
   (b) What is the forced response?
   (c) Determine the steady-state response.

*5. Consider the forced-oscillations IVP $y'' + \omega^2 y = k \cos at, \qquad y(0) = y'(0) = 0.$

The solution is

$$y = \frac{k}{\omega^2 - a^2}(\cos at - \cos \omega t).$$

If there are positive integers m_0 and n_0 such that $a = (m_0/n_0)\omega$, then the circular frequencies a and ω are commensurate. (a) If $n_0 = 1$, so that $a = m_0\omega$, show that the fundamental period of $y(t)$ is $2\pi/\omega$, which is equal to $m_0(2\pi/a)$. Make a sketch of $y(t)$ for $m_0 = 3$. It is called a subharmonic oscillation because its fundamental period is m_0 times the fundamental period $2\pi/a$ of the forcing function. Make a sketch of $y(t)$ for $n_0 = 3$. (b) If $m_0 = 1$, so that $a = \omega/n_0$, show that the fundamental period of $y(t)$ is $n_0(2\pi/\omega)$, which is the same as the fundamental period of the forcing function. Although the forcing function and $y(t)$ are both harmonic functions of the same frequency, they differ because the forcing function is a single harmonic function while $y(t)$ is a linear combination of two simple harmonic motions with different fundamental frequencies. In this case $y(t)$ is called an ultraharmonic oscillation. Make a sketch of a typical ultraharmonic oscillation.

6. Determine the forced response of a pendulum of length l that is started from rest, $\theta(0) = \theta'(0) = 0$, with a forcing function

$$r(t) = \sum_{n=1}^{N} b_n \sin n\pi t,$$

where b_n are specified constants.

*7. An LC electric circuit is subjected to an impressed voltage $E(t) = E_0(\sin t + \sin at)$, where E_0 is a given constant. If a is not commensurate with 1, then $E(t)$ is a quasi-periodic function. The DE for the charge $y(t)$ in the circuit is

$$y'' + \omega^2 y = \frac{E}{L}, \qquad \omega^2 = (LC)^{-1}.$$

Find the forced response. Discuss the forced response for a and ω in the following ranges: $a \ll \omega^2$; $a \ll 1$; $1 \ll a \ll \omega^2$; a near 1; $a = 1$; $a = \omega$. When do "beats" and resonance occur?

8. The small amplitude forced oscillations of a pendulum with damping is described by the DE $\theta'' + 2s\theta' + \omega^2\theta = r(t)/ml$ where $s > 0$ is the damping coefficient and ω is the frequency of the free vibrations of the pendulum in the absence of damping.

(a) Show that if $s < \omega$ (small damping) then

$$\theta_c = e^{-st}(c_1 \cos qt + c_2 \sin qt)$$

where $q = (\omega^2 - s^2)^{1/2}$. Observe that θ_c is a transient since $s > 0$.

(b) If $r/ml = \cos at$ show that

$$\theta_p = \frac{1}{z}\left[\frac{2as}{z}\sin at + \frac{\omega^2 - a^2}{z}\cos at\right]$$

where $z = [(\omega^2 - a^2)^2 + 4s^2a^2]^{1/2}$. This is a particular solution even when the forcing frequency a is equal to the natural frequency ω. Thus θ_p remains bounded for all t and the effect of damping is to eliminate the unbounded resonance discussed in this section.

(c) Express θ_p in (b) in phase amplitude form. Show that $1/z$ is the amplitude of θ_p.

(d) Sketch curves of $1/z$ as a function of a for fixed ω. Find the maximum value of these curves. Observe that as $s \to 0$ the maximum value of $1/z$ approaches infinity. This corresponds to the undamped resonance discussed earlier.

8.4 THE ELECTRO-MECHANICAL ANALOGY

In Appendix I.3 we used Kirchhoff's Law to show that the charge on the capacitor $y(t)$ satisfies the DE

$$Ly'' + Ry' + \frac{1}{C}y = E(t). \tag{1}$$

Here the constants R, L, and C are the resistance, inductance, and capacitance respectively, and $E(t)$ is the applied voltage. Equation (1) is of the same form as the DE for the forced oscillations of a simple pendulum,

$$m\theta'' + \mu\theta' + \left(\frac{mg}{l}\right)\theta = \frac{r(t)}{l}. \tag{2}$$

Here l is the length of the pendulum, m is the mass, μ is the coefficient of friction, g is the gravitational constant, and $r(t)$ is the applied force. If we identify terms in Eqs. (1) and (2) as indicated in the following table, then the equations are identical, and hence they must have the same solutions. There is, therefore, an analogy between the mechanical oscillations of a pendulum and the flow of charge in an electric circuit.

TABLE 8.1 Electro-Mechanical Analogy

Mechanical System		*Electrical System*	
Mass	m	Inductance	L
Damping coefficient	μ	Resistance	R
Length/weight	l/mg	Capacitance	C
Applied force	$\frac{r(t)}{l}$	Applied voltage	$E(t)$
Displacement	θ	Charge	$y(t)$
Velocity	θ'	Current	$y'(t)$

This analogy between electrical and mechanical systems enables us to solve problems for one system and use these results to study the analogous system. In practice it is easier and cheaper to build electric circuits and thus many mechanical oscillation problems are studied experimentally by treating the analogous electric circuit problem. This is the principle behind the analog computer. Such computers are particularly useful in studying the effects of changes in the parameters (length, mass, friction coefficient, etc.) of the system on the solution.

In many applications of electric circuits, such as radio or television sets, a capacitor or inductor is varied in order to achieve the maximum amplitude for a fixed frequency applied voltage. The applied voltage is supplied, for example, by the transmitter of the radio or television station.

The ideas of tuning and resonance are closely related. In discussing resonance, we usually consider a fixed system (for the pendulum, the mass, length, and friction coefficient are fixed) and consider the maximum amplitude of the steady-state response as the forcing frequency is varied. In tuning, the forcing frequency is fixed and we try to choose the system parameters (inductance or capacitance for the RCL circuit) to maximize the amplitude of the steady-state solution. Mechanical systems are usually designed to avoid resonances while in many electrical systems resonances are used to achieve a desired effect.

Exercises

1. Show that if $R = 0$, then the solution of the DE (1) with $E(t) \equiv 0$ is a periodic function. This result implies that there is no mechanism for dissipating the electrical energy in the circuit and thus the charge and the current vary periodically. The capacitor and the inductor do not dissipate energy. A mechanical analogy is a pendulum without friction. As the pendulum moves, the energy is transferred periodically from potential to kinetic energy. The capacitor acts to store electrical energy and the energy in an inductor is analogous to kinetic energy.

2. Discuss the phenomena of beats and resonance for an LC circuit ($R = 0$) by using the electro-mechanical analogy and the analysis of Section 8.3-1. Interpret the results in terms of the circuit parameters.

3. The motion of a mass m on a spring when the mass is acted upon by an external force $F(t)$ and a damping force proportional to its velocity is given by

$$my'' + \mu y' + ky = F(t)$$

where k is the spring constant and μ is the damping constant. (See Exercises 7–10 of Section 5.3.) Discuss the analogy between this system and the electric circuit.

4. Use the results of Exercise 8 of Section 8.3 and the electro-mechanical analogy to discuss tuning for an RCL circuit with an electromotive force $E(t) = \cos at$. Show that if $a = (LC)^{-1/2}$, then the circuit is tuned.

chapter 9

THE nTH-ORDER LINEAR EQUATION WITH CONSTANT COEFFICIENTS

9.1 INTRODUCTION

In this chapter we shall extend the theory and solution techniques that were developed in the previous chapters for second-order equations to the nth-order linear DE

$$y^{(n)} + b_{n-1}y^{(n-1)} + \cdots + b_1y' + b_0y = r(x). \tag{1}$$

Here the constant coefficients, $b_0, b_1, \ldots, b_{n-1}$, are specified real numbers and the inhomogeneous term $r(x)$ is a specified function. We shall use the notation

$$y^{(m)} = \frac{d^my}{dx^m}$$

for the derivatives of y of order $m \geq 4$. It is convenient to retain the "prime" notation for derivatives of order $m \leq 3$. The homogeneous DE corresponding to Eq. (1) is

$$y^{(n)} + b_{n-1}y^{(n-1)} + \cdots + b_1y' + b_0y = 0. \tag{2}$$

The IVP is to find a solution $y(x)$ of Eq. (1) that satisfies the initial conditions

$$y(x_0) = a_0, \quad y'(x_0) = a_1, \ldots, \quad y^{(n-1)}(x_0) = a_{n-1}, \tag{3}$$

where x_0 is a specified point and $a_0, a_1, \ldots, a_{n-1}$ are the given initial values. We observe that for the nth-order equation, n initial conditions must be satisfied. Hence, in analogy with the first- and second-order DEs, we expect that a general solution of Eq. (2) is a linear combination of n solutions whose Wronskian is not zero. The Wronskian and the related concept of linear independence of n solutions will be discussed in Section 9.4.

As in the second-order case, the basic technique for constructing general solutions is to first determine exponential solutions $y = e^{mx}$ of homogeneous DE (2). In principle, there is no difficulty in obtaining a general solution for higher-order equations in this way and using it to solve IVPs. However, computational difficulties can arise, as we shall see.

In analogy with the first-order and second-order DEs, we define a general solution of inhomogeneous DE (1) by

$$y = y_c + y_p,$$

where y_c and y_p are the complementary and particular solutions, respectively. A complementary solution is a general solution of the homogeneous DE. A particular solution is any solution of the inhomogeneous DE. Techniques for solving the homogeneous DE are discussed in Sections 9.2 through 9.4. Methods to determine particular solutions are developed in Section 9.5.

9.2 EXPONENTIAL SOLUTIONS

In analogy with the second-order DE with constant coefficients, we seek solutions of the homogeneous DE

$$y^{(n)} + b_{n-1}y^{(n-1)} + \cdots + b_1 y' + b_0 y = 0 \tag{1}$$

in the form $y = e^{mx}$, where m is a constant. Then by substituting $y = e^{mx}$ into Eq. (1), we obtain

$$(m^n + b_{n-1}m^{n-1} + \cdots + b_1 m + b_0)e^{mx} = 0.$$

Since $e^{mx} \neq 0$ for any mx, we conclude from this equation that DE (1) has the exponential solution e^{mx} if m satisfies the characteristic equation

$$p(m) = m^n + b_{n-1}m^{n-1} + \cdots + b_1 m + b_0 = 0. \tag{2}$$

We know from algebra that this algebraic equation has n roots. They are called the *characteristic roots*. The roots may be all distinct, or some of the roots may be repeated. In contrast to the case of a quadratic equation

$(n = 2)$, if $n > 2$, a repeated root may be repeated more than twice. For example, $m = 1$ is a triple root of $m^3 - 3m^2 + 3m - 1 = 0$ because $m^3 - 3m^2 + 3m - 1 = (m - 1)^3$. Since the coefficients $b_0, b_1, \ldots, b_{n-1}$ in Eq. (2) are all real numbers, complex roots occur in conjugate pairs. That is, if m is a complex root, then its complex conjugate is also a root. Complex roots may be repeated if $n \geq 4$.

The primary difficulty in constructing a solution of Eq. (1) is to determine the characteristic roots if $n \geq 3$. If $n = 3$ (cubic equation) or $n = 4$ (quartic equation), then there are explicit formulas for the roots. However, they are complicated and frequently inconvenient to apply. For $n \geq 5$, there are no explicit solution formulas for Eq. (2). Thus, numerical methods are frequently employed in practice to approximate the characteristic roots. However, in many of the examples and exercises in this chapter at least one of the characteristic roots is an integer. The integer roots can usually be found by guessing or by using synthetic division.

If $m = m_i$ is a characteristic root, then $y = e^{m_i x}$ is an exponential solution of DE (1). If m_i is a complex number, then $e^{m_i x}$ is a complex-valued exponential solution. If there are n distinct characteristic roots $m_1, m_2, \ldots, m_n$, then $e^{m_1 x}, e^{m_2 x}, \ldots, e^{m_n x}$ are exponential solutions. In addition, the linear combination

$$y = c_1 e^{m_1 x} + c_2 e^{m_2 x} + \cdots + c_n e^{m_n x} \tag{3}$$

is also a solution for all values of the arbitrary constants $c_1, c_2, \ldots, c_n$.

The main objective in this section and the next is to illustrate the techniques of constructing general solutions and using them to solve IVPs. A precise definition of a general solution will be given in Section 9.4. There we show that when the characteristic roots are distinct, then Eq. (3) is a general solution.

EXAMPLE 1

Find a general solution of the DE

$$y''' - 4y'' + y' + 6y = 0.$$

solution. We look for an exponential solution by substituting $y = e^{mx}$ into the DE. This gives the characteristic equation

$$m^3 - 4m^2 + m + 6 = 0. \tag{4}$$

Using synthetic division or simply trying several integer values, we see that $m = 2$ is a root of this cubic equation because $m^3 - 4m^2 + m + 6 = (2)^3 - 4(2)^2 + 2 + 6 = 0$. Thus $(m - 2)$ is a factor of the characteristic polynomial $m^3 - 4m^2 + m + 6$. Dividing the characteristic polynomial by $(m - 2)$, we obtain the quotient $m^2 - 2m - 3$. Thus Eq. (4) can be written as

$$m^3 - 4m^2 + m + 6 = (m - 2)(m^2 - 2m - 3) = 0.$$

Hence, the other two roots must satisfy the quadratic equation $m^2 - 2m - 3 = 0$. The roots of this equation are $m = -1$ and $m = 3$, and thus the three characteristic roots are $m = 3$, $m = 2$, and $m = -1$. The corresponding exponential solutions are e^{3x}, e^{2x}, and e^{-x}. Since the characteristic roots are distinct, the linear combination of solutions

$$y = c_1e^{3x} + c_2e^{2x} + c_3e^{-x}$$

is a general solution.

EXAMPLE 2

Find a general solution of

$$y^{(4)} - 2y'' - 3y = 0.$$

solution. The characteristic equation corresponding to the DE is

$$m^4 - 2m^2 - 3 = 0.$$

This is a fourth-degree algebraic equation. Since it contains only even powers of the unknown m, it is a quadratic equation for m^2. If we let $q = m^2$, then the characteristic equation is $q^2 - 2q - 3 = 0$. The roots of this equation are $q = m^2 = 3$ and $q = m^2 = -1$. Solving these equations for m yields the four distinct characteristic roots, $m = \pm i$ and $m = \pm\sqrt{3}$. Then

$$e^{ix}, \qquad e^{-ix}, \qquad e^{\sqrt{3}x}, \qquad \text{and} \qquad e^{-\sqrt{3}x}$$

are solutions of the DE. The solutions e^{ix}, e^{-ix} are complex-valued. Thus we have a complex-valued general solution in the form

$$y = A_1e^{ix} + A_2e^{-ix} + A_3e^{\sqrt{3}x} + A_4e^{-\sqrt{3}x}$$

where the arbitrary constants A_1, A_2, A_3, and A_4 may be complex numbers. However, we recall that e^{ix} and e^{-ix} can be expressed as a linear combination of $\sin x$ and $\cos x$. Thus

$$y = c_1 \cos x + c_2 \sin x + c_3e^{\sqrt{3}x} + c_4e^{-\sqrt{3}x}$$

is an alternate (real-valued) form for a general solution.

EXAMPLE 3

Solve the IVP

$$y''' - 6y'' + 11y' - 6y = 0$$

$$y(0) = 1,\ y'(0) = 0,\ y''(0) = 0.$$

solution. The characteristic roots corresponding to the DE are $m = 1, 2, 3$. Since they are distinct, a general solution is the linear combination

$$y = c_1e^x + c_2e^{2x} + c_3e^{3x}.$$

Applying the initial condition, we find that the constants c_1, c_2, and c_3 must satisfy

$$c_1 + c_2 + c_3 = 1$$
$$c_1 + 2c_2 + 3c_3 = 0$$
$$c_1 + 4c_2 + 9c_3 = 0.$$

The solution of this system of three linear algebraic equations in three unknowns is, $c_1 = 3$, $c_2 = -3$, and $c_3 = 1$. Consequently, the solution of the IVP is

$$y = 3e^x - 3e^{2x} + e^{3x}.$$

In solving the characteristic equation corresponding to certain higher-order DEs we must find roots of complex numbers. For example, the DE $y^{(4)} + y = 0$ has the characteristic equation $m^4 + 1 = 0$. We can solve this for m^2 to get $m^2 = \pm\sqrt{-1} = \pm i$. To determine m we must evaluate $\sqrt{i}$ and $\sqrt{-i}$. The determination of the roots of complex numbers is discussed in Exercise 8 of Appendix II and Exercise 3 of this section.

Exercises

1. Find a general solution of each of the following DEs.
   (a) $y''' - y'' - 2y' + 2y = 0$
   (b) $y''' + 3y'' - 20y' + 6y = 0$
   (c) $y''' + y' = 0$
   (d) $y''' - y'' + 9y' - 9y = 0$
   (e) $y''' - 2y'' - 3y' + 10y = 0$
   (f) $y^{(6)} + 4y^{(4)} - y'' - 4y = 0$

2. Solve the following IVPs.
   (a) $y''' + 9y' = 0, \quad y(0) = y'(0) = 0, y''(0) = 1$
   (b) $y''' + 7y'' - y' - 7 = 0, \quad y(0) = 1, y'(0) = 0, y''(0) = 2$
   (c) $y''' - 2y'' + 3y' - 6y = 0, \quad y(0) = 1, y'(0) = 1, y''(0) = 0$
   (d) $y^{(4)} - y = 0, \quad y(0) = y'(0) = y''(0) = 0, y'''(0) = 1$

3. Use DeMoivre's formula (see Exercise 8 of Appendix II) to determine the characteristic roots for the following DEs, and find a general solution.
   (a) $y^{(4)} + y = 0$
   (b) $y''' - y = 0$
   (c) $y^{(6)} + 64y = 0$
   (d) $y^{(4)} + 2y'' + 2y = 0$

9.3 REPEATED ROOTS

We shall now show how to construct a general solution of the nth-order homogeneous DE when one or more roots of the characteristic equation

$$p(m) = m^n + b_{n-1}m^{n-1} + \cdots + b_1 m + b_0 = 0 \tag{1}$$

are repeated. We first recall the definition of the multiplicity of a repeated root.

definition

A root $m = m_1$ is of multiplicity k if $(m - m_1)^k$ is a factor of $p(m)$, but $(m - m_1)^{k+1}$ is not a factor.

Thus the multiplicity of a simple root (nonrepeated root) is $k = 1$. If $m = m_1$ is a repeated root of Eq. (1), then we can rewrite Eq. (1) as

$$p(m) = (m - m_1)^k p_1(m) = 0,$$

where $p_1(m)$ is a polynomial of degree $n - k$.

A basic result from algebra is that if $m_1, m_2, \ldots, m_j$, where $j \leq n$, are the distinct roots of Eq. (1) and $k_1, k_2, \ldots, k_j$ are the corresponding multiplicities of these roots, then $k_1 + k_2 + \cdots + k_j = n$, and

$$p(m) = (m - m_1)^{k_1}(m - m_2)^{k_2} \cdots (m - m_j)^{k_j}.$$

EXAMPLE 1

Find all the roots of the equation

$$p(m) = m^6 + m^5 - m^4 - m^3 = 0,$$

and determine their multiplicities.

solution. The equation is of sixth degree and thus has six roots. We can factor out m^3 and rewrite it as

$$p(m) = m^3(m^3 + m^2 - m - 1) = 0.$$

We thus see that $m = 0$ is a root of multiplicity $k = 3$ because m^3 is a factor of p and m^4 is not a factor. The remaining three roots are solutions of the cubic equation $m^3 + m^2 - m - 1 = 0$. It is easy to verify that $m = 1$ is a root of this equation. Then we can factor the cubic equation and obtain

$$m^3 + m^2 - m - 1 = (m - 1)(m^2 + 2m + 1) = (m - 1)(m + 1)^2 = 0.$$

The original sixth-degree equation can be written in the factored form

$$m^3(m - 1)(m + 1)^2 = 0.$$

We conclude that in addition to the root $m = 0$ of multiplicity three, $m = 1$ is a simple root (multiplicity one), and $m = -1$ is a root of multiplicity two (double root).

We recall from Section 4.4 that if the characteristic roots are repeated for the second-order DE ($m_1 = m_2$), then a general solution is a linear combination of the exponential solution $e^{m_1 x}$ and $xe^{m_1 x}$. The generalization of this result to nth-order DEs when m_1 is a characteristic root of multiplicity k is summarized in Theorem 1.

theorem 1

If m_1 is a characteristic root of multiplicity k, then k solutions corresponding to this root are

$$e^{m_1x},\ xe^{m_1x},\ x^2e^{m_1x},\ \ldots,\ x^{k-1}e^{m_1x}.$$

A proof of the theorem simply consists of substituting each of these functions into the DE and showing that it is a solution. Therefore the proof will not be given. The theorem is illustrated below in Example 2. Theorem 1 is valid when m_1 is a complex number and e^{m_1x} a complex-valued exponential solution. However, it is useful to restate this result in terms of real solutions in the following theorem.

theorem 2

If $m = \alpha + i\beta$ is a complex characteristic root of multiplicity k, then so is its conjugate $m = \alpha - i\beta$. The real solutions corresponding to these complex conjugate roots are

$$e^{\alpha x}\cos\beta x,\ xe^{\alpha x}\cos\beta x,\ x^2e^{\alpha x}\cos\beta x,\ \ldots,\ x^{k-1}e^{\alpha x}\cos\beta x$$

and

$$e^{\alpha x}\sin\beta x,\ xe^{\alpha x}\sin\beta x,\ x^2e^{\alpha x}\sin\beta x,\ \ldots,\ x^{k-1}e^{\alpha x}\sin\beta x.$$

EXAMPLE 2

Find a general solution of

$$y''' - 3y'' + 3y' - y = 0.$$

solution. The reader can verify that $m = 1$ is the only characteristic root and it is of multiplicity $k = 3$. Therefore the only exponential solution is e^x. The solutions corresponding to this root are e^x, xe^x, and x^2e^x and

$$y = c_1e^x + c_2xe^x + c_3x^2e^x,$$

is a general solution, as we shall verify in Section 9.4.

EXAMPLE 3

Find a general solution of

$$y^{(4)} + 2y'' + y = 0.$$

solution. The characteristic equation is $m^4 + 2m^2 + 1 = 0$. Factoring this, we find that $(m^2 + 1)^2 = 0$ or, equivalently, $(m - i)^2(m + i)^2 = 0$. Thus, $m = i$ and $m = -i$ are characteristic roots of multiplicity $k = 2$. The exponential solutions corresponding to these roots are e^{ix} and e^{-ix}. The equivalent real solutions are $\sin x$ and $\cos x$. Since the roots are of multiplicity $k = 2$, $x \sin x$, and $x \cos x$ are also solutions. We conclude that

$$y = c_1 \sin x + c_2x \sin x + c_3 \cos x + c_4x \cos x$$

is a solution for all values of c_1, c_2, c_3, and c_4. In the next section we show that this is a general solution.

Exercises

1. Find candidates for general solutions of the following DEs.
   (a) $y''' + 3y'' - 4y = 0$
   (b) $y''' + 6y'' + 12y' + 8y = 0$
   (c) $y^{(4)} + 2y''' + y'' = 0$
   (d) $y^{(4)} + 6y'' + 9y = 0$
   (e) $y^{(6)} + 3y^{(4)} + 3y'' + y = 0$
   (f) $y^{(6)} + 8y^{(5)} + 8y^{(4)} = 0$

2. Solve the following IVPs.
   (a) $y''' + 3y'' + 3y' + y = 0, \quad y(0) = 1, y'(0) = y''(0) = 0$
   (b) $y^{(4)} + 8y'' + 16y = 0, \quad y(0) = y'(0) = y''(0) = 0, y'''(0) = 1$
   (c) $y^{(4)} - 4y'' = 0, \quad y(0) = 1, y'(0) = 0, y''(0) = 0, y'''(0) = 0$

3. (a) Show that the DE $y^{(4)} - 2\beta^2 y'' + \beta^4 y = 0$, $\beta =$ constant has solutions $e^{\beta x}$, $e^{-\beta x}$, $xe^{\beta x}$, and $xe^{-\beta x}$. (b) Using (a) and the fact that a linear combination of solutions is a solution, show that $\sinh \beta x$, $\cosh \beta x$, $x \sinh \beta x$, and $x \cosh \beta x$ are also solutions. (c) Verify directly that $\sinh \beta x$ and $x \cosh \beta x$ are solutions of the DE.

4. In Exercise 3 the characteristic equation was $m^4 - 2\beta^2 m^2 + \beta^4 = (m^2 - \beta^2)^2 = 0$, and a candidate for general solution was

$$y = (c_1 + c_2 x) \sinh \beta x + (c_3 + c_4 x) \cosh \beta x.$$

   Generalize this by giving a candidate for a general solution of any DE with characteristic equation of the form $(m^2 - \beta^2)^q = 0$, for some positive integer q.

5. Solve the DE $y^{(4)} - 4y''' + 6y'' - 4y' + y = 0$. Suppose that a linear constant-coefficient homogeneous DE of order p has the characteristic equation $(m - a)^p = 0$ for some integer p. Determine a general solution of the DE.

6. Show that $m = 1$ is a characteristic root of multiplicity three of the DE, $y''' - 3y'' + 3y' - y = 0$. Thus $y = e^x$ is a solution. Use variation of parameters (reduction of order) to find other solutions.

9.4 GENERAL SOLUTIONS AND THE INITIAL VALUE PROBLEM

The IVP for the nth-order linear homogeneous DE

$$y^{(n)} + b_{n-1}y^{(n-1)} + \cdots + b_1 y' + b_0 y = 0 \tag{1}$$

is to find solutions of Eq. (1) that satisfy the initial conditions

$$y(x_0) = a_0, \; y'(x_0) = a_1, \; \ldots, \; y^{(n-1)}(x_0) = a_{n-1}. \tag{2}$$

If $y_1(x), y_2(x), \ldots, y_n(x)$ are n solutions of Eq. (1), then

$$y = c_1y_1(x) + c_2y_2(x) + \cdots + c_ny_n(x) \tag{3}$$

is a solution of Eq. (1) for all values of the arbitrary constants $c_1, c_2, \ldots, c_n$ because Eq. (1) is a linear homogeneous DE. In order for Eq. (3) to qualify as a general solution of the DE, it must be possible to find a set of values of the constants such that Eq. (3) satisfies the initial conditions given in Eq. (2) for every set of initial values $(a_0, a_1, \ldots, a_{n-1})$. That is, it must be possible to solve the IVP with the solution given by Eq. (3) for any choice of the initial values. For the second-order equation ($n = 2$) we showed in Section 4.5 that this is always possible if the Wronskian of the two functions y_1 and y_2 in the general solution is not zero or, equivalently, if y_1 and y_2 are linearly independent functions.

A similar result, which we shall now derive, holds for nth-order equations. We evaluate the solution given in Eq. (3) and its derivatives at $x = x_0$ and then substitute the results into the initial conditions given in Eq. (2). This yields

$$\begin{aligned} c_1y_1(x_0) + c_2y_2(x_0) + \cdots + c_ny_n(x_0) &= a_0 \\ c_1y_1'(x_0) + c_2y_2'(x_0) + \cdots + c_ny_n'(x_0) &= a_1 \\ &\cdots \\ c_1y_1^{(n-1)}(x_0) + c_2y_2^{(n-1)}(x_0) + \cdots + c_ny_n^{(n-1)}(x_0) &= a_{n-1}. \end{aligned} \tag{4}$$

This is a system of n linear algebraic equations for the n unknowns $(c_1, \ldots, c_n)$. The coefficients in the algebraic equations are obtained from the values of $y_1, y_2, \ldots, y_n$ and their derivatives evaluated at $x = x_0$.

A basic result in the theory of linear algebraic equations is that a system of n linear inhomogeneous algebraic equations has a unique solution if and only if the determinant of the coefficients of the equations is not zero. The determinant of the coefficients of Eq. (4), which we denote by $W(x_0)$, is given by

$$W(x_0) = \begin{vmatrix} y_1(x_0) & y_2(x_0) & \cdots & y_n(x_0) \\ y_1'(x_0) & \cdot & \cdots & y_n'(x_0) \\ \cdot & \cdot & \cdots & \cdot \\ \cdot & \cdot & \cdots & \cdot \\ \cdot & \cdot & \cdots & \cdot \\ y_1^{(n-1)}(x_0) & y_2^{(n-1)}(x_0) & \cdots & y_n^{(n-1)}(x_0) \end{vmatrix} \tag{5}$$

Thus algebraic equations (4) have a unique solution if and only if $W(x_0) \neq 0$. This shows that the IVP has a solution for any choice of the initial values if there are n solutions $y_1, y_2, \ldots, y_n$ of DE (1) such that

$$W(x_0) \neq 0.$$

If $W(x_0) = 0$, then it is not possible to solve Eq. (4) for all choices of the initial data. (See Exercise 5.)

In analogy with the second-order equation we call W the Wronskian of $y_1, y_2, \ldots, y_n$. The reader should show that for $n = 2$, the definition given in Eq. (5) of the Wronskian coincides with the definition of the Wronskian for the second-order equation given in Eq. (7) of Section 4.5.

For the second-order equation we have shown that if $W \neq 0$ for a single value of x, then $W \neq 0$ for all values of x. The same result is valid for the nth-order equation. The demonstration of this fact is discussed in Exercises 7 and 8.

We now give the following definition of a general solution.

definition

A general solution of DE (1) is a linear combination

$$y = c_1y_1 + c_2y_2 + \cdots + c_ny_n$$

of n solutions $y_1, y_2, \ldots, y_n$ of the DE for which $W(x) \neq 0$ for at least one value of x.

We can prove that all solutions of DE (1) are contained in a general solution by following the same reasoning that was used for the second-order DE in Section 4.5. (See Exercise 9.)

EXAMPLE 1

Show that

$$y = (c_1 + c_2x + c_3x^2)e^x \tag{6}$$

is a general solution of

$$y''' - 3y'' + 3y' - y = 0. \tag{7}$$

solution. We have already shown in Example 2 of Section 9.3 that $y_1 = e^x$, $y_2 = xe^x$, and $y_3 = x^2e^x$ are solutions of DE (7). To show that Eq. (6) is a general solution, we must demonstrate that the Wronskian of y_1, y_2, and y_3 is not equal to zero for one value of x. For simplicity, we choose $x = 0$. We evaluate the derivatives of y_1, y_2, y_3 at $x = 0$ and substitute them into Eq. (5) with $n = 3$ and $x_0 = 0$. This gives

$$W(0) = \begin{vmatrix} 1 & 0 & 0 \\ 1 & 1 & 0 \\ 1 & 2 & 2 \end{vmatrix} = 2.$$

Since $W(0) \neq 0$, Eq. (6) gives a general solution.

In analogy with the second-order DE, n functions $y_1(x), y_2(x), \ldots,$ $y_n(x)$, which are defined on an interval $\alpha < x < \beta$, are *linearly dependent* if there are constants $c_1, c_2, \ldots, c_n$, not all zero, such that

$$c_1y_1(x) + c_2y_2(x) + \cdots + c_ny_n(x) = 0 \tag{8}$$

for all x in the interval. If the functions are not linearly dependent, then they are said to be *linearly independent*. Furthermore, it is possible to demonstrate the equivalence of linear independence of n solutions and the condition $W(x) \neq 0$, as we did for the second-order DE.

A general solution of the nth-order DE in Eq. (1) is given by a linear combination of all the solutions corresponding to each characteristic root (see Theorem 1 of Section 9.3). This result is summarized in the next theorem.

theorem

Let $m_1, m_2, \ldots, m_i$ denote the characteristic roots of DE (1) and $k_1, k_2, \ldots, k_i$ their respective multiplicities ($k_1 + k_2 + \cdots + k_i = n$). Then a general solution is

$$y = e^{m_1x}p_1(x) + e^{m_2x}p_2(x) + \cdots + e^{m_ix}p_i(x),$$

where $p_j(x)$, $j = 1, 2, \ldots, i$ are the polynomials $p_j(x) = c_{1j} + c_{2j}x + \cdots + c_{k_jj}x^{k_j-1}$ of degree $k_j - 1$, with $c_{1j}, c_{2j}, \ldots, c_{kj}$ as arbitrary constants.

The proof of this theorem is discussed in more advanced texts.

EXAMPLE 2

Suppose that $m_1 = 2$, $m_2 = -3$, and $m_3 = \frac{1}{2}$ are the distinct characteristic roots of a DE and, furthermore, that the corresponding multiplicities are $k_1 = 2$, $k_2 = 3$, and $k_3 = 1$. Find a general solution.

solution. Since $k_1 + k_2 + k_3 = 6$, the DE must be of sixth order. The part of the general solution corresponding to m_i is $e^{m_ix}p_i(x)$, for $i = 1, 2, 3$. Since $k_1 = 2$, $k_2 = 3$, and $k_3 = 1$, we have $p_1(x) = c_{11} + c_{21}x$, $p_2(x) = c_{12} + c_{22}x + c_{32}x^2$, and $p_3 = c_{13}$. Thus a general solution is

$$y = (c_{11} + c_{21}x)e^{2x} + (c_{12} + c_{22}x + c_{32}x^2)e^{-3x} + c_{13}e^{x/2}.$$

Exercises

1. Find a general solution of each of the following DEs. Prove that it is a general solution by computing the appropriate Wronskian.
   (a) $y''' + 4y' = 0$
   (b) $y''' + 2y'' - y' - 2y = 0$
   (c) $y^{(4)} + 2y''' + y'' = 0$
   (d) $y^{(5)} = 0$
2. We have seen in Sections 9.2 and 9.3 that the following equations have the indicated solutions. In each case compute the Wronskian of these solutions and verify that the solutions are linearly independent.

(a) $y''' - 4y'' + y' + 6y = 0, \qquad e^{3x}, e^{2x}, e^{-x}$
(b) $y^{(4)} + 2y'' + y = 0, \qquad \sin x, \cos x, x \sin x, x \cos x$
(c) $y^{(4)} - 2y'' - 3y = 0, \qquad \sin x, \cos x, e^{\sqrt{3}x}, e^{-\sqrt{3}x}$

3. (a) Show that the solutions 1, e^x, e^{-x}, and $1 + e^x$ of the DE $y^{(4)} - y'' = 0$ are linearly dependent by finding appropriate constants c_1, c_2, c_3, and c_4 so that Eq. (8) is satisfied.
   (b) Proceed as in (a) for the solutions $\cos x$, $\sin x$, $x + \sin x$, x, and 1 of the DE $y^{(4)} + y'' = 0$.

4. The distinct characteristic roots m_i and their multiplicities k_i for a given DE are $m_1 = 1$ $(k_1 = 4)$, $m_2 = i\sqrt{3}$ $(k_2 = 2)$, $m_3 = 2 + i3$ $(k_3 = 2)$. Find a general solution.

5. Consider the IVP

$$y''' - y' = 0, \qquad y(0) = a_0, y'(0) = a_1, y''(0) = a_2.$$

   (a) Verify that the Wronskian of the three solutions $y_1 = e^x$, $y_2 = 1$, and $y_3 = 1 + e^x$ is zero. The three solutions are thus linearly dependent, and they cannot comprise a general solution.
   (b) Verify that the linear combination $y = c_1y_1 + c_2y_2 + c_3y_3$ can be used to solve the IVP with $a_0 = 2$, $a_1 = 0$, and $a_2 = 0$, but that it *cannot* solve the IVP with $a_0 = a_1 = 0$ and $a_2 = 1$.
   *(c) Find the most general initial values for which the linear combination in (b) solves the IVP.

6. Consider the determinant of functions $f_{ij}(x)$:

$$D(x) = \begin{vmatrix} f_{11} & f_{12} \\ f_{21} & f_{22} \end{vmatrix} = f_{11}f_{22} - f_{12}f_{21}.$$

Show by direct differentiation that

$$D' = \frac{dD}{dx} = \begin{vmatrix} f_{11}' & f_{12}' \\ f_{21} & f_{22} \end{vmatrix} + \begin{vmatrix} f_{11} & f_{12} \\ f_{21}' & f_{22}' \end{vmatrix}.$$

Observe that D' is the sum of two 2×2 determinants each of which results from differentiating one row of D. This can be generalized to $n \times n$ determinants. Then D' is the sum of n, $n \times n$ determinants each of which results from differentiating one row of D.

*7. Show that if y_1, y_2, and y_3 are three solutions of the DE $y''' + b_2y'' + b_1y' + b_0y = 0$, then their Wronskian satisfies the first-order linear constant-coefficient DE

$$W' + b_2W = 0.$$

(*Hint:* Use Exercise 6 to show that

$$W' = \begin{vmatrix} y_1 & y_2 & y_3 \\ y_1' & y_2' & y_3' \\ y_1''' & y_2''' & y_3''' \end{vmatrix}$$

Use the DE to eliminate all third-order derivatives from W'.)

*8. Show that the Wronskian $W(x)$ of n solutions of the DE (1) satisfies $W' + b_{n-1}W = 0$. (See Exercise 7.) Use this result to show that if the Wronskian is not zero at one value of x, then it is not zero for all values of x.

*9. Prove that every solution of DE (1) is contained in a general solution. (*Hint:* See Theorem 2, Section 4.5 and assume that the IVP has a unique solution.)

10. Show that the Wronskian of the n functions $e^{m_1x}, e^{m_2x}, \ldots, e^{m_nx}$ is never equal to zero if $m_1 \neq m_2 \neq \cdots \neq m_n$.

9.5 THE METHOD OF UNDETERMINED COEFFICIENTS

The method of undetermined coefficients, which was developed in Section 7.2 for the second-order DE, can be applied to find particular solutions of the inhomogeneous DE,

$$y^{(n)} + b_{n-1}y^{(n-1)} + \cdots + b_1y' + b_0y = r(x) \tag{1}$$

when $r(x)$ is given by

$$r(x) = (k_0 + k_1x + \cdots + k_mx^m)e^{\alpha x}\sin\beta x \tag{2}$$

or by

$$r(x) = (k_0 + k_1x + \cdots + k_mx^m)e^{\alpha x}\cos\beta x. \tag{3}$$

Then it can be shown that the appropriate choice[1] for a particular solution y_p is

$$y_p = (B_0 + B_1x + \cdots + B_mx^m)e^{\alpha x}\cos\beta x + (D_0 + D_1x + \cdots + D_mx^m)e^{\alpha x}\sin\beta x. \tag{4}$$

This choice is correct if no term in Eq. (4) is a solution of the homogeneous equation. We shall not attempt to justify Eq. (4) or its modifications because of the similarity with the second-order DE. The following example illustrates the technique. (Following the example we shall discuss the necessary modification of Eq. (4) when it contains terms that are solutions of the homogeneous equation.)

[1] See line 5 of Table 7.1.

EXAMPLE 1

Find a particular solution of

$$y''' - 4y'' + y' + 6y = 4 \sin 2x. \tag{5}$$

solution. A general solution of the homogeneous DE corresponding to Eq. (5) was obtained in Example 1 of Section 9.2 as

$$y_c = c_1e^{3x} + c_2e^{2x} + c_3e^{-x}.$$

Since $r(x) = 4 \sin 2x$, we suppose, by analogy with the second-order DE, that an appropriate choice for y_p is

$$y = B \cos 2x + D \sin 2x. \tag{6}$$

Since no term of Eq. (6) is in y_c, no alteration of this choice is necessary. To determine B and D, we require that Eq. (6) satisfy DE (5). This gives

$$(22D + 6B) \sin 2x + (-6D + 22B) \cos 2x = 4 \sin 2x.$$

Equating coefficients of sin $2x$ and cos $2x$ on both sides of this equation, we obtain

$$22D + 6B = 4 \quad \text{and} \quad -6D + 22B = 0.$$

The solution of these algebraic equations is $B = \frac{3}{65}$ and $D = \frac{11}{65}$, and thus

$$y_p = \tfrac{3}{65} \cos 2x + \tfrac{11}{65} \sin 2x.$$

If any term in the candidate for y_p given in Eq (4) is a solution of the homogeneous equation corresponding to DE (1), then the candidate must be modified. In the second-order case we saw that the original choice for y_p must be multiplied by x, and this product then taken as a candidate for y_p. If any term in the new choice is contained in the complementary solution, then we must multiply by x again or, equivalently, take x^2y_p.

In the second-order case this procedure will always provide the appropriate form for y_p. For higher-order equations x^2y_p may still contain terms that are solutions of the homogeneous equation. Then x^2y_p will not be an appropriate choice. It is then necessary to seek a particular solution in the form x^sy_p, where s is some integer ≥ 3, as we shall see in the following example.

EXAMPLE 2

Find a particular solution of

$$y''' - 3y'' + 3y' - y = 3e^x. \tag{7}$$

solution. A complementary solution was found in Example 2 of Section 9.3 to be

$$y_c = c_1e^x + c_2xe^x + c_3x^2e^x. \tag{8}$$

Thus $r(x) = 3e^x$ is a solution of the homogeneous DE. So are xe^x and x^2e^x, as we see from Eq. (8). However, x^3e^x is not a solution of the homogeneous DE, and hence we shall try

$$y_p = Bx^3e^x.$$

Substituting this into DE (7), we find that

$$6Be^x = 3e^x, \qquad \text{or } B = \tfrac{1}{2}.$$

Thus $y_p = \frac{1}{2}x^3e^x$ is a particular solution. It is in the form of x^3 times the candidate Be^x that is suggested by Eq. (4).

The foregoing example suggests the general rule: If any term in the choice for y_p given in (4) is a solution of the homogeneous equation, multiply y_p by x^s, where s is the smallest positive integer, such that no term in x^sy_p is a solution of the homogeneous DE.

The method of variation of parameters for determining particular solutions of the second-order DE, which was presented in Section 7.3, can be extended to the higher-order DE. This is discussed in Exercises 8-10.

Exercises

1. Find a general solution of each of the following DEs.
   (a) $y''' - 4y'' + y' + 6y = 3e^x$ (b) $y''' - y'' - 2y' + 2y = 5e^x$
   (c) $y''' + 4y' = \sin x$ (d) $y''' + 7y'' - y' - 7y = x^2 + 2x$
   (e) $y^{(4)} + y'' = x^2 + 2x$
   (f) $y''' - 6y'' + y' - 6y = \sin x + 2$
   (g) $y''' + 5y'' - 3y' - 15y = xe^x$ (h) $y^{(4)} - y = \cos x$
2. Solve the following IVPs.
   (a) $y''' - 2y'' - y' + 2y = e^{-2x}$, $y(0) = y'(0) = y''(0) = 0$
   (b) $y''' + y'' = 1$, $y(0) = 1, y'(0) = y''(0) = 0$
   (c) $y^{(4)} - 5y'' + 4y = \sin x$, $y(0) = y'(0) = 0, y''(0) = 1, y'''(0) = 2$
3. For each of the following DEs determine an appropriate guess for a particular solution to be used in the method of undetermined coefficients. Do not determine the unknown constants.
   (a) $y^{(4)} + 2y'' + y = \cos x + e^x$
   (b) $y^{(5)} - 2y^{(4)} + 2y''' = 2e^x \cos x + 2$
   (c) $y''' - 3y'' + 3y' - y = x^3 + e^x$ (d) $y^{(4)} + y'' = x^3 + e^x$
   (e) $y^{(4)} - y'' = x^3e^x$
   (f) $y''' + 3y'' + 3y' + y = x^2e^{-x} \sin x$
4. Consider the DE $y''' + 2y'' + 2y' = 3e^{-ax}$, where a is a constant. Find y_p for each value of a.
5. Show that $y_p = Bx^3e^x$ is an appropriate form for a particular solution of $y''' - 3y'' + 3y' - y = 3e^x$ by seeking y_p in the form $y_p = f(x)e^x$ and determining f.

6. (a) Consider the DE

$$y''' + b_2y'' + b_1y' + b_0y = k_0 + k_1x + \cdots + k_jx^j$$

where b_0, b_1, b_2, and k_i, for $i = 0, \ldots, j$ are known constants. Show that the choice $y_p = B_0 + B_1x + \cdots + B_jx^j$ will not give a particular solution if $b_0 = 0$. If $b_0 = 0$, observe that $m = 0$ is a characteristic root, and thus $y = B_0$ is a solution of the homogeneous equation.

(b) Show that if $b_0 = 0$, then the polynomial of degree $j + 1$, $y_p = B_0x + B_1x^2 + \cdots + B_jx^{j+1}$, is an appropriate choice unless $b_1 = 0$. Observe that if $b_0 = b_1 = 0$, then $y = c_1 + c_2x$, with c_1 and c_2 arbitrary constants, is a solution of the corresponding homogeneous equation.

(c) Discuss the cases $b_2 \neq 0$, $b_0 = b_1 = 0$, and $b_2 = b_1 = b_0 = 0$.

7. (a) Consider the DE, $y''' + b_2y'' + b_1y' + b_0y = e^{ax}P_j(x)$, where the polynomial $P_j(x) = k_0 + k_1x + \cdots + k_jx^j$. Show that the change of variables $y = v(x)e^{ax}$ reduces the DE to a DE for $v(x)$ that is of the same form as the DE in Exercise 6(a).

(b) Use the results of Exercise 6 to find the appropriate form for y_p. Consider the case where e^{ax} is a solution of the corresponding homogeneous DE and the case where e^{ax} is not such a solution.

(c) Generalize your results to the nth-order DE with $r(x) = e^{ax}P_j(x)$.

The method of variation of parameters for finding particular solutions of higher order equations is discussed in Exercises 8–11.

8. Find a particular solution of $y''' + y' = \tan x$ for x in the interval $(-\pi/2, \pi/2)$, by proceeding as follows: (a) Show that the complementary solution is $y_c = c_1 + c_2 \cos x + c_3 \sin x$. (b) Look for a particular solution in the form

$$y_p = u(x) + v(x) \cos x + w(x) \sin x.$$

Show that if we require that only first derivatives of the unknown functions u, v, and w should appear in the final equation, then u, v, and w must satisfy

$$u' + v' \cos x + w' \sin x = 0$$

$$-v' \sin x + w' \cos x = 0$$

$$-v' \cos x - w' \sin x = \tan x.$$

(c) Solve the three DEs derived in (b) as algebraic equations for u', v', and w' and then integrate the results directly to obtain u, v, and w.

9. Find a particular solution of each of the following DEs by variation of parameters.

(a) $y''' + y' = \cot x$, for $0 < x < \pi$

(b) $y''' + 2y'' = x^2$

(c) $y''' + 4y'' - y' - 4y = e^x$

(d) $y''' + 2y'' - y' - 2y = \dfrac{3e^{-2x}}{1 + e^x}$

*10. Let

$$y_c = c_1y_1 + \cdots + c_ny_n$$

be a general solution of the homogeneous equation corresponding to Eq. (1). It can be shown, by proceeding as in Exercise 8, that

$$y_p = u_1(x)y_1 + \cdots + u_n(x)y_n,$$

where

$$u_i = \int \frac{D_i(x)}{W(x)}\,dx, \qquad i = 1, 2, \ldots, n.$$

Here $W(x)$ is the Wronskian of the solutions y_i, for $i = 1, \ldots, n$, and D_i is the determinant formed by replacing the ith column of W by the column

$$\begin{pmatrix} 0 \\ 0 \\ \cdot \\ \cdot \\ \cdot \\ r(x) \end{pmatrix}.$$

Verify this result for $n = 3$.

9.6 APPLICATIONS

Higher-order DEs occur in a wide variety of applications. They appear in IVPs, BVPs, and eigenvalue problems. Typical applications are given in the following subsections.

9.6-1 Asymptotic stability; the Routh-Hurwitz test

When all solutions of the DE

$$y^{(n)} + b_{n-1}y^{(n-1)} + \cdots + b_1y' + b_0y = 0 \tag{1}$$

approach zero as $t \to \infty$, we say that the DE is *asymptotically stable*. The concept is of importance in applications of DEs to physical problems.

It follows easily from the theorem of Section 9.4 that if all the characteristic roots of Eq. (1) have negative real parts, then the DE is asymptotically stable. It is also possible to establish the converse of this result.

A convenient test for determining whether all the characteristic roots have negative real parts was discovered by Routh and Hurwitz independently. It is called the Routh-Hurwitz test.[2] Applying this test we can see whether or not the nth-order DE is asymptotically stable without actually solving it. For a third-order DE the Routh-Hurwitz test for asymptotic stability is that the following three quantities are positive:

$$b_2, \qquad \begin{vmatrix} b_2 & 1 \\ b_0 & b_1 \end{vmatrix}, \qquad \begin{vmatrix} b_2 & 1 & 0 \\ b_0 & b_1 & b_2 \\ 0 & 0 & b_0 \end{vmatrix}.$$

Exercises

1. Use the Routh-Hurwitz test to determine which of the following DEs are asymptotically stable. Verify your answer by solving the DE.
   (a) $y''' + 3y'' + 3y' + y = 0$ (b) $y''' + 6y'' + 2y' = 0$
   (c) $y''' + 2y'' - y' - 2y = 0$

2. The flow of chemically reacting mixtures of gases plays a fundamental role in studying such diverse problems as the solar atmosphere and the atmosphere of other stars, and the gas flow in the combustion chamber of a rocket engine. It can be shown that for certain types of gases the propagation of small disturbances through the gas as time t varies is described by the DE $y''' + ay'' + by' + cy = 0$, where the given constants a, b, and c are all positive. The independent variable $y(t)$ is proportional to the gas pressure. The coefficients a, b, and c are related to the physical properties and the temperature of the gas. In particular, the constants b and c are usually called the frozen and equilibrium sound speeds of the gas, respectively. From the physical properties, it is known that $b > c$. If the DE is asymptotically stable, then all disturbances to the gas will eventually disappear because they are dissipated by the chemical reactions. If the DE is not asymptotically stable, then there are disturbances which do not decay as $t \to \infty$. Then shock waves may form in the gas. (a) Use the Routh-Hurwitz test to determine conditions on the constants a, b, and c for which the DE is asymptotically stable. (b) Is the DE asymptotically stable if $c = 0$?

9.6-2 Deflections of elastic beams

One of the most commonly used elements in structures such as aircraft, buildings, ships, and bridges is the elastic beam. A beam is a long slender rod that is supported at either or both ends and is subjected to

[2] For a detailed discussion of this test see, "Stability and Asymptotic Behavior of Differential Equations" by W. A. Coppel, D. C. Heath, 1965.

external forces or displacements along its length and possibly at its ends. Beams are used for girders and for wall and floor supports in building construction.

A modern airplane contains thousands of beams in its fuselage, wing, and tail structures. The rails of a railroad track are examples of beams that are supported along their entire length by the roadbed (the ties are considered as parts of the roadbed). Then the beam is said to rest on an elastic foundation.

Because of the importance of beams in construction, their deformations have been extensively studied since the earliest days of scientific inquiry when Greek and Roman architects compiled data from experiments. The modern engineering theory of beams was developed in the eighteenth century, principally through the efforts of Euler and the Bernoullis. Galileo and Coulomb had made earlier contributions to the theory.

In this theory the deformations of the beam are described by a boundary-value problem for a fourth-order ordinary differential equation for the vertical displacement $y(x)$ of the axis of the beam at the point x; see Figure 9.1. The beam is deformed by a vertical force per unit length of the beam, $q(x)$, which acts on the surface of the beam as shown in the figure. We refer to $q(x)$ as the load. The beam is assumed to deform in the plane of the figure. Then the differential equation is

$$EIy^{(4)} = q. \tag{1}$$

Here $q(x)$ is an inhomogeneous term, E is Young's elastic modulus for the beam material, and I is the moment of inertia of the beam's cross section about an axis perpendicular to the plane of the paper in Figure 9.1.

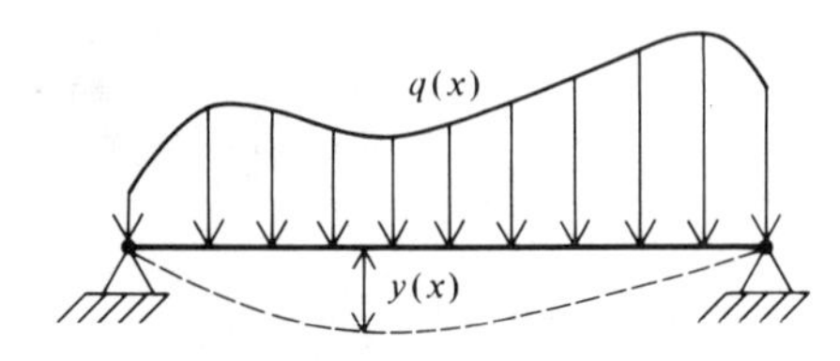

Figure 9.1

The boundary conditions that y must satisfy are obtained by considering the forces and/or displacements that are applied to the ends $x = 0$ and $x = L$ of the beam. Typical boundary conditions are:

The simply supported beam,

$$y(0) = 0,\ y(L) = 0,\ y''(0) = 0,\ y''(L) = 0. \tag{2}$$

The clamped beam,

$$y(0) = 0, y(L) = 0, y'(0) = 0, y'(L) = 0. \tag{3}$$

The condition $y(0) = 0$ means that the end of the beam cannot deflect vertically. The condition $y'(0) = 0$ means that the slope of the beam's axis at $x = 0$ is zero; that is, the axis is horizontal at the end. The clamped beam is frequently called a built-in beam, as illustrated in Figure 9.2a. The condition $y''(0) = 0$ implies that the end is free to rotate about an axis perpendicular to the plane of deformation and passing through the end of the beam. The simply supported end is sometimes called a pinned end, as illustrated in Figure 9.2b.

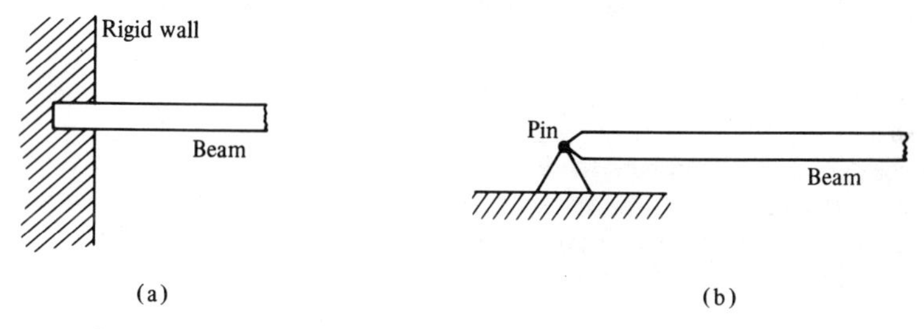

Figure 9.2

If the beam is resting on an elastic foundation, then the DE must be modified. We get

$$EIy^{(4)} + ky = q, \tag{4}$$

where k is called the spring constant of the elastic foundation. If the beam is subjected to an axial force T, then the DE is

$$EIy^{(4)} - Ty'' = q. \tag{5}$$

Here T is positive if the force acts to stretch the beam.

EXAMPLE

A uniform load

$$q(x) = q_0 = \text{constant}$$

is applied to the surface of a simply supported beam of length L. Determine the deflection of the beam.

solution. The BVP describing the deflection of the beam is

$$EIy^{(4)} = q_0,$$

$$y(0) = y''(0) = 0 \quad \text{and} \quad y(L) = y''(L) = 0.$$

A general solution of the DE can easily be obtained by integrating it directly four times. However, we shall use the techniques developed in this chapter to determine a general solution. The characteristic equation corresponding to the homogeneous DE $y^{(4)} = 0$ is $m^4 = 0$. Thus $m = 0$ is a root of multiplicity four, and the complementary solution is

$$y_c = c_1 + c_2x + c_3x^2 + c_4x^3.$$

To determine a particular solution by the method of undetermined coefficients, we note that $r(x) = q_0/EI$ is a constant, and thus it is a solution of the homogeneous equation. Since $x^s r(x)$, for $s = 1$, 2, and 3 are also solutions of the homogeneous DE, an appropriate choice for y_p is

$$y_p = Bx^4.$$

To determine B we substitute this choice into the inhomogeneous DE. Then a general solution is $y_c + y_p$, or

$$y(x) = c_1 + c_2x + c_3x^2 + c_4x^3 + \frac{q_0}{EI}\frac{x^4}{4!}. \tag{6}$$

Application of the boundary conditions yields four algebraic equations for the four arbitrary constants in Eq. (6). By solving these algebraic equations and substituting the results into Eq. (6) (we leave the details as an exercise for the reader), we find that the deflection of the beam is

$$y(x) = \frac{q_0}{12EI}\left(\frac{L^3}{2}x - Lx^3 + \frac{x^4}{2}\right). \tag{7}$$

Since the load q is symmetric about the center $x = L/2$ of the beam, it is intuitively clear that the maximum deflection must occur at $x = L/2$. We can also establish this by considering y' and y'', and showing that $x = L/2$ is the only maximum of Eq. (7) in the interval $0 \leq x \leq L$. We leave this as an exercise for the reader. Thus we have

$$y_{\max} = y\left(\frac{L}{2}\right) = \frac{q_0}{12EI}\left(\frac{L^4}{4} - \frac{L^4}{8} + \frac{L^4}{32}\right) = \frac{5q_0L^4}{32EI}. \tag{8}$$

We observe from Eq. (8) that the maximum deflection is inversely proportional to EI. As EI increases, the maximum deflection decreases. That is why EI is called the bending rigidity of the beam. As L increases, $y_{\max}$ increases. This is consistent with our intuition that the beam becomes more flexible as its length increases, and hence it deflects more for the same lateral load per unit length. Furthermore $y_{\max}$ is proportional to the magnitude q_0 of the applied lateral load.

If the ends of the beam are clamped instead of simply supported, then we determine the deflection of the beam by using the general solution given in Eq. (6) and evaluating the constants from the clamped boundary conditions given in Eq. (3). Then after some calculation, we can show that the maximum deflection of the clamped beam, which occurs at $x = L/2$, is

$$y_{\max} = y\left(\frac{L}{2}\right) = \frac{5q_0L^4}{96EI} = \frac{1}{3}\left(\frac{5q_0L^4}{32EI}\right) = \frac{1}{3}y^s{}_{\max}, \tag{9}$$

where y^s_{max} is the maximum deflection of the simply supported beam given in Eq. (8). This result shows that the clamped beam has a smaller deflection than the simply supported beam for the same load. This follows because the ends of a clamped beam are not permitted to rotate and hence the clamped beam is stiffer.

Exercises

1. A simply supported beam of length $L = 12$ ft is subjected to a load $q(x) = \sin(\pi x/12)$. Determine the deflection of the beam. Determine the location and magnitude of the maximum deflection.
2. Repeat Exercise 1 for $q(x) = q_0 \cos(\pi x/12)$, where q_0 is a constant.
3. (a) Determine the maximum deflection of the clamped beam subjected to a uniform load.
   (b) A clamped beam of length $L = 50$ ft is subjected to a load $q(x) = (25 - x)^2$. Determine the magnitude and location of the maximum deflection of the beam.
4. Determine the deflection of a clamped beam with the load $q(x) = x^2$ and compare the results with those of Exercise 3.
5. A beam that is clamped at one end and completely free at the other is called a cantilevered beam. It can be shown that the boundary conditions at the free end $x = L$ are $y''(L) = y'''(L) = 0$. The boundary conditions at the clamped end are $y(0) = y'(0) = 0$. Suppose that $q(x) = q_0 =$ constant. Determine the deflection of the beam, and the location and magnitude of the maximum deflection.
6. Verify that the maximum value of $y(x)$ given in Eq. (8) occurs at $x = L/2$.
7. An axially loaded beam of length $L = 10$ ft is subjected to a load $q(x) = x$, applied on the surface of the beam, and an axial force $T = 100$ lb. Determine the deflection of the beam if the ends are simply supported and $EI = 100$.
8. Repeat Exercise 7 for clamped ends.
9. Solve $EIy^{(4)} - Ty'' = q_0$, where EI, T, and q_0 are positive constants such that $T = EI$ and $y(0) = y''(0) = y(1) = y''(1) = 0$.
10. A beam on an elastic foundation is subjected to a uniform load of q_0 pounds per foot. If $k^2/EI = 0.04$, find a general solution of the DE that describes the deflection of the beam.

9.6-3 The vibrating beam

In many beam problems the forces or loads are moving or changing with time. This causes the beam to vibrate. For example, some of the loads acting on a monorail track are its own weight, the weight of the moving cars, and thermal forces. The weight of the rail itself is a constant load, but the latter two loads are time-dependent. In the analysis of the effect of time-dependent loads on the deflection of a simply supported beam, the following eigenvalue problem occurs:

Determine the values of the constant λ such that the EVP

$$y^{(4)} - \lambda y = 0, \tag{1}$$

$$y(0) = y''(0) = 0,\ y(L) = y''(L) = 0, \tag{2}$$

has nonzero solutions. The quantity λ in Eq. (1) is given by

$$\lambda = \frac{Am}{EI}\,\omega^2 \tag{3}$$

where A is the cross-sectional area of the beam, m is the mass per unit volume of the beam's material, and ω is the frequency of free vibration of the beam. We have already discussed the free vibrations of the pendulum in Section 5.3. Thus for each eigenvalue of the EVP, ω obtained from Eq. (3) gives a corresponding frequency of free vibration of the beam. As we shall see, there are an infinite number of such frequencies. They are called the natural frequencies of the beam. The eigenfunctions, $y(x)$, give the shapes of the beam as it oscillates at the corresponding natural frequencies.

If $\lambda > 0$, then the characteristic roots are $m_1 = B$, $m_2 = -B$, $m_3 = iB$, and $m_4 = -iB$, where $B = \lambda^{1/4}$. A general solution is,

$$y = c_1 \cosh Bx + c_2 \sinh Bx + c_3 \cos Bx + c_4 \sin Bx. \tag{4}$$

The boundary conditions at $x = 0$, which are given in Eq. (2), yield the two algebraic equations

$$c_1 + c_3 = 0 \qquad \text{and} \qquad c_1 - c_3 = 0$$

Thus we have $c_1 = c_3 = 0$. Substituting this result into Eq. (4) and applying the boundary conditions at $x = L$, we get

$$c_2 \sinh BL + c_4 \sin BL = 0$$

$$c_2 \sinh BL - c_4 \sin BL = 0. \tag{5}$$

By adding these two equations we find that $c_2 \sinh BL = 0$. Since $B \neq 0$ and $\sinh A \neq 0$ for $A \neq 0$, we conclude that $c_2 = 0$. Thus Eq. (5) implies that

$$c_4 \sin BL = 0. \tag{6}$$

We conclude from Eq. (6) that $\sin BL = 0$ and hence that $B = n\pi/L$,

$n = \pm 1, \pm 2, \ldots$. The eigenvalues are then

$$\lambda = B^4 = \left(\frac{n\pi}{L}\right)^4, \qquad n = 1, 2, \ldots.$$

Since c_4 is an arbitrary constant, the eigenfunctions are

$$y = y_n(x) = A_n \sin \frac{n\pi x}{L}, \qquad n = 1, 2, 3, \ldots.$$

Here we have replaced the arbitrary constant c_4 by A_n in order to emphasize that the constant can be different for different values of n.

The reader should verify that there are no eigenvalues with $\lambda \leq 0$. (See Exercise 3.)

Exercises

*1. Find the eigenvalues and eigenfunctions for the vibrating clamped beam,

$$y^{(4)} - \lambda y = 0, \qquad y(0) = y'(0) = 0, \; y(L) = y'(L) = 0.$$

2. Find the eigenvalues and eigenfunctions of:
(a) $y^{(4)} - \lambda y = 0, \qquad y(0) = y'(0) = 0, \; y''(L) = y'''(L) = 0$
(b) $y^{(3)} + \lambda y' = 0, \qquad y(0) = y''(0) = 0, \; y(1) = 0$
(c) $y^{(4)} + \lambda y'' + y = 0, \qquad y(0) = y''(0) = y(1) = y''(1) = 0$
(*Hint:* The eigenfunctions are proportional to $\sin m\pi x$, $m = 1, 2, \ldots$.)

3. Verify that the EVP given by Eqs. (1) and (2) has no negative eigenvalues and that $\lambda = 0$ is not an eigenvalue.

chapter 10

SYSTEMS OF LINEAR DIFFERENTIAL EQUATIONS WITH CONSTANT COEFFICIENTS

10.1 INTRODUCTION

We have seen in the discussion of species interactions given in Section 5.2 and the resource conservation problem in Section 8.2 that mathematical models need not be formulated in terms of a single differential equation. The natural formulation may require the simultaneous solution of a system of two or more DEs, such as

$$y' = ay + bz + r_1(x) \tag{1}$$

$$z' = cy + dz + r_2(x) \tag{2}$$

in two unknown functions $y(x)$ and $z(x)$. The prescribed constants a, b, c, and d are the coefficients of the DEs, and the prescribed functions $r_1(x)$ and $r_2(x)$ are the inhomogeneous terms. Equations (1) and (2) are a system of two first-order DEs. Applications of DEs to the biological, social, and economic sciences usually lead to such systems. For simplicity of presentation, we shall primarily consider the system given by Eqs. (1) and (2), although more general systems will be studied.

A solution of the system is a pair of functions $[y(x), z(x)]$ that simultaneously satisfy Eqs. (1) and (2). Sometimes we shall refer to $[y(x), z(x)]$ as a solution pair.

A more general system of two first-order linear DEs is one in which both equations contain derivatives of both unknown functions $y(x)$ and $z(x)$. Frequently these systems can be transformed into the form of Eqs. (1) and (2) as we discuss in Exercise 7. Thus we shall call any system of the form given by Eqs. (1) and (2), where y' appears only in the first DE and z' only in the second DE, a *standard system.*

The theory and technique of solving single DEs with constant coefficients can be extended to systems of first-order equations with constant coefficients. The concepts of general solutions, Wronskians, and linear independence carry over to systems. The mathematical theory of systems is most efficiently presented by using the ideas of linear algebra and, in particular, the theory of matrices. However, we assume that the reader is unfamiliar with this theory. Therefore, we shall only give a brief introduction to systems, concentrating mainly on standard systems and emphasizing applications.

We shall now demonstrate that every second-order linear DE with constant coefficients

$$y'' + 2by' + cy = r(x) \tag{3}$$

is equivalent to a standard system of first-order DEs. By equivalent we mean that any solution of either one is also a solution of the other. We introduce the auxiliary dependent variable $z = y'$. This gives $z' = y''$. By substituting these expressions for y' and y'' into Eq. (3) we get

$$z' + 2bz + cy = r(x).$$

Consequently, we have the standard system

$$y' = z \tag{4a}$$

$$z' = -cy - 2bz + r(x). \tag{4b}$$

It is a special case of Eqs. (1) and (2). Thus we have shown that every solution of second-order DE (3) is a solution of first-order system (4). To establish the converse, namely that every solution of the system (4) provides a solution of Eq. (3), we assume that we have a solution $[y(x), z(x)]$ of the system. Then by substituting $z = y'$ from Eq. (4a) into Eq. (4b), we get

$$y'' = -cy - 2by' + r(x),$$

which is precisely second-order DE (3). Thus every solution of the system (4) gives a solution of Eq. (3). Consequently, Eqs. (3) and (4) are equivalent.

This demonstration of equivalence of the single second-order DE and the standard system suggests one of the principal techniques for solving standard systems. It is called the *method of elimination*. One of the unknowns, either y or z, is eliminated from Eqs. (1) and (2) by differentiation and algebraic manipulation. Then we obtain a second-order DE with constant coefficients for the remaining unknown. This resulting DE can, of course, be solved by the techniques developed in Chapters 4 through 7. The method of elimination is discussed[1] in Section 10.2.

The equivalence also suggests that a general solution of the homogeneous DEs corresponding to Eqs. (1) and (2) should contain two arbitrary constants. We shall see that a general solution is a pair of functions,

$$y(x) = c_1 y_1(x) + c_2 y_2(x) \tag{5}$$

$$z(x) = c_1 z_1(x) + c_2 z_2(x), \tag{6}$$

where c_1 and c_2 are arbitrary constants and the pairs $[y_1(x), z_1(x)]$ and $[y_2(x), z_2(x)]$ are solutions of the homogeneous system. General solutions will be defined and discussed in Section 10.3.

Let the pair $[y_p(x), z_p(x)]$ be any solution of the inhomogeneous system given by Eqs. (1) and (2) and let $[y_c(x), z_c(x)]$ be a general solution of the corresponding homogeneous system. Then in analogy with a single DE, a general solution of the inhomogeneous system given by Eqs. (1) and (2) is defined by

$$y(x) = y_c(x) + y_p(x)$$

and

$$z(x) = z_c(x) + z_p(x).$$

We refer to the pairs $[y_c(x), z_c(x)]$ and $[y_p(x), z_p(x)]$ as a complementary solution and a particular solution, respectively.

The methods of variation of parameters and undetermined coefficients, which we used to find a particular solution for a single inhomogeneous DE, can be extended to inhomogeneous systems. We discuss these techniques in the exercises of Section 10.3. However, in the text we use the method of elimination to find a general solution of the inhomogeneous system directly.

[1] A natural generalization of the techniques we employed to solve the homogeneous, second-order DE is to seek exponential solutions of systems of first-order DEs. Since this procedure is most efficiently applied by using matricies, it shall not be presented. Pertinent references are: "Ordinary Differential Equations" by I. G. Petrovski (Prentice-Hall, Inc., 1966); and "Differential Equations" by H. Hochstadt, (Holt, Rinehart and Winston, 1964).

Exercises

1. Write an equivalent standard system for each of the following DEs. Obtain the solution of the system by solving the given equation.
(a) $y'' + y = \sin x$ (b) $y'' + 6y' + 8y = 0$
(c) $y''' + 3y'' + 3y' + y = x$
(*Hint:* Let $w = y''$.)

2. Show that a general nth-order linear equation

$$y^{(n)} + b_{n-1}(x)y^{(n-1)} + b_{n-2}(x)y^{(n-2)} + \cdots + b_1(x)y' + b_0(x)y = r(x)$$

can be written as a standard system of n first-order equations with n unknowns.

3. Show that the system

$$y' = ay + bz + r_1(x)$$
$$z' = cy + dz + r_2(x)$$

can always be reduced to a single second-order equation for either y or z.

4. Prove in general that if $[y_1, z_1]$ is a solution of the homogeneous system

$$y' = cy + bz$$
$$z' = cy + dz,$$

then $[c_1y_1, c_1z_1]$ is a solution for all values of the constant c_1.

5. Show that if $[y, z]$ is any solution of the homogeneous system corresponding to Eqs. (1) and (2) in the text and $[y_p, z_p]$ is a particular solution of the system, the sum $[y + y_p, z + z_p]$ is a solution.

6. Show that the results of Exercises 4 and 5 are valid for the linear system with variable coefficients

$$y' = A(x)y + B(x)z + r_1(x)$$
$$z' = C(x)y + D(x)z + r_2(x).$$

7. (a) Show that the system

$$ay' + bz' + cy + dz = r_1$$
$$a_1y' + b_1z' + c_1y + d_1z = r_2,$$

can be written in standard form provided the constants a, b, a_1, and b_1 satisfy the condition $ab_1 - a_1b \neq 0$. If $ab_1 - a_1b = 0$, then the system is said to be degenerate. (*Hint:* Successively eliminate y' and z' from these equations.)
(b) Degenerate systems of two equations are not equivalent to a single second-order equation in general. For example, show that the system

$$y' + z' - y = x$$
$$y' + z' = 0$$

has solutions $y = -x$, $z = x + c$, where c is arbitrary. These are the only solutions, and thus there is only one arbitrary constant instead of the two we would expect if the system were equivalent to a single second-order DE.

8. Introduce the auxiliary variables $v = y'$ and $w = z'$ and rewrite the system

$$z'' + 3z' + y' + 2z + 4y = x$$

$$y'' + 3y' + 4z + 7y = \sin x$$

as a system of four first-order equations.

10.2 METHOD OF ELIMINATION

The method of elimination for the solution of systems of two first-order linear DEs in two unknowns, $y(x)$ and $z(x)$, is similar to the method of elimination for solving linear algebraic equations. We shall illustrate this technique in the following examples. Because the manipulations can be lengthy, the method of elimination is usually not employed for systems of more than three DEs. The method was already used in Sections 5.2 and 8.2.

EXAMPLE 1

Find a general solution of

$$y' = y - 3z \tag{1}$$

$$z' = 2y - 4z. \tag{2}$$

solution. We can eliminate either y or z from these equations. We choose to eliminate z. To do this, we first solve Eq. (1) for z and obtain

$$z = \frac{1}{3}(y - y'). \tag{3}$$

This equation expresses z in terms of y and y'. To eliminate z and z' from Eq. (2), we differentiate Eq. (3) and substitute the result and Eq. (3) into Eq. (2). After collecting terms, this gives

$$y'' + 3y' + 2y = 0. \tag{4}$$

Thus by eliminating z and z' from Eq. (2) we have reduced the problem to a single second-order DE with constant coefficients for the remaining unknown y. This shows that if DEs (1) and (2) have a solution $[y(x), z(x)]$, then y must satisfy Eq. (4) and z is given by Eq. (3). A general solution of Eq. (4) is

$$y = c_1e^{-x} + c_2e^{-2x}. \tag{5}$$

The corresponding function z is obtained by substituting this solution for $y(x)$ into

Eq. (3). This gives

$$z = \frac{2}{3} c_1 e^{-x} + c_2 e^{-2x}. \tag{6}$$

Thus we have shown that if Eqs. (1) and (2) have a solution, it is given by Eqs. (5) and (6). It is easy to verify by direct substitution in Eqs. (1) and (2) that the pair $[y(x), z(x)]$ given by Eqs. (5) and (6) is a solution for all values of the arbitrary constants c_1 and c_2.

If we set $c_1 = 1$ and $c_2 = 0$ in Eqs. (5) and (6), then we conclude that the pair $[y_1(x), z_1(x)]$, where $y_1 = e^{-x}$ and $z_1 = (2/3)e^{-x}$, is a solution. Similarly, if we set $c_1 = 0$ and $c_2 = 1$ we obtain the solution pair $[y_2(x), z_2(x)]$, where $y_2 = e^{-2x}$ and $z_2 = e^{-2x}$. Thus Eqs. (5) and (6) show that y is a linear combination of y_1 and y_2 and that z is a linear combination of z_1 and z_2. The pair $[y(x), z(x)]$ is a general solution, as we shall verify in the next section.

If y and y' are eliminated from Eqs. (1) and (2) instead of z and z', it can easily be shown that the solution still is given by Eqs. (5) and (6). Thus it is immaterial which of the variables is eliminated. The choice of which variable to eliminate is usually dictated by calculational simplicity.

EXAMPLE 2

Solve the IVP that consists of the system given in Eqs. (1) and (2) and the initial conditions

$$y(0) = 1, \qquad z(0) = \frac{1}{3}.$$

solution. In Example 1, we showed that Eqs. (5) and (6) give a solution for all values of c_1 and c_2. To solve the IVP we choose c_1 and c_2 so that this solution satisfies the initial conditions. Thus, substituting $x = 0$ in Eqs. (5) and (6), and using the initial conditions, yields

$$y(0) = c_1 + c_2 = 1$$

$$z(0) = \frac{2}{3} c_1 + c_2 = \frac{1}{3}.$$

By solving these algebraic equations and substituting the results into Eqs. (5) and (6), we obtain a solution of the IVP as,

$$y = 2e^{-x} - e^{-2x}, \qquad z = \frac{4}{3} e^{-x} - e^{-2x}.$$

The method of elimination can be readily applied to a system of two first-order inhomogeneous DEs, (see Exercise 2).

Exercises

1. Solve each of the following systems.

(a) $y' = 2y + 3z$
$z' = \frac{1}{3}y + 2z$

(b) $y' = 3y + 3z$
$z' = y + z$

(c) $y' = 2y - z$
$z' = y + 2z$

(d) $y' = 2y + z$
$z' = -8y + 2z$

(e) $y' = 3y - z$
$z' = y + z$

(f) $y' = 3y - 4z$
$z' = 8y + 2z$

(g) $y' + 3z' + 2y - z = 0$
$2y' - z' - 2y - 3z = 0$
(*Hint:* Eliminate z' or y'.)

(h) $3y' + 4z' - 2y = 0$
$y' + 3z' + y - 2z = 0$

(i) $y' = -3y - 2w$
$z' = 2z$
$w' = 3z + 2w$

(j) $y' = y + z$
$z' = y - 2z + w$
$w' = -2y + z - w$

2. Solve the following inhomogeneous systems. In each case determine y_c and y_p.

(a) $y' = 2y + 4z + x$
$z' = 3y - z + 1$

(b) $y' = 3y - z + e^{2x}$
$z' = y + 4z + \sin x$

(c) $y' - y + 6z = x^2$
$z' + y - 2z = 1 + x$

(d) $y' = az + r_1(x)$
$z' = by + r_2(x)$
(Here a and b are prescribed constants and r_1 and r_2 are prescribed continuous functions.)

3. Solve the following IVPs.

(a) $y' = 6y + z, \quad y(0) = 1$
$z' = 2y - z, \quad z(0) = 0$

(b) $y' = 2y + 2z, \quad y(1) = 1$
$z' = -y + 2z, \quad z(1) = -1$

(c) $y' = 2y - z, \quad y(0) = 1$
$z' = y + z, \quad z(0) = 6$

4. Solve the system

$$y' = ay - z$$

$$z' = y + dz,$$

where a and d are any constants. Discuss the behavior of the solution as $x \to \infty$.

10.3 GENERAL SOLUTIONS AND THE INITIAL VALUE PROBLEM

The IVP for the homogeneous system

$$y' = ay + bz \tag{1}$$

$$z' = cy + dz \tag{2}$$

consists of finding a solution pair $[y(x), z(x)]$ of Eqs. (1) and (2) that satisfies the initial conditions

$$y(x_0) = a_0,\ z(x_0) = b_0. \tag{3}$$

In the preceding section we solved an IVP by first determining a solution in the form

$$y = c_1y_1(x) + c_2y_2(x) \tag{4}$$

$$z = c_1z_1(x) + c_2z_2(x) \tag{5}$$

where the pairs $[y_1, z_1]$ and $[y_2, z_2]$ are solutions of the DEs. Then the initial conditions gave two algebraic equations to determine the two constants c_1 and c_2.

We shall give a definition of a general solution by considering the IVP. It is convenient to first define a linear combination of two pairs of functions.

definition 1

If $[y_1(x), z_1(x)]$ and $[y_2(x), z_2(x)]$ are two pairs of functions, then the pair $[y(x), z(x)]$ given by

$$y = c_1y_1 + c_2y_2$$

$$z = c_1z_1 + c_2z_2,$$

where c_1 and c_2 are any constants, is called a linear combination of the given pairs.

By analogy with the case of the single second-order DE and from the examples in the preceding section, we would expect that the superposition property is valid for solutions of the system given by Eqs. (1) and (2). This is contained in the following theorem.

theorem

Any linear combination of two solutions $[y_1, z_1]$ and $[y_2, z_2]$ of the homogeneous system (1) and (2) is also a solution.

The proof follows directly by substituting the linear combination into the DEs. We leave the details as an exercise for the reader.

We shall now give the definition of a general solution. The solution given in Eqs. (4) and (5) is a general solution if we can determine c_1 and c_2 so that Eqs. (4) and (5) satisfy the initial conditions given by Eq. (3) for all values of a_0 and b_0. Thus we set $x = x_0$, $y = a_0$, and $z = b_0$ in Eqs. (4) and (5). This gives

$$c_1y_1(x_0) + c_2y_2(x_0) = a_0 \tag{6}$$

and

$$c_1z_1(x_0) + c_2z_2(x_0) = b_0. \tag{7}$$

This is a system of two linear algebraic equations for c_1 and c_2. The coefficients in these equations are known numbers given by the values of $[y_1, z_1]$ and $[y_2, z_2]$ at $x = x_0$.

We solve these algebraic equations and get,

$$W(x_0)c_1 = a_0 z_2(x_0) - b_0 y_2(x_0) \tag{8}$$

and

$$W(x_0)c_2 = -a_0 z_1(x_0) + b_0 y_1(x_0) \tag{9}$$

where $W(x_0)$ is the value at $x = x_0$ of the function

$$W(x) = y_1(x)z_2(x) - y_2(x)z_1(x). \tag{10}$$

If

$$W(x_0) \neq 0, \tag{11}$$

then we can uniquely solve Eqs. (8) and (9) for c_1 and c_2. Thus the solution given by Eqs. (4) and (5) will satisfy the initial conditions for all values of a_0 and b_0. If $W(x_0) = 0$, then the left sides of Eqs. (8) and (9) are zero for all values of c_1 and c_2 and, consequently, the initial conditions can be satisfied only for special values of a_0 and b_0.

The quantity $W(x)$ is called the Wronskian of the solutions $[y_1, z_1]$ and $[y_2, z_2]$. In analogy with the second-order DE, we can show (see Exercise 2) that $W(x)$ is either identically zero or never zero. Thus we define a general solution as follows:

definition

A general solution of a standard system is a linear combination

$$y = c_1 y_1 + c_2 y_2$$

$$z = c_1 z_1 + c_2 z_2$$

where $[y_1, z_1]$ and $[y_2, z_2]$ are two solution pairs for which $W(x) \neq 0$ for at least one value of x.

Thus we have shown that if a general solution can be determined, then the IVP can be solved for all initial values a_0 and b_0. Furthermore, we can show, as for the second-order DE, that there is only one solution of the IVP (see Exercise 5). This implies that every solution of DEs (1) and (2) is contained in a general solution.

EXAMPLE

Show that

$$y = c_1 e^{-x} + c_2 e^{-2x} \qquad \text{and} \qquad z = \frac{2}{3} c_1 e^{-x} + c_2 e^{-2x} \tag{12}$$

is a general solution of the system

$$y' = y - 3z$$

$$z' = 2y - 4z.$$

solution. We have already shown in Example 1 of Section 10.2 that the pair given by Eq. (12) is a linear combination of the solution pairs $[y_1, z_1]$ and $[y_2, z_2]$, where

$$y_1 = e^{-x}, \qquad y_2 = e^{-2x}$$

$$z_1 = \frac{2}{3}e^{-x}, \qquad z_2 = e^{-2x}.$$

Therefore, it is a general solution if the Wronskian of $[y_1, z_1]$ and $[y_2, z_2]$ is not equal to zero. Thus we have

$$W = y_1 z_2 - y_2 z_1 = \frac{1}{3}e^{-3x} \neq 0.$$

Since $W \neq 0$, the pair $[y, z]$ given by Eq. (12) is a general solution.

Exercises

1. Find a general solution of the following systems. Prove that your answer is a general solution by using the Wronskian.
   (a) $y' = 3y + 2z$, $z' = -5y + z$
   (b) $y' = -3y + 4z$, $z' = -y + z$
   (c) $y' = -bz$, $z' = 4y$, b any constant
   (d) $y' = ay - z$, $z' = y + 2z$, a any constant

2. Prove that the Wronskian $W(x)$ of any two solution pairs $[y_1, z_1]$, $[y_2, z_2]$ of the system

$$y' = ay + bz$$
$$z' = cy + dz \tag{i}$$

   satisfies the DE

$$W' = (a + d)W.$$

   Conclude that $W(x)$ is either identically zero or never zero. (*Hint:* Differentiate the definition of W given in Eq. (10) to determine W'.)

3. Consider the system

$$y' + z' - 2y - 2z = 0$$

$$y' + 2z' - 3y - 2z = 0.$$

   Find a general solution and verify that the appropriate Wronskian is nonzero. The definition of general solution is the same as for the standard system.

4. (a) Derive Eqs. (8) and (9) from Eqs. (6) and (7). (b) Suppose $W(x_0) = 0$ in Eqs. (8) and (9). Discuss possible initial conditions for which a solution to the IVP can still be found. In all cases observe that if there is more than one solution to the IVP, then there are infinitely many solutions.

5. Prove that the IVP

$$y' = ay + bz, \qquad y(x_0) = a_0$$
$$z' = cy + dz, \qquad z(x_0) = b_0$$

has no more than one solution. (*Hint:* Use elimination to reduce this IVP to an IVP for a second-order DE. Then use the uniqueness theorem of Section 4.6.)

6. Prove that every solution of DEs (1) and (2) is contained in a general solution. (*Hint:* Proceed analogously to Theorem 2 of Section 4.5 and use the uniqueness result of Exercise 5.)

Exercises 7 through 9 discuss the method of variation of parameters for determining particular solutions of inhomogeneous standard systems.

7. Consider the system

$$y' = z + x$$
$$z' = -y + 1.$$

(a) Verify that a general solution of the homogeneous system is

$$y = c_1 \sin x + c_2 \cos x$$
$$z = c_1 \cos x - c_2 \sin x.$$

(b) Vary the constants c_1 and c_2 and determine a particular solution in the form

$$y = u(x) \sin x + v(x) \cos x$$
$$z = u(x) \cos x - v(x) \sin x.$$

8. Proceed as in Exercise 7 to find a general solution of the following systems.
(a) $y' = 2y - z + 6e^x$
$z' = 3y - 2z + 4e^x$
(b) $y' = 3y - 7z$
$z' = y - 3z + \tan 4x, \qquad 0 < x < \pi/4$
(c) $y' = 6y - z + e^x$
$z' = 3y + 2z + 1$

The method of undetermined coefficients, which was developed for a single equation, can be extended to inhomogeneous systems.

9. (a) Determine a particular solution of the system

$$y' = y + 4z + e^x$$
$$z' = y + z + 2e^x$$

by the method of undetermined coefficients. (*Hint:* An appropriate form for a particular solution is $[B_1e^x, B_2e^x]$.)
(b) Show that the system

$$y' = y + 4z + e^{ax}$$
$$z' = y + z + 2e^{ax}$$

has a particular solution in the form $[B_1e^{ax}, B_2e^{ax}]$ except when $a = -1$ or $a = 3$.

(c) Show that the system

$$y' = y + 4z + e^{-x}$$

$$z' = y + z + 2e^{-x}$$

does not have a solution in the form $[B_1xe^{-x}, B_2xe^{-x}]$ even though $[e^{-x}, -\frac{1}{2}e^{-x}]$ is a solution of the homogeneous system. Verify that the appropriate form for $[y_p, z_p]$ is $[(B_1 + B_2x)e^{-x}, (B_3 + B_4x)e^{-x}] = [B_1e^{-x}, B_3e^{-x}] + [B_2xe^{-x}, B_4xe^{-x}]$.

10. (a) Find a particular solution of the system

$$y' = -z + \sin ax,$$

$$z' = y + \sin ax, \qquad \text{with } a \neq 1,\ a \neq -1.$$

(b) If $a = 1$, find the appropriate form for a particular solution.

11. Find a particular solution for each of the following systems by the method of undetermined coefficients.

(a) $y' = -z + e^{-x}\cos 4x$
$z' = y + 6e^{-x}\sin 4x$

(b) $y' = 3y + z + x$
$z' = 2y + 2z + \sin x$

10.4 COUPLED PENDULUMS; NORMAL MODES

In Figure 10.1 we show two identical pendulums of mass m and length l, suspended from two pivots that lie on the same horizontal line. The angular displacements of the pendulums are $\theta(t)$ and $\psi(t)$. The masses are coupled by a spring with a spring constant K. The spring is in its natural state (unextended) when both pendulums are in their vertical position $\theta = \psi = 0$. For small-amplitude motions without damping, we can show by applying Newton's law to each mass that the equations of motion for the coupled pendulums are

$$\theta'' + \omega^2\theta = -k(\theta - \psi) \tag{1}$$

and

$$\psi'' + \omega^2\psi = k(\theta - \psi), \tag{2}$$

where ω and k are defined by

$$\omega = \left(\frac{g}{l}\right)^{1/2} \quad \text{and} \quad k = \frac{K}{m} \geq 0.$$

The pendulums are given initial angular displacements θ_0 and ψ_0 and released from rest, so the initial angular velocities are zero. Then the IVP for the oscillations of the coupled pendulums is to determine solutions

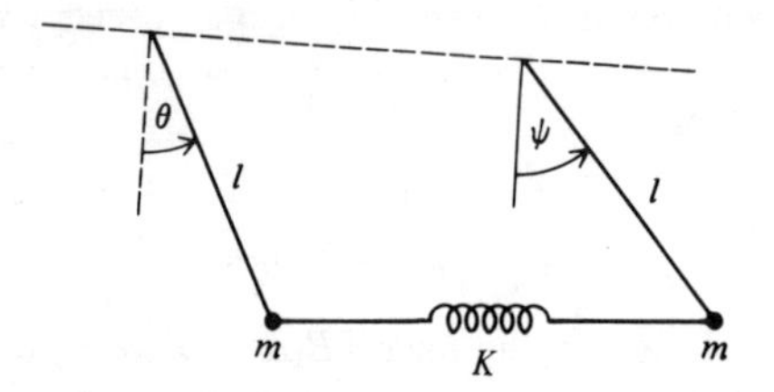

Figure 10.1

of DEs (1) and (2) that satisfy the initial conditions:

$$\theta(0) = \theta_0, \qquad \theta'(0) = 0, \qquad \psi(0) = \psi_0, \qquad \psi'(0) = 0. \tag{3}$$

Equations (1) and (2) are a coupled system of two second-order DEs for the two unknowns $\theta(t)$ and $\psi(t)$. We observe that if $k = 0$ (the coupling spring is removed), then the DEs uncouple and become $\theta'' + \omega^2\theta = 0$ and $\psi'' + \omega^2\psi = 0$. These are precisely the DEs for the vibrations of each pendulum separately, as we would expect. The free vibrations of a single pendulum are discussed in Section 5.3.

By defining the new dependent variables $u = \theta'$ and $v = \psi'$, it is possible to reduce the system given by Eqs. (1) and (2) to a standard system of four first-order DEs. This system can then be solved by the method of elimination. However, the algebraic manipulations that are required are quite lengthy. We shall therefore solve the IVP by using a trick that employs the special form of DEs (1) and (2). The idea underlying the trick is the basis of the "normal mode" method for analyzing coupled systems of oscillators. This method is described in more advanced texts, since it requires knowledge of matrix theory for its proper presentation.

We observe that the coefficients in the left sides of DEs (1) and (2) are identical and that the right sides differ by only a sign. Thus by first adding Eqs. (1) and (2) and then subtracting them, we get

$$\theta'' + \psi'' + \omega^2(\theta + \psi) = 0$$

and

$$\theta'' - \psi'' + \omega^2(\theta - \psi) = -2k(\theta - \psi).$$

Since $\theta'' \pm \psi'' = (\theta \pm \psi)''$, the variables θ and ψ in these equations appear only in the form $\theta \pm \psi$. Thus we define new dependent variables $y(t)$ and $z(t)$ by

$$y = \theta + \psi \qquad \text{and} \qquad z = \theta - \psi. \tag{4}$$

Then the DEs can be written as

$$y'' + \omega^2 y = 0 \tag{5}$$

and

$$z'' + (\omega^2 + 2k)z = 0. \tag{6}$$

In the y and z variables, DEs (5) and (6) are uncoupled. Hence they may be solved independently of each other. General solutions of Eqs. (5) and (6) are

$$y = c_1 \cos \omega t + c_2 \sin \omega t \tag{7}$$

and

$$z = c_3 \cos pt + c_4 \sin pt, \qquad p = (\omega^2 + 2k)^{1/2} > 0. \tag{8}$$

The corresponding expressions for θ and ψ are obtained first by solving the two equations in Eq. (4) for θ and ψ in terms of y and z. This gives

$$2\theta = y + z \qquad \text{and} \qquad 2\psi = y - z.$$

Then, by inserting Eqs. (7) and (8) into these equations, we get a general solution

$$2\theta = c_1 \cos \omega t + c_2 \sin \omega t + c_3 \cos pt + c_4 \sin pt \tag{9}$$

and

$$2\psi = c_1 \cos \omega t + c_2 \sin \omega t - c_3 \cos pt - c_4 \sin pt. \tag{10}$$

The four constants $c_1, \ldots, c_4$ are determined by requiring Eqs. (9) and (10) to satisfy the four initial conditions given in Eq. (3). This leads to four algebraic equations for $c_1, \ldots, c_4$. By substituting the solution of these equations into Eqs. (9) and (10), we get the solution of the IVP as

$$\theta = \tfrac{1}{2}[(\theta_0 + \psi_0) \cos \omega t + (\theta_0 - \psi_0) \cos pt] \tag{11}$$

and

$$\psi = \tfrac{1}{2}[(\theta_0 + \psi_0) \cos \omega t - (\theta_0 - \psi_0) \cos pt]. \tag{12}$$

The solution pair $[\theta, \psi]$ is a linear combination of periodic functions of fundamental periods $2\pi/\omega$ and $2\pi/p$, respectively. Of course, $\cos \omega t$ and $\cos pt$ are also periodic functions, with periods $2m\pi/\omega$ and $2n\pi/p$ for any integers m and n. Thus $\cos \omega t$ and $\cos pt$ are periodic functions with a common period if and only if $2m\pi/\omega = 2n\pi/p$; that is, if

$$p = \frac{n}{m}\,\omega, \qquad \text{for some integers } m \text{ and } n. \tag{13}$$

If there are integers m and n such that Eq. (13) is true, then the circular frequencies are said to be commensurate and the pendulums oscillate periodically with period $2\pi/\omega = (n/m)2\pi/p$; see the discussion in Section 8.3-1. We note that $n > m$, since $p > \omega$. If the circular frequencies are incommensurate, then the motion is not periodic. It is a more complicated quasi-periodic motion. If p is near ω, which from the definition of p in Eq. (8) means that k is small (a weak spring), then "beats" will be obtained. We discuss this in Exercise 5.

For further interpretation of the solution we consider the special initial conditions $\theta_0 = \psi_0$. Both pendulums are then raised to the same initial angular displacement. Then the solution is

$$\theta = \psi = \theta_0 \cos \omega t. \tag{14}$$

Thus both pendulums oscillate in unison (in phase), with the same circular frequency ω. The spring remains unextended throughout the motion.

If $\theta_0 = -\psi_0$, as we show in Figure 10.2, then the solution is

$$\theta = -\psi = \theta_0 \cos pt. \tag{15}$$

In this motion the pendulums are exactly out of phase. That is, when the right pendulum is moving to the right with an angular displacement $\psi(t)$, the left pendulum is moving to the left with an angular displacement $-\psi(t)$.

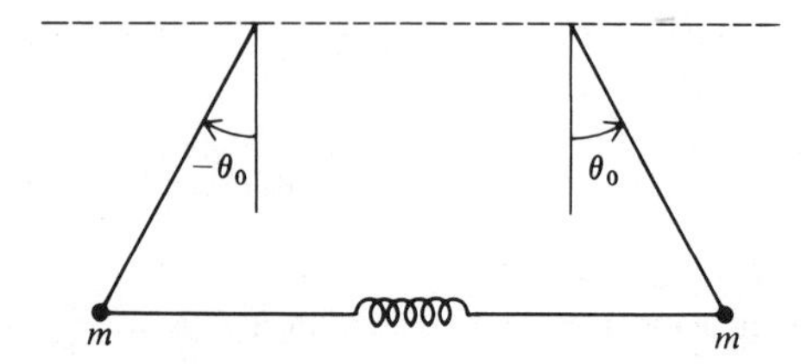

Figure 10.2

The special solutions given by Eqs. (14) and (15) are called the normal modes of vibration of the pendulum, because if $\theta_0 \neq \pm\psi_0$, then as we see from Eqs. (11) and (12), the general motion is a linear combination of these normal modes.

Exercises

1. Suppose that the coupled pendulums of Figure 10.1 are 2 ft long, that the spring constant is $k = 5$ lb/ft, and that $m = 2$ slugs. (a) If the masses are given initial displacements θ_0 and ψ_0, respectively, determine their subsequent motion. (b) Is it a periodic motion? (*Hint:* See whether the condition given by Eq. (13) is satisfied for some choice of n and m.) (c) Determine the normal modes of vibration.

2. Repeat Exercise 1 for $l = 16$ ft, $m = 5$ slugs, and $k = 5$ lb/ft.

3. Suppose that the two pendulums discussed in the text are started from the vertically hanging position with initial angular velocities $\theta'(0) = \theta_1$ and $\psi'(0) = \psi_1$, respectively. Start with the general solution given in Eqs. (9) and (10) and answer the following. (a) Determine the subsequent motion of the masses. (b) Give conditions for the motion to be periodic. (c) Show that the normal modes of vibration can be used to describe this motion.

4. Show that the solution for the motion of the coupled pendulums for any initial conditions can be written in terms of normal modes of vibrations.

*5. Let $\psi_0 = 0$ in Eqs. (11) and (12) and use the trigonometric identities

$$\cos a + \cos b = 2 \cos \left(\frac{a - b}{2}\right) \cos \left(\frac{a + b}{2}\right)$$

and

$$\cos a - \cos b = 2 \sin \left(\frac{b - a}{2}\right) \sin \left(\frac{b + a}{2}\right)$$

to rewrite the solution as

$$\theta = \theta_0 \cos \left(\frac{\omega - p}{2}\right) t \cos \left(\frac{\omega + p}{2}\right) t$$

and

$$\psi = \psi_0 \sin \left(\frac{p - \omega}{2}\right) t \sin \left(\frac{p + \omega}{2}\right) t.$$

Assuming that p is almost equal to ω, show that beats occur (see Section 8.3-1). Verify that in this case the motion can be described as follows. Initially the pendulum on the left vibrates, whereas the one on the right remains almost stationary. As t increases, the pendulum on the right moves more, but the amplitude of the one on the left decreases. At $t = \pi/(p - \omega)$ the pendulum on the left is stationary while the one on the right is moving. For larger values of t the amplitude of the motion of the right pendulum decreases and the left increases. This motion is repeated, and thus serves as an illustration of the phenomenon of beats.

6. If a resisting, or damping, force that is proportional to the velocity of the masses acts to impede the motion of each mass, then the appropriate system is

$$\theta'' + s\theta' + \omega^2\theta = -k(\theta - \psi)$$
$$\psi'' + s\psi' + \omega^2\psi = k(\theta - \psi)$$

where s is a constant. Show that if $s > 0$, then all solutions of the system approach zero as $t \to \infty$. The resisting force dissipates the initial energy of the system.

7. Suppose that each of the masses in a coupled pendulum without damping are acted upon by a force $R(t) = R_0 \cos at$. Here R_0 and a are constants. (a) Find the subsequent motion of each mass if the system is initially at rest. (b) Determine the resonant frequencies. *(c) When is the motion periodic?

10.5 TWO-LOOP ELECTRIC NETWORKS

The simple *RCL* electric circuit that we have studied is frequently an inadequate model for the description of electrical phenomena that are encountered in many applications. For example, the operation of an electric transformer cannot be described by means of a single *RCL* circuit. The analysis of more complicated devices, such as a television receiver or the complex electric circuit models of the brain and the nervous system, require more sophisticated models. A similar comment applies to the regional and nationwide electric power networks.

A generalization of the simple *RCL* circuit is to consider a number of interconnected *RCL* circuits. This is called an electric network. We consider the typical example of a two-loop (or two-mesh) electric network shown in Figure 10.3. The points n_1 and n_2 in the figure are called the junction points. They show where the circuit loops are joined. At both junction points there are three circuit branches. We denote the currents in the three branches by $i_1(t)$, $i_2(t)$, and $i_3(t)$, respectively. We choose the positive directions as shown in the figure. The quantity $E(t)$ is an electromotive force applied to the left-hand loop of the network.

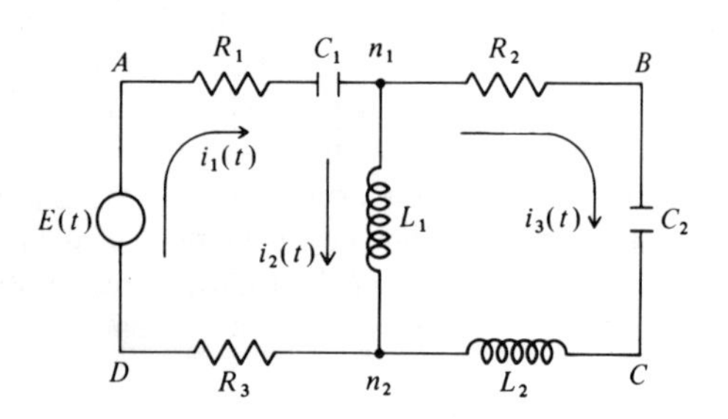

Figure 10.3

We recall that the voltage drops across resistors, inductors, and capacitors respectively are given by

$$Ri_R, \qquad Li_L', \qquad \text{and} \qquad \frac{1}{C}\int^t i_C(z)\,dz,$$

where i_R, i_L, and i_C are the respective currents in the resistor, inductor, and capacitor. Furthermore, Kirchhoff's two circuit laws for multiloop electric circuits, which are, in effect, laws for the conservation of charge and energy in a circuit, can be written respectively as

1. The algebraic sum of all the currents flowing into or out of a junction point is zero for all t. By convention a current is taken as positive when it flows into a junction and negative when it flows out.

2. The algebraic sum of all voltage drops around a closed loop is zero for all t.

The second law was used in Appendix I.3 to derive the DE for the simple *RCL* circuit. There are three closed loops in the circuit sketched in Figure 10.3. They are the two "small loops," given by An_1n_2D and n_1BCn_2, and the "big loop," given by $ABCD$.

Application of Kirchhoff's first law to either of the junctions in Figure 10.3 gives

$$i_1 - i_2 - i_3 = 0, \tag{1}$$

because, for example, for junction n_1, the current i_1 is flowing into the junction and the currents i_2 and i_3 are flowing out. We now apply Kirchhoff's second law to the loop An_1n_2D. This gives

$$R_1i_1 + \frac{1}{C_1}\int_0^t i_1(z)\,dz + L_1i_2' + R_3i_1 = E. \tag{2}$$

Similarly, for the loop n_1BCn_2 we get, from the second law,

$$R_2i_3 + \frac{1}{C_2}\int^t i_3(z)\,dz + L_2i_3' - L_1i_2' = 0. \tag{3}$$

Equations (1), (2), and (3) are three equations to determine the three unknown currents i_1, i_2, and i_3. A fourth equation relating the three currents is obtained by applying Kirchhoff's second law to the loop $ABCD$. However, this equation is redundant because it is a linear combination of Eqs. (2) and (3). Consequently, we need only consider Eqs. (1), (2), and (3).

A special case occurs when $1/C_1 = 1/C_3 = 0$, that is, when there are no capacitors in the network. Then by using Eq. (1) to eliminate i_1, we find that the DEs for the two-loop capacitanceless circuit are

$$L_1i_2' + (R_1 + R_3)i_2 + (R_1 + R_3)i_3 = E \tag{4}$$

and

$$L_2i_3' + R_2i_3 - L_1i_2' = 0. \tag{5}$$

The DEs given in Eqs. (4) and (5) are a system of two coupled first-order DEs. They are not in standard form. They can be solved by directly applying the elimination method, or they can first be transformed into standard form (see Exercise 7 of Section 10.1) and then solved by elimination.

Exercises

1. Suppose that the applied emf in the circuit of Figure 10.4 is a constant E_0. (a) Determine the currents i_1, i_2, and i_3. (b) Find the steady-state currents.

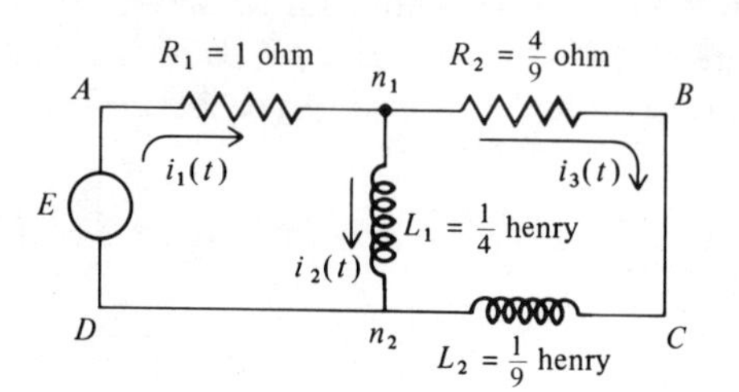

Figure 10.4

2. Consider the circuit in Figure 10.4, with $R_2 = 0$, $E(t) = \sin t$ and all other parameters as indicated. (a) Determine the currents i_1, i_2, and i_3. (b) Does the current i_3 have a transient part?

3. (a) Derive Kirchhoff's laws for the circuit shown in Figure 10.3, with $L_1 = L_2 = 0$. (b) Observe that the circuit is equivalent to a single RC circuit. (c) Give the values of R and C for the equivalent RC circuit.

4. A circuit for a simple transformer, which is widely used in electric devices, is shown in the Figure 10.5. A typical example is the transformer in model electric trains. The two circuits interact due to magnetic coupling of the inductances. The current passing through each inductance produces an effect on the other inductance, and thereby on the current in the other circuit. The appropriate system to describe the transformer is

$$L_1 i_1' - M i_2' + R_1 i_1 = E(t)$$

$$-M i_1' + L_2 i_2' + R_2 i_2 = 0, \tag{i}$$

where M is called the mutual inductance. If $M = 0$, the two circuits do not interact. Typical initial conditions are

$$i_1(0) = i_2(0) = 0. \tag{ii}$$

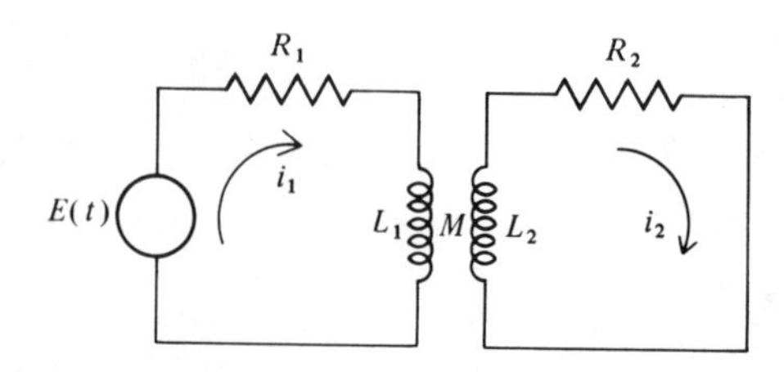

Figure 10.5

Let $E(t) = \sin at$ and assume that $R_1 = R_2 = 0$. This is a lossless transformer, because there are no resistors present to dissipate energy. (a) Solve the IVP given by Eqs. (i) and (ii) in this case. (b) Is there a resonant value for the frequency of the applied emf? (c) Find the ratios i_1/i_2 and V_1/V_2, where V_1 and V_2 are the voltage drops across the inductors L_1 and L_2, respectively. (d) Assume that $M^2 - L_1L_2 = 0$. This is called an ideal transformer. Specialize your results in (a), (b), and (c) to this case.

5. Given $L_1 = 1$, $L_2 = 4M = \sqrt{2}$, $R_1 = R_2 = 3$, and $E(t) = \sin t$ determine the current in the loops of the transformer of Figure 10.5. These values of the circuit parameters were chosen to simplify algebraic computations. They do not represent realistic values for actual transformers.

chapter 11

SECOND-ORDER LINEAR DIFFERENTIAL EQUATIONS WITH VARIABLE COEFFICIENTS

11.1 INTRODUCTION

In many of the applications of differential equations the coefficients in these equations are not constants. For example, the differential equation for the RCL electric circuit is

$$Ly'' + Ry' + \frac{1}{C}y = E(t), \tag{1}$$

where $y(t)$ is the charge. The inductance, resistance, and capacitance of the circuit may vary with time. Then the coefficients L, R, and C in Eq. (1) are prescribed functions of the independent variable t. The variation may be deliberately imposed or it may result from heating of the circuit elements by the current. Such circuits are called time-varying electric circuits. For chemical and biochemical kinetic systems, variable coefficients represent time-varying reaction rates.

Ordinary differential equations with variable coefficients also occur in the solution of partial differential equations. A typical physical problem in which such equations arise is the vibrations of the head of a drum. The drumhead is treated as a tightly stretched, circular membrane. It can be shown that the frequencies of vibration and the corresponding shapes of a circular membrane are obtained from the solutions of

$$y'' + \frac{1}{x}y' + \left(k^2 - \frac{n^2}{x^2}\right)y = 0. \qquad (2)$$

Here, x is the radial distance from the center of the membrane, the constant k is proportional to the vibration frequency, and n is a nonnegative integer. This DE, which has variable coefficients, is called Bessel's equation of order n.

Other typical DEs with variable coefficients that occur in applications are the Cauchy-Euler equation

$$y'' + \frac{b_1}{x}y' + \frac{b_2}{x^2}y = 0, \qquad b_1,\ b_2 \text{ constants} \qquad (3)$$

and Legendre's equation

$$y'' - \frac{2x}{1-x^2}y' + \frac{k}{1-x^2}y = 0, \qquad k \text{ constant.} \qquad (4)$$

These two equations occur, for example, in studying the temperature distribution generated by a heat source such as the sun or a nuclear reactor.

We shall consider the general second-order linear DE in the form[1]

$$y'' + P(x)y' + Q(x)y = r(x). \qquad (5)$$

The coefficients $P(x)$ and $Q(x)$ and the inhomogeneous term $r(x)$ are prescribed functions. The corresponding homogeneous DE is

$$y'' + P(x)y' + Q(x)y = 0. \qquad (6)$$

When the coefficients $P(x)$ and $Q(x)$ are constants, as they were in Chapters 4 through 8, we saw that any solution $y(x)$ of the DE satisfied it for all values of x. (Recall that for a second-order DE this implies that the solution $y(x)$ has a second derivative for all values of x.) This is not

[1] A more general form is

$$a_0(x)y'' + a_1(x)y' + a_2(x)y = R(x).$$

This can always be reduced to the form of Eq. (5) by dividing by $a_0(x)$. Then we have

$$P(x) = \frac{a_1(x)}{a_0(x)}, \qquad Q(x) = \frac{a_2(x)}{a_0(x)}, \qquad \text{and} \qquad r(x) = \frac{R(x)}{a_0(x)}.$$

Thus $P(x)$, $Q(x)$ and $r(x)$ may be discontinuous for the values of x at which $a_0(x) = 0$.

the case for DE (6) for all choices of $P(x)$ and $Q(x)$. In fact we have already seen that for the first-order DE with a variable coefficient, $y' + q(x)y = 0$, if $q(x)$ is discontinuous at a point x_1, then the solution y might not be defined at $x = x_1$. We also found that if we restrict $q(x)$ to be a continuous function on an interval (α, β), then all solutions of the DE have continuous first derivatives, at least on (α, β). (See Section 2.2, Exercises 6 and 7.) Similar results are valid for DE (6).

In more advanced texts the following result is established:

theorem

If $P(x)$ and $Q(x)$ are continuous on an interval (α, β), then all solutions of the DE given in Eq. (6) are continuous and have continuous second derivatives on this interval.

In particular, if $P(x)$ and $Q(x)$ are continuous for all values of x, then all solutions are valid for all values of x. However, in many DEs of interest $P(x)$ and $Q(x)$ are not continuous for all x, as we see, for example, in Bessel's equation, given by Eq. (2). In Eq. (2), $P(x) = 1/x$ and $Q(x) = k^2 - (n^2/x^2)$ are not continuous at $x = 0$. In Legendre's equation, $P(x) = -2x/(1 - x^2)$ and $Q(x) = k/(1 - x^2)$ are not continuous at $x = \pm 1$. Thus in solving DEs with variable coefficients it is important to remember that solutions may not be defined for all values of x. The interval on which we seek a solution of the DE is usually determined by the specification of initial or boundary conditions.

When P and Q are constants, homogeneous DE (6) always has at least one exponential solution $y = e^{mx}$, where m is a constant. In the exercises we shall show that it has an exponential solution only for a limited class of variable coefficients $P(x)$ and $Q(x)$. In general, it does not possess an exponential solution or, in fact, solutions that are elementary functions. This accounts for the greater difficulty in solving DEs with variable coefficients. Thus approximate methods and computers are used frequently to solve these equations. Approximate methods are discussed in Chapter 15.

Exercises

1. Show that $y = e^x$ is the only exponential solution of $y'' + xy' - (1 + x)y = 0$ by inserting $y = e^{mx}$, for m a constant, into this equation. Show that m must satisfy

$$m^2 + xm - (1 + x) = 0, \qquad \text{for all } x.$$

2. Find an exponential solution of each of the following DEs if it exists.

(a) $y'' - \dfrac{4(x+1)}{2x+1}y' + \dfrac{4}{2x+1}y = 0, \qquad \text{for } x > -\dfrac{1}{2}$

(b) $y'' - \dfrac{2x+5}{x+2}y' + \dfrac{2}{x+2}y = 0, \qquad \text{for } x > -2$

3. Show that if $y = e^{mx}$ is a solution of Eq. (6) for some constant m, then m must satisfy the quadratic equation $m^2 + P(x)m + Q(x) = 0$ for all values of x in the interval of definition of the DE. Show that if the quadratic equation has two constant roots, then $P(x)$ and $Q(x)$ are both constants. Obtain conditions relating $P(x)$ and $Q(x)$ that ensure that the DE has one exponential solution.

4. Show that the substitution

$$y(x) = v(x) \exp\left[-\frac{1}{2}\int^x P(z)\,dz\right]$$

transforms the DE in Eq. (5) into the form $v'' + q(x)v = R(x)$. If $q(x)$ is a constant, then the DE can be easily solved.

5. Determine the discontinuities of $P(x)$ and $Q(x)$ if any, in each of the following DEs.
(a) $y'' + 3xy' + (\sin x)y = 0$
(b) $y'' + 3(\cos x)y' + (\tan x)y = 0$
(c) $e^x y'' + 3xy' + x^2 y = 0$
(d) $(1 + x^2)y'' + 3y' + xy = 0$

6. (a) Solve the DE $xy'' + y' = 0$, by rewriting it as $(xy')' = 0$.
(b) Observe that $P(x) = 1/x$ is discontinuous at $x = 0$. Discuss the behavior of your solution as $x \to 0$.
(c) Solve the BVP $xy'' + y' = 0$, $y(x_0) = a_0$, $y(x_1) = a_1$. This BVP describes the temperature distribution in a ring between two concentric circles with constant temperatures a_0 and a_1 on the inner and outer circles, respectively. Observe that the solution has continuous second derivatives if $x > 0$.
(d) Solve the BVP of (c) with the boundary condition at x_1 replaced by $y'(x_1) = 0$. This corresponds to insulating the outer circle $x = x_1$ and fixing the temperature on the inner circle. Insulating the surface means that the heat flow across the surface, which is proportional to the temperature gradient y', must be zero.

7. In each of the following DEs show that $P(x)$ and $Q(x)$ are not continuous for all x. Show that the given functions are solutions. Hence the solutions are continuous and have continuous second derivatives for all x. Thus continuity of $P(x)$ and $Q(x)$ is a sufficient condition, but not a necessary one, for a solution to exist for all x.
(a) $x^2y'' - 3xy' + 3y = 0, \qquad y = c_1x + c_2x^3$
(b) $(1 - x)y'' - xy' + y = 0, \qquad y = c_1x + c_2e^x$

11.2 THE INITIAL VALUE PROBLEM

The IVP for the homogeneous DE

$$y'' + P(x)y' + Q(x)y = 0 \tag{1}$$

is to find a function $y(x)$ with a continuous second derivative that satisfies the initial conditions,

$$y(x_0) = a_0 \qquad \text{and} \qquad y'(x_0) = a_1. \tag{2}$$

We can solve the IVP by using a general solution. The definition of a general solution is the same as for DEs with constant coefficients.

definition

A general solution $y(x)$ on the interval (α, β) is a linear combination

$$y(x) = c_1y_1(x) + c_2y_2(x)$$

of two solutions, $y_1(x)$ and $y_2(x)$, of the DE whose Wronskian $W(x) \neq 0$ on (α, β). The Wronskian is defined by

$$W(x) = y_1(x)y_2'(x) - y_1'(x)y_2(x). \tag{3}$$

We showed in the previous chapters for DEs with constant coefficients that if $W(x) \neq 0$ at one point, then $W(x) \neq 0$ for any x. A similar result is valid for DE (1), with appropriate restrictions on the coefficients $P(x)$ and $Q(x)$. Specifically, we have the following theorem:

theorem

If the coefficients $P(x)$ and $Q(x)$ are continuous functions in the open interval (α, β), then the Wronskian of two solutions of homogeneous DE (1) is either zero for every x in (α, β) or it is not zero for any x in (α, β).

The proof of this theorem is similar to the proof of Theorem 1 of Section 4.5, and therefore it will not be given (see Exercise 2).

For coefficients $P(x)$ and $Q(x)$ that satisfy the conditions of the theorem, linear independence and linear dependence of two solutions of the DE given in Eq. (1) are equivalent to the conditions $W \neq 0$ and $W = 0$, respectively. This corresponds to the discussion in Section 4.5-1.

A general solution of the inhomogeneous DE

$$y'' + P(x)y' + Q(x)y = r(x) \tag{4}$$

is given by

$$y(x) = y_c(x) + y_p(x),$$

where the complementary solution $y_c(x)$ is a general solution of the homogeneous DE and the particular solution $y_p(x)$ is any solution of the inhomogeneous DE. If $y_c(x)$ can be determined, then we can find a particular solution by the method of variation of parameters, as we did in Section 7.3 for the constant-coefficient case. In fact we can show (see Exercise 5) that if $y_1(x)$ and $y_2(x)$ are linearly independent solutions of the homogeneous DE, then a particular solution is given by

$$y_p(x) = -y_1(x) \int^x \frac{y_2(z)r(z)}{W(z)}\, dz + y_2(x) \int^x \frac{y_1(z)r(z)}{W(z)}\, dz. \tag{5}$$

In order to determine general solutions and use them to solve IVPs and BVPs, we require two linearly independent solutions of the homogeneous DE. Consequently, in this chapter and in Chapters 12 and 13 we shall only consider techniques for constructing these solutions of the homogeneous DE. Little reference will be made to solving initial and boundary value problems or to determining particular solutions.

Exercises

1. Show that $y_1 = x$ and $y_2 = x^2$ are solutions of

$$y'' - \frac{2}{x}y' + \frac{2}{x^2}y = 0.$$

   Observe that the coefficients $P(x) = -2/x$ and $Q(x) = 2/x^2$ are not continuous at $x = 0$. However, the two solutions y_1 and y_2 are continuous and have continuous second derivatives at $x = 0$. Show that the Wronskian of y_1 and y_2 is $W(x) = x^2$. Here $W(0) = 0$, but W is not zero for any other value of x. This does not violate the theorem in the text, since $x = 0$ is the point of discontinuity of the coefficients and therefore it is not included in the interval (α, β).

*2. Prove the theorem in the text. (*Hint:* See Section 4.5.) In particular, show where the continuity of $P(x)$ and $Q(x)$ is used in the proof.

3. Use the method of variation of parameters to show that $y_p = \frac{1}{2}x^3$ is a particular solution of

$$y'' - \frac{2}{x}y' + \frac{2}{x^2}y = x,$$

   given that $y_1 = x$ and $y_2 = x^2$ are solutions of the homogeneous DE.

4. Show that the Wronskian of the solutions $y_1 = e^x$ and $y_2 = e^x x^3$ of the DE, $xy'' - 2(x+1)y' + (x+2)y = 0$ is not equal to zero. Use the method of variation of parameters to show that $y_p = e^{2x}$ is a particular solution of $xy'' - 2(x+1)y' + (x+2)y = (x-2)e^{2x}$.

5. Let y_1 and y_2 be two linearly independent solutions of DE (1). By the method

of variation of parameters, we assume that

$$y_p(x) = u(x)y_1(x) + v(x)y_2(x)$$

is a particular solution. Then show that $u(x)$ and $v(x)$ satisfy the conditions

$$u'y_1 + v'y_2 = 0$$
$$u'y_1' + v'y_2' = r.$$

Solve these equations for u' and v' and then show that a particular solution is given by Eq. (5).

6. Prove that if y_c and y_p are complementary and particular solutions of the DE given in Eq. (4), then $y = y_c + y_p$ is a solution.

11.3 REDUCTION OF ORDER

To obtain a general solution of the homogeneous DE

$$y'' + P(x)y' + Q(x)y = 0 \tag{1}$$

we must determine two solutions of Eq. (1) whose Wronskian is not zero. As we shall see in this and in subsequent chapters, it is sometimes possible either to guess one solution of the DE or to obtain one solution by techniques to be discussed. We shall show in this section that when one solution is known, it is possible to obtain the second solution by using the method of reduction of order. This procedure has been used already in Section 4.4. We illustrate it in the following example.

EXAMPLE

Find a general solution of the DE

$$xy'' - 2(x + 1)y' + (x + 2)y = 0, \qquad \text{for } x > 0.$$

solution. It is easy to verify by substitution into the DE that $y = e^x$ is a solution. Since the DE is linear and homogeneous, $y = ce^x$ is also a solution for all values of the constant c. We look for a second, linearly independent solution by varying the constant c. Thus we replace c by a function $u(x)$, which is to be determined, and we seek a solution in the form

$$y = u(x)e^x. \tag{2}$$

We substitute Eq. (2) and its derivatives

$$y' = u'e^x + ue^x \qquad \text{and} \qquad y'' = u''e^x + 2u'e^x + ue^x$$

into the DE. Since $e^x \neq 0$, we conclude from the resulting equation that u must satisfy the DE

$$xu'' - 2u' = 0.$$

We set $v(x) = u'(x)$ in this equation and obtain the first-order DE

$$xv' - 2v = 0, \qquad \text{for } x > 0. \tag{3}$$

A general solution of Eq. (3) is $v = bx^2$, where b is an arbitrary constant. Then $v(x) = u'(x) = bx^2$, and integration of this equation yields

$$u(x) = \frac{bx^3}{3} + b_1.$$

Here b_1 is another arbitrary constant. Using this value of u in Eq. (2) and defining $c_1 = b_1$ and $c_2 = b/3$, we conclude that

$$y = c_1e^x + c_2x^3e^x. \tag{4}$$

It is a linear combination of the known solution $y_1 = e^x$ and the function $y_2 = x^3e^x$. The reader should verify by direct substitution that y_2 is a solution of the DE. Since the DE is linear, it then follows that y given by Eq. (4) is a solution for all values of the arbitrary constants c_1 and c_2. To show that it is a general solution, we calculate the Wronskian of y_1 and y_2. It is given by

$$\begin{aligned} W(x) &= y_1y_2' - y_2y_1' \\ &= e^x(3x^2e^x + x^3e^x) - x^3e^x(e^x) = 3x^2e^{2x}. \end{aligned}$$

The Wronskian is not zero for all $x > 0$. Hence y as given by Eq. (4) is a general solution of the DE for $x > 0$.

The name, reduction of order, results from the fact that the second solution is determined by solving a first-order DE, such as Eq. (3), for v, rather than a second-order DE.

The technique used in the example can be applied to DE (1). Specifically, we can show that if $P(x)$ is a continuous function on the interval (α, β) and if $y = f(x)$ is a known solution of DE (1) that is not zero for any x in (α, β), then a second, linearly independent solution can always be found by proceeding as we did in the foregoing example. The method is discussed further in Exercise 5. Reduction of order may also work if $f(x)$ is zero at one or more distinct points. This is illustrated in Exercise 6.

Exercises

Use the method of reduction of order in Exercises 1 through 3.

1. Show that $y = e^{2x}$ is a solution of the DE

$$y'' - \frac{4(x+1)}{2x+1}y' + \frac{4}{2x+1}y = 0,$$

and then find a general solution for $x > -\frac{1}{2}$.

2. Show that e^{2x} is a solution of the DE

$$y'' - \frac{2x+5}{x+2}y' + \frac{2}{x+2}y = 0,$$

and then find a general solution for $x > -2$.

3. Show that $y = x$ is a solution of the DE

$$y'' - \frac{x+2}{x}y' + \frac{x+2}{x^2}y = 0$$

and then find a general solution for $x > 0$.

*4. Consider the DE $y'' + Q(x)y = 0$. Assume that one solution $y = f(x)$ is known. (a) Derive a formula for the second solution. (b) Give conditions on f that will ensure that your result in (a) provides a second, linearly independent solution even if f vanishes at a point.

5. (a) Consider the DE given in Eq. (1) with $P(x)$ a continuous function on an interval (α, β). Use reduction of order to show that if $y_1 = f(x)$ is a known solution of the DE and $f(x)$ is nonzero for all x in (α, β), then a second solution y_2 is given by

$$y_2 = f(x) \int^x f^{-2}(z_1) \exp\left[-\int^{z_1} P(z)\,dz\right] dz_1.$$

Where is the continuity of $P(x)$ used in your derivation? Where is the condition $f(x) \neq 0$ used?

(b) Verify that the Wronskian of y_1 and y_2 is nonzero for all x in (α, β).

6. Show that $y_1 = f(x) = x$ is a solution of the Tchebycheff equation $(1 - x^2)y'' - xy' + y = 0$. Determine a second solution in the interval $(-1, 1)$ by reduction of order. Since $f(0) = 0$, consider the intervals $(-1, 0)$ and $(0, 1)$ separately. Show that the resulting function $y = u(x)f(x)$ has continuous second derivatives and is a solution of the Tchebycheff equation in $(-1, 1)$. Determine a general solution.

11.4 THE CAUCHY-EULER EQUATION

The Cauchy-Euler equation is

$$y'' + \frac{b_1}{x}y' + \frac{b_2}{x^2}y = 0,$$

where b_1 and b_2 are specified constants. It is a special case of the general second-order DE with the variable coefficients $P(x) = b_1/x$ and $Q(x) = b_2/x^2$. Since these coefficients are infinite at $x = 0$, we shall analyze this equation in the interval $x > 0$. The solution for $x < 0$ can be obtained directly from the solution with $x > 0$. (See Exercise 3.) By multiplying both sides of this DE by x^2 we get the alternative form of the Cauchy-Euler equation

$$x^2y'' + b_1xy' + b_2y = 0. \tag{1}$$

From Eq. (1) we observe the distinguishing feature of the Cauchy-Euler equation: The coefficient of the nth derivative of y, for $n = 0, 1, 2$, is proportional to x^n. The zeroth derivative of y is equal to the function y itself.

We devote a section to the Cauchy-Euler equation because it occurs frequently in applications and because it is useful in some of the analysis in Chapter 13.

We now show how to find a general solution of DE (1). First we consider the special case,

$$x^2y'' + b_1xy' = 0, \qquad b_1 \neq 1. \tag{2}$$

By setting $v = y'$ we reduce this equation to the first-order linear DE $xv' + b_1v = 0$. A general solution of this equation is easily found to be

$$v = cx^{-b_1},$$

where c is an arbitrary constant. To determine y we integrate $y' = v$ and obtain

$$y = c_1x^{1-b_1} + c_2, \tag{3}$$

where $c_1 = c/(1 - b_1)$ and c_2 are arbitrary constants. Letting $c_1 = 1$, $c_2 = 0$, and $c_1 = 0$, $c_2 = 1$ we obtain the two solutions

$$y_1 = x^{1-b_1} \qquad \text{and} \qquad y_2 = 1. \tag{4}$$

By evaluating the Wronskian of y_1 and y_2, the reader can verify that y_1 and y_2 are linearly independent for $x > 0$; and thus y as given by Eq. (3) is a general solution of Eq. (2) for $x > 0$. Both y_1 and y_2 are of the form $y = x^m$, $m =$ constant, because $1 = x^0$.

If $b_1 = 1$ in DE (2), then $v = cx^{-1}$ and

$$y = c \ln x + c_2 \tag{5}$$

is a general solution.

Since there are solutions of Eq. (2) in the form x^m, it is reasonable to look for solutions to the full Eq. (1) in the same form. We substitute $y = x^m$ into Eq. (1). This yields

$$x^2[m(m - 1)x^{m-2}] + b_1x[mx^{m-1}] + b_2x^m = 0.$$

Simplifying, we find that

$$x^m[m^2 + (b_1 - 1)m + b_2] = 0.$$

The bracketed term in this equation must equal zero, because $x^m \neq 0$ for $x > 0$. Thus in order for $y = x^m$ to be a solution of Eq. (1), m must satisfy the quadratic equation

$$m^2 + (b_1 - 1)m + b_2 = 0. \tag{6}$$

If the roots of this equation are real and distinct, then we obtain two solutions x^{m_1} and x^{m_2}, as we illustrated in Eq. (4) for Eq. (2) with $b_1 \neq 1$. If Eq. (6) has a repeated root, then there is only one solution of the form x^m, as we illustrated in Eq. (5). A second solution is obtained in the form $y = u(x)x^m$ by reduction of order. This gives (see Exercise 4) $y_2 = x^m \ln x$, and hence a general solution is

$$y = c_1 x^m + c_2 x^m \ln x. \tag{7}$$

If the coefficients in the Cauchy-Euler equation are such that the exponents m are complex conjugate numbers, then the solutions are of the form $y = x^m$, where m is a complex number. Thus we must define the function x^m for complex numbers m. This can be done by using the complex valued exponential function. This is pursued further in Exercise 5. The result of this analysis is that when

$$m_1 = \alpha + i\beta \qquad \text{and} \qquad m_2 = \alpha - i\beta, \qquad \text{for } \beta \neq 0,$$

then a general solution of the Cauchy-Euler equation is

$$y = c_1 x^\alpha \cos(\beta \ln x) + c_2 x^\alpha \sin(\beta \ln x). \tag{8}$$

Exercises

1. Find a general solution of each of the following DEs.
   (a) $x^2y'' - 2xy' + 2y = 0$ (b) $x^2y'' - 4xy' + 6y = x$
   (c) $x^2y'' - 5xy' + 5y = 0$ (d) $2x^2y'' - xy' + y = 0$

2. Find a general solution of each of the following DEs using reduction of order.
   (a) $2x^2y'' - xy' + y = 0$ (b) $4x^2y'' + y = 0$
   (c) $4x^2y'' + 8xy' + y = 0$

3. Show that the transformation $x = -t$, $z(t) = y(x)$, leaves DE (1) invariant. That is, in the new variables the DE is

$$t^2 \frac{d^2z}{dt^2} + b_1 t \frac{dz}{dt} + b_2 z = 0.$$

   Since $x < 0$ corresponds to $t > 0$, show that the solutions for $x < 0$ are obtained from the solutions with $x > 0$ by replacing x by $-x$.

4. Use reduction of order to establish Eq. (7) when the coefficients in the Cauchy-Euler equation satisfy $(b_1 - 1)^2 = 4b_2$.

5. To derive the general solution given in Eq. (8) for $m = \alpha + i\beta$ use the following definition:

$$x^m = x^\alpha e^{i\beta \ln x}.$$

Apply Euler's Formula to verify that

$$x^\alpha \cos(\beta \ln x) \quad \text{and} \quad x^\alpha \sin(\beta \ln x)$$

are real-valued solutions of the Cauchy-Euler equation. Show that the Wronskian of these two solutions is not zero.

6. Find a general solution of the following DEs.
   (a) $x^2y'' + 9xy' + y = 0$
   (b) $x^2y'' + xy' + y = 0$
   (c) $x^2y'' - 5xy' + 25y = 0$

Another method of solving the Cauchy-Euler equation is to reduce it to a DE with constant coefficients, as is discussed in Exercise 7.

7. Change the independent variable x to t by the transformation $t = \ln x$ in DE (1) and show that it is reduced to

$$z'' + (b_1 - 1)z' + b_2 z = 0.$$

Obtain general solutions of this DE in the three cases when its characteristic roots m_1 and m_2 are real and distinct, real and equal, and complex conjugates. Then use the transformation $t = \ln x$ to obtain general solutions in each case.

*8. Derive the change of variable used in Exercise (7) starting from the general transformation $t = g(x)$. Determine g so that the transformed equation will have constant coefficients.

chapter 12

POWER SERIES SOLUTIONS

12.1 INTRODUCTION

In this chapter we shall show how to solve second-order linear differential equations with variable coefficients by using power series. We recall from calculus that the Taylor's series representation about $x = x_0$ of a function $f(x)$ is

$$f(x) = f(x_0) + \frac{f'(x_0)}{1}(x - x_0) + \frac{f''(x_0)}{2\cdot 1}(x - x_0)^2 + \frac{f'''(x_0)}{3\cdot 2\cdot 1}(x - x_0)^3 + \cdots.$$

It is a series in powers of $x - x_0$; that is, it is a power series representation. The three dots indicate that there are an infinite number of additional terms in the series. We shall use this notation throughout the chapter.

Solutions of linear homogeneous differential equations with constant coefficients always can be represented by power series. We illustrate this for the simple DE

$$y' - y = 0. \tag{1}$$

A general solution is $y = ce^x$, where c is an arbitrary constant. We know from calculus that e^x can be represented by the Taylor series, with $x_0 = 0$,

$$e^x = 1 + x + \frac{x^2}{2\cdot 1} + \frac{x^3}{3\cdot 2\cdot 1} + \cdots.$$

Thus, the general solution can be written as

$$y = c\left(1 + x + \frac{x^2}{2\cdot 1} + \frac{x^3}{3\cdot 2\cdot 1} + \cdots\right). \tag{2}$$

This is a power series representation of the solution (in powers of x). We refer to it as a *power series solution.*

Our objective is not to expand known solutions in power series but rather to obtain power series solutions directly from the differential equation in cases where the solutions are not known explicitly. To introduce the technique we consider DE(1) and show how to evaluate the first few terms in a power series solution in powers of x directly from the differential equation. A power series solution in powers of $(x - x_0)$ can be obtained by a similar procedure.

Thus we seek a solution in the form,

$$y = b_0 + b_1x + b_2x^2 + b_3x^3 + \cdots, \tag{3}$$

where the unknown coefficients $b_0, b_1, \ldots$ are constants. They are determined by requiring the power series given in Eq. (3) to be a solution of DE (1). In order to substitute Eq. (3) into the differential equation, we must first calculate its derivative. We assume that y' is obtained by differentiating the series in Eq. (3), term by term. This gives

$$y' = b_1 + 2b_2x + 3b_3x^2 + \cdots. \tag{4}$$

We substitute the series given by Eqs. (3) and (4) into DE (1) and get

$$(b_1 + 2b_2x + 3b_3x^2 + \cdots) - (b_0 + b_1x + b_2x^2 + \cdots) = 0. \tag{5}$$

Collecting like powers of x, we rewrite Eq. (5) as

$$(b_1 - b_0) + (2b_2 - b_1)x + (3b_3 - b_2)x^2 + \cdots = 0. \tag{6}$$

Hence, Eq. (3) is a solution of DE (1) if the left side of Eq. (6) is equal to zero. It is obvious that if the coefficient of each power of x in Eq. (6) is zero, then the left side of Eq. (6) is zero for all values of x; that is,

$$b_1 - b_0 = 0, \qquad 2b_2 - b_1 = 0, \qquad 3b_3 - b_2 = 0, \qquad \cdots. \tag{7}$$

It follows from Eq. (7) that

$$b_1 = b_0, \qquad b_2 = \frac{b_1}{2} = \frac{b_0}{1\cdot 2}, \quad \cdots.$$

Thus we can express each coefficient $b_1, b_2, \ldots$ in terms of the first coefficient b_0. The coefficient b_0 is not determined, and hence it is an arbitrary constant. By substituting these values into the series given by Eq. (3), we obtain the power series solution

$$y = b_0 + b_0 x + \frac{b_0 x^2}{2} + \cdots = b_0\left(1 + x + \frac{x^2}{2} + \cdots\right). \tag{8}$$

This is equivalent to the general solution given by Eq. (2), since b_0 is an arbitrary constant.

Thus we have shown formally how the power series solution given in Eq. (2) is obtained directly from the DE. However, in our analysis we have not considered several important mathematical questions, such as the convergence of the power series to the solution and the validity of the term-by-term differentiation that was used in Eq. (4) to determine the derivative of the power series solution. In order to resolve these questions, we shall review in the next section some relevant mathematical results about power series.

Exercises

1. Find the first four terms in a Taylor series expansion for the following functions about the given point.
   (a) $\sin x, \quad x = 0$ (b) $\sin x, \quad x = \pi/2$
   (c) $e^{x^2}, \quad x = 1$ (d) $x^2 e^x, \quad x = 0$
   (e) $\dfrac{1}{1-x}, \quad x = 0$ (f) $\dfrac{1}{1+x^2}, \quad x = 0$

2. Determine the first four nonzero terms in a power series representation in powers of x of a general solution for each of the following DEs in two ways: First, find an explicit general solution and expand it in a Taylor series in powers of x; second, substitute Eq. (3) directly into the DE and determine the unknown coefficients b_n.
   (a) $y' + 3x^2 y = 0$ (b) $y' + (1 + x)y = 0$
   (c) $y' + (1 + x + x^2)y = 0$ (d) $y'' + 2y' = 0$

3. (a) Show directly from the DE that every derivative of a solution of the DE $y' - y = 0$ can be determined in terms of $y(0)$. (b) Observe that the result of

(a) implies that if the solution $y(x)$ has a Taylor series

$$y(x) = y(0) + y'(0)x + \frac{y''(0)}{2}x^2 + \cdots,$$

then all the coefficients, $y'(0)$, $y''(0)$, . . ., in this series can be determined in terms of $y(0)$. Hence show that this gives the power series solution in Eq. (8).

4. Generalize the result of Exercise 3 to show that all derivatives of y of order two or greater can be determined in terms of $y(0)$ and $y'(0)$ for each of the following second-order DEs.
   (a) $y'' + y' + y = 0$ (b) $y'' + xy' + x^2y = 0$.

5. Proceeding as in Exercises 3 and 4, determine a power series solution for the following IVPs.
   (a) $y' + 2x^2y = 0, \quad y(0) = 2$
   (b) $y'' + y' = 0, \quad y(0) = 0, y'(0) = 1$
   (c) $y'' - y = 0, \quad y(0) = y'(0) = 2$

6. (a) Show that the DE $y' + (1/x)y = 0$ has no power series solution by looking for a solution in the form of Eq. (3). Observe that the coefficient of y is infinite at $x = 0$. (b) Determine an explicit solution of the DE and verify the result of (a).

7. Find a power series solution of the DE (1) in powers of $(x - x_0)$. Verify that this solution is an expansion of the general solution ce^x in powers of $(x - x_0)$ and not a new solution.

8. Find a power series solution of the DE $y' + xy = 0$ in powers of $(x - 1)$ directly from the DE. (*Hint:* Expand the coefficient x in a Taylor series about $x = 1$.)

12.2 REVIEW OF POWER SERIES: CONVERGENCE

We shall now present the basic mathematical results about power series that are needed for our subsequent work. Some or all of the material is usually studied in calculus. Consequently, the results are stated without proof.

We begin by defining a power series.

definition 1

An infinite series of the form

$$b_0 + b_1(x - x_0) + b_2(x - x_0)^2 + \cdots + b_n(x - x_0)^n + \cdots \tag{1}$$

is called a power series in powers of $(x - x_0)$.

The real numbers $b_0, b_1, b_2, \ldots, b_n, \ldots$ are called the coefficients of the power series. A power series is a finite power series (polynomial) if there is a positive integer N such that $b_n = 0$ for all $n > N$.

The series

$$b_0 + b_1 x + b_2 x^2 + \cdots + b_n x^n + \cdots \tag{2}$$

is called a power series in x. It is a special case of series (1) that is obtained by setting $x_0 = 0$. By shifting the origin from $x = 0$ to $x = x_0$ by the transformation $x = x_0 + z$, we can always express a power series in $(x - x_0)$ as a power series in z. Thus, we shall be primarily concerned with power series of the form given in Eq. (2).

We shall employ the summation notation for power series. This notation is convenient for the manipulation of the series. Thus we denote a power series in x by

$$b_0 + b_1 x + b_2 x^2 + \cdots + b_n x^n + \cdots = \sum_{n=0}^{\infty} b_n x^n. \tag{3}$$

For example,[1]

$$1 + x + \frac{x^2}{2!} + \frac{x^3}{3!} + \cdots \frac{x^n}{n!} + \cdots = \sum_{n=0}^{\infty} \frac{x^n}{n!}.$$

definition 2

The mth partial sum $S_m(x)$, for $m = 1, \ldots,$ of the series given by Eq. (2) is the sum of the first m terms; that is,

$$S_m(x) = \sum_{n=0}^{m-1} b_n x^n.$$

definition 3

A power series converges at a point x if the sequence of its partial sums $S_1(x), S_2(x), \ldots$ converges as $m \to \infty$; that is, the series converges if

$$\lim_{m \to \infty} S_m(x)$$

exists. When this limit exists it is called the sum $S(x)$ of the power series at the point x. If this limit does not exist, then the series is said to diverge.

The convergence of a power series for a fixed value of x is the same as the convergence of infinite series of numbers. The interval of values of x for which a power series converges is described in the following theorem.

[1] We recall that factorial n, which is written as $n!$, is defined by $n! = n(n-1) \cdot (n-2) \cdots 2 \cdot 1$. Furthermore, we have $0! = 1$.

theorem

If the power series

$$\sum_{n=0}^{\infty} b_n x^n$$

converges for some value of $x \neq 0$, then either the series converges for all values of x, or there is a real number R (>0) such that the series converges for all $|x| < R$ and it diverges for all $|x| > R$. Then R is called the *radius of convergence* of the power series and the interval $(-R, R)$ is the *interval of convergence* of the series.

The convergence of the series at the ends of the interval, $x = \pm R$, depends upon the specific power series that is being analyzed. If the power series converges for all values of x, then we say $R = \infty$. Clearly, every power series such as Eq. (3) converges for $x = 0$ and the sum at $x = 0$ is $S(0) = b_0$. If a power series converges only at $x = 0$, then $R = 0$.

The radius of convergence of a power series can frequently be established by using the *ratio test.* We consider a power series for which $b_n \neq 0$, for $n = 0, 1, \ldots$. We form the ratio of any two successive terms:

$$\frac{b_{n+1}x^{n+1}}{b_n x^n} = \frac{b_{n+1}x}{b_n}.$$

The limit of the absolute value of this ratio as $n \to \infty$, if it exists, is denoted by L. Then we have

$$L = \lim_{n\to\infty} \left| \frac{b_{n+1}x}{b_n} \right| = \lim_{n\to\infty} \left| \frac{b_{n+1}}{b_n} \right| |x| = |x| \lim_{n\to\infty} \left| \frac{b_{n+1}}{b_n} \right|. \tag{4}$$

The ratio test states that the power series converges at the value x if $L < 1$ and it diverges if $L > 1$. If $L = 1$, the test provides no information. Hence we conclude from Eq. (4) that the power series converges for all values of x that satisfy

$$|x| < \lim_{n\to\infty} \left| \frac{b_n}{b_{n+1}} \right|$$

and it diverges for all values of x that satisfy

$$|x| > \lim_{n\to\infty} \left| \frac{b_n}{b_{n+1}} \right|.$$

Therefore the radius of convergence of the power series is

$$R = \lim_{n\to\infty} \left| \frac{b_n}{b_{n+1}} \right|, \tag{5}$$

provided that this limit exists. If the limit is infinite, then $R = \infty$.

EXAMPLE

Find the radius of convergence of the power series

$$\sum_{n=1}^{\infty} \frac{x^n}{n}.$$

solution. Since $b_n = 1/n$ for $n = 1, 2, \ldots$, then we have from Eq. (5) that the radius of convergence

$$R = \lim_{n\to\infty} \left| \frac{b_n}{b_{n+1}} \right| = \lim_{n\to\infty} \frac{n+1}{n} = 1.$$

Thus the series converges for all values of x for which $|x| < 1$ and it diverges for all $|x| > 1$.

A convergent power series in powers of x or $x - x_0$, defines a function $f(x)$ for all x in its interval of convergence. For example, the series in the Example is the Taylor series expansion in powers of x of the function $f(x) = \ln|1 - x|$. Furthermore, the value of the function at any point x_1 in the interval of convergence is the sum of the power series at the point x_1.

Exercises

1. Find the radius of convergence of the following power series.

(a) $\displaystyle\sum_{n=0}^{\infty} \frac{x^n}{(n+1)(n+2)}$ (b) $\displaystyle\sum_{n=0}^{\infty} \frac{x^n}{n!}$

(c) $\displaystyle\sum_{n=0}^{\infty} \frac{(n!)^2}{(2n)!} x^n$ (d) $\displaystyle\sum_{n=1}^{\infty} \frac{(-1)^{n-1}}{n} x^n$

(e) $\displaystyle\sum_{n=0}^{\infty} \frac{(-1)^n x^{2n}}{(2n)!}$ (f) $\displaystyle\sum_{n=0}^{\infty} n^n x^n$

12.2-1 Manipulation

In the application of power series to the solution of DEs it is necessary to add two power series, to multiply power series by a power x^p and to differentiate power series. We consider two functions $f(x)$ and $g(x)$, which are defined by the power series

$$f(x) = \sum_{n=0}^{\infty} b_n x^n \quad \text{and} \quad g(x) = \sum_{n=0}^{\infty} d_n x^n. \tag{1}$$

The coefficients b_n and d_n, for $n = 0, 1, \ldots$, are known numbers. We assume that both series in Eq. (1) are convergent for $|x| < R$. Then we have

$$cx^p f(x) = cx^p \sum_{n=0}^{\infty} b_n x^n = \sum_{n=0}^{\infty} cb_n x^{n+p}, \tag{2}$$

where c is any constant, and

$$f(x) + g(x) = \sum_{n=0}^{\infty} (b_n + d_n) x^n. \tag{3}$$

The results stated in Eqs. (2) and (3) can be established by using Definition 3 in Section 12.2 for the convergence of power series. Equation (3) shows that the sum of two power series is a power series whose coefficients are the term-by-term sum of the coefficients in the original power series.

We now state, without proof, the differentiation theorem for power series.

theorem 1

Let $f(x)$ denote the series

$$f(x) = b_0 + b_1 x + b_2 x^2 + \cdots = \sum_{n=0}^{\infty} b_n x^n, \tag{4}$$

which converges in the interval $|x| < R$. Then $f'(x)$ is obtained by term-by-term differentiation of the series in Eq. (4). That is,

$$f'(x) = b_1 + 2b_2 x + \cdots = \sum_{n=1}^{\infty} nb_n x^{n-1}. \tag{5}$$

This series converges in the same interval $|x| < R$ as Eq. (4). Similarly, we have

$$f''(x) = 2b_2 + \cdots = \sum_{n=2}^{\infty} n(n-1)b_n x^{n-2}, \tag{6}$$

and it converges in $|x| < R$.

The indices n in the sums given by Eqs. (4), (5), and (6) are dummy indices, similar to dummy variables of integration in definite integrals. The sums do not depend on n; they depend only on x. That is,

$$\sum_{n=0}^{\infty} b_n x^n = \sum_{m=0}^{\infty} b_m x^m.$$

We are frequently required to add two power series, such as Eqs. (4)

and (5). That is, we must consider sums such as

$$S = \sum_{n=0}^{\infty} b_n x^n + \sum_{n=1}^{\infty} n b_n x^{n-1}, \tag{7}$$

where the powers of x are indexed differently in each sum. We wish to make transformations of the dummy summation indices so that in the resulting sums the powers of x are the same for the two sums. Then, according to Eq. (3), the sum of the two series is obtained by term-by-term addition of the coefficients. For example, in the second sum in Eq. (7) we define a transformation from the index n to the index m by $n = m + 1$. Then the second sum is given by

$$\sum_{n=1}^{\infty} n b_n x^{n-1} = \sum_{m=0}^{\infty} (m+1) b_{m+1} x^m.$$

The range of summation in the transformed sum is from $m = 0$ to $m = \infty$, because by the substitution $n = m + 1$, we have $m = 0$ when $n = 1$ and $m = \infty$ when $n = \infty$. We substitute this result in Eq. (7) and change the dummy index of summation in the first sum to m by the substitution $n = m$. This gives

$$S = \sum_{m=0}^{\infty} b_m x^m + \sum_{m=0}^{\infty} (m+1) b_{m+1} x^m = \sum_{m=0}^{\infty} [b_m + (m+1) b_{m+1}] x^m.$$

Finally, we give an important theorem about power series that will be used in the following sections.

theorem 2

If

$$\sum_{n=0}^{\infty} b_n x^n = 0$$

for each x in the interval (α, β), then each of the coefficients $b_n = 0$, for $n = 0, 1, \ldots$.

A proof of this result is discussed in Exercise 5.

Exercises

1. Find $f'(x)$ and $f''(x)$ for the following series. Find the largest interval on which differentiation is valid.

(a) $f(x) = \sum_{n=0}^{\infty} \dfrac{x^n}{n!}$ (b) $f(x) = \sum_{n=0}^{\infty} \dfrac{x^n}{2^n n!}$

(c) $f(x) = \sum_{n=1}^{\infty} \frac{(-1)^{n-1}x^n}{n}$ (d) $f(x) = \sum_{n=0}^{\infty} n!\,x^n$

2. Show that $(\sin x)' = \cos x$ and $(\cos x)' = -\sin x$ by using the Taylor series expansions for $\sin x$ and $\cos x$ in powers of x.

3. Recall that Euler's formula is given by $e^{ix} = \cos x + i \sin x$. Show that the Taylor series for the functions on the right and left sides of this identity are equal. In expanding e^{ix} proceed as if i were a real constant.

4. Add the following power series termwise and write the result as a single series.

(a) $\sum_{n=0}^{\infty} 3x^n + \sum_{n=2}^{\infty} n(n-1)x^{n-2}$ (b) $\sum_{n=2}^{\infty} 6n(n-1)x^{n-2} + \sum_{n=1}^{\infty} 2nx^{n-1}$

(c) $\sum_{n=1}^{\infty} 3nx^{n-1} + \sum_{n=2}^{\infty} n(n-1)x^{n-2} + \sum_{n=0}^{\infty} x^n$

5. Use Theorem 1 to prove that

$$\sum_{n=0}^{\infty} b_n x^n = 0 \qquad \text{for all } |x| < R,$$

implies that $b_n = 0$ for all n. (*Hint:* Let

$$f(x) = \sum_{n=0}^{\infty} b_n x^n.$$

Then, since $f(0) = 0$, observe that $b_0 = 0$. Differentiate $f(x)$ and conclude that $b_1 = 0$, etc.).

6. If the functions $f(x)$ and $g(x)$ are defined by the power series given in Eq. (1), then show by multiplying the series term by term and collecting terms with the same powers of x, that

$$f(x)g(x) = \sum_{n=0}^{\infty} c_n x^n, \qquad \text{where } c_n = \sum_{k=0}^{n} d_k b_{n-k}.$$

7. Find the product of the series

$$f(x) = \sum_{n=0}^{\infty} x^n \qquad \text{and} \qquad g(x) = \sum_{n=0}^{\infty} (-1)^n x^n,$$

for $|x| < 1$.

12.3 THE POWER SERIES METHOD

In this section we study how to obtain power series solutions of DEs with variable coefficients. However, we shall first analyze the simple constant-coefficient DE $y' - y = 0$ to illustrate the techniques. The first few terms in the power series solution in powers of x of this equation were obtained previously in Section 12.1. We shall show how to obtain a general expression for any term and study the convergence of the resulting series.

EXAMPLE

Find a power series solution in powers of x of the DE

$$y' - y = 0. \tag{1}$$

solution. We assume that DE (1) has a power series solution in the form

$$y = b_0 + b_1x + b_2x^2 + \cdots = \sum_{n=0}^{\infty} b_nx^n, \tag{2}$$

which converges for $|x| < R$ for some $R > 0$. The coefficients are determined by substituting into the DE. From Theorem 1 of Section 12.2-1 we know that the power series given by Eq. (2) can be differentiated term by term in its interval of convergence. Thus differentiation of Eq. (2) gives

$$y' = b_1 + 2b_2x + 3b_3x^2 + \cdots = \sum_{n=1}^{\infty} nb_nx^{n-1}.$$

Substituting these series for y and y' in Eq. (1) and using the summation notation, we obtain

$$\sum_{n=1}^{\infty} nb_nx^{n-1} - \sum_{n=0}^{\infty} b_nx^n = 0. \tag{3}$$

In order to subtract these two series term by term we first collect terms in this equation with the same powers of x. Thus we rewrite the first sum in Eq. (3) by changing the dummy index of summation n to the index m by the substitution $n = m + 1$. Then the range of summation in the first sum will be from $m = 0$ because by the substitution $n = m + 1$, $m = 0$ when $n = 1$. Since n is also a dummy index of summation in the second sum, we can change its name to m with no other change in the second sum. Thus Eq. (3) is reduced to

$$\sum_{m=0}^{\infty} (m + 1)b_{m+1}x^m - \sum_{m=0}^{\infty} b_mx^m = 0.$$

We now combine the two sums and get

$$\sum_{m=0}^{\infty} [(m + 1)b_{m+1} - b_m]x^m = 0.$$

Since this equation must be satisfied for all values of x, it follows from Theorem 2 of Section 12.2-1 that the coefficient of each power of x must vanish. Thus we get

$$b_{m+1} = \frac{b_m}{m + 1}, \qquad \text{for } m = 0, 1, \ldots. \tag{4}$$

This equation relating the coefficients b_m and b_{m+1} is called the *recurrence relation* for the coefficients. It shows that every coefficient b_{m+1} is proportional to the previous coefficient. We can determine the radius of convergence of the power series by use of this recurrence relation. From Eq. (5) of Section 12.2 we have

$$R = \lim_{m\to\infty} \left| \frac{b_m}{b_{m+1}} \right| = \lim_{m\to\infty} (m+1) = \infty . \tag{5}$$

Thus the series converges for all values of x.

We now use the recurrence relation of Eq. (4) to express every coefficient in terms of b_0. By setting $m = 0$, $m = 1$, and $m = 2$ in Eq. (4), we get

$$b_1 = b_0, \qquad b_2 = \frac{b_1}{2} = \frac{b_0}{2}, \qquad b_3 = \frac{b_2}{3} = \frac{b_0}{3\cdot 2}.$$

This suggests that, in general,

$$b_n = \frac{b_0}{n!}, \qquad n = 0, 1, \ldots. \tag{6}$$

A proof of this result for arbitrary n follows from the recurrence relation given by Eq. (4) and the technique of mathematical induction. We leave the proof as an exercise for the reader.

Inserting the coefficient given in Eq. (6) into Eq. (2), we get

$$y = b_0 \sum_{n=0}^{\infty} \frac{x^n}{n!}, \tag{7}$$

which is precisely the power series representation of b_0e^x. The constant b_0 is arbitrary.

This example illustrates the techniques of the power series method. The procedure can be classified into three major steps:

1. Obtain the recurrence relation for the coefficients by inserting the assumed power series solution into the DE.
2. Evaluate the radius of convergence of the series by using the ratio test or some other technique.
3. Obtain a general formula for the nth coefficient.

Step 1 is the most important. Once the recurrence relation is known, then any desired number of terms in the series can be calculated. In practice, digital computers are now routinely used to calculate the coefficients from the recurrence relation. The formula obtained from Step 3 is useful in deriving properties of the function represented by the series. It is normally difficult to obtain this formula.

In the Example, we have obtained a power series solution in powers of x. Other power series solutions can be obtained by expanding about another point $x_0 \neq 0$. The choice of the point x_0 is usually determined by the specification of initial or boundary conditions. For example, it is almost essential to expand the solution about the initial point of an IVP, because in an IVP we are given the solution and its first derivative at the initial point and we wish to study how the solution evolves from this initial data as x deviates from the initial point. If we expand about x_0 and x_0 is not the initial point, then it may be difficult to satisfy the initial conditions (see Exercise 3). Unless it is stipulated otherwise, we shall take $x_0 = 0$ throughout this and the following chapter. This corresponds to studying an IVP for the DE with $x_0 = 0$ as the initial point.

Exercises

1. For each of the following DEs determine the recurrence relation, the radius of convergence, and the general terms for a power series solution in powers of x.
   (a) $y' + 2y = 0$ (b) $y' - xy = 0$
   (c) $y' + (1 + x^2)y = 0$
   (d) $y' + x^N y = 0$ (where N is a positive integer)
2. Prove that Eq. (6) is valid by using the principle of mathematical induction.
3. (a) Solve the IVP $y' + y = 0$, $y(0) = 3$ by using the power series solution given by Eq. (7). (b) The power series solution used in (a) was in powers of x. Thus it is an expansion about the point $x_0 = 0$. Try to solve the IVP in (a) by using a power series solution in powers of $x - 1$.

12.3-1 Second-order equations

In the following example we apply the power series techniques that were developed in Section 12.3 to a second-order DE with variable coefficients. The limitations of the power series method will be discussed in Section 12.4.

EXAMPLE

Find a power series solution in powers of x of

$$y'' - xy' - y = 0$$

in the interval $-\infty < x < \infty$. Then find a general solution.

solution. We assume that the DE has a power series solution

$$y = \sum_{n=0}^{\infty} b_n x^n. \tag{1}$$

We differentiate this series term by term and obtain

$$y' = \sum_{n=1}^{\infty} n b_n x^{n-1}, \quad \text{and} \quad y'' = \sum_{n=2}^{\infty} n(n-1) b_n x^{n-2}. \tag{2}$$

By substituting these series for y, y', and y'' into the DE we obtain

$$\sum_{n=2}^{\infty} n(n-1)b_n x^{n-2} - x\sum_{n=1}^{\infty} nb_n x^{n-1} - \sum_{n=0}^{\infty} b_n x^n = 0. \tag{3}$$

We wish to combine these three sums into a single sum and then collect terms with the same powers of x in the resulting equation. To do this we first transform the dummy index of summation in the first sum from n to m by the transformation $n = m + 2$. This gives

$$\sum_{n=2}^{\infty} n(n-1)b_n x^{n-2} = \sum_{m=0}^{\infty} (m+2)(m+1)b_{m+2}x^m.$$

We substitute this into Eq. (3) and then bring the factor x in the middle sum in Eq. (3) under the summation sign. This gives

$$\sum_{m=0}^{\infty} (m+2)(m+1)b_{m+2}x^m - \sum_{n=1}^{\infty} nb_n x^n - \sum_{n=0}^{\infty} b_n x^n = 0. \tag{4}$$

Since $nb_n x = 0$ for $n = 0$, we can rewrite the second sum in Eq. (4) as

$$\sum_{n=1}^{\infty} nb_n x^n = \sum_{n=0}^{\infty} nb_n x^n.$$

We now change the name of the summation variables from n to m in this sum and in the third sum in Eq. (4). Then by combining the resulting three sums, we get

$$\sum_{m=0}^{\infty} [(m+2)(m+1)b_{m+2} - (m+1)b_m]x^m = 0.$$

Since this equation is valid for all x in some interval containing $x = 0$, the coefficient of each power must be zero. Thus we get

$$b_{m+2} = \frac{(m+1)b_m}{(m+2)(m+1)} = \frac{b_m}{m+2}, \qquad m = 0, 1, \ldots. \tag{5}$$

Equation (5) is the recurrence relation for the coefficients b_m. It shows that any coefficient b_{m+2} is proportional to b_m. Hence

$$b_2 = \frac{b_0}{2}, \qquad b_3 = \frac{b_1}{3}, \qquad b_4 = \frac{b_2}{4}, \qquad b_5 = \frac{b_3}{5},$$

and so forth. We conclude that the even and odd indexed coefficients are independent and we can decompose the series in Eq. (1) into two sums:

$$y = y_{\text{even}} + y_{\text{odd}}, \tag{6}$$

where y_{even} and y_{odd} are defined by

$$y_{\text{even}} = \sum_{n=0,2,4,\ldots}^{\infty} b_n x^n \tag{7}$$

and

$$y_{\text{odd}} = \sum_{n=1,3,5,\ldots}^{\infty} b_n x^n. \tag{8}$$

This completes Step 1 of the procedure outlined in Section 12.3.

We now determine the radius of convergence of these series by the ratio test. Thus we must study the ratio of any two successive terms. For the series in Eq. (7) we have by using the recurrence relation given in Eq. (5) with $m = 2n$ that

$$L = \lim_{n\to\infty} \left| \frac{b_{2n+2}x^{2n+2}}{b_{2n}x^{2n}} \right| = x^2 \lim_{n\to\infty} \left| \frac{1}{2n+2} \right| = 0.$$

Since $L = 0$ for all x, the series converges for all x and hence $R = \infty$. Similarly it can be shown that the series for y_{odd} in Eq. (8) converges for all x.

This completes Step 2. We now proceed with Step 3 and determine the general term in the series given in Eqs. (7) and (8) by using the recurrence relations. By setting $n = 0, 2, 4, 6, \ldots$ in Eq. (5) we can solve for the first few even coefficients in terms of b_0. Thus we get

$$b_2 = \frac{b_0}{2}, \qquad b_4 = \frac{b_2}{4} = \frac{b_0}{4\cdot 2}, \qquad b_6 = \frac{b_4}{6} = \frac{b_0}{6\cdot 4\cdot 2}.$$

This suggests in general that

$$b_{2p} = \frac{b_0}{2p(2p-2)(2p-4)\cdots 4\cdot 2}.$$

This result can be proved by using mathematical induction. Each factor in the denominator has a common factor of 2 and there are p factors. So this equation becomes

$$b_{2p} = \frac{b_0}{2^p p!}. \tag{9}$$

Thus we have shown that every coefficient b_n, with n an even integer, is proportional to b_0.

The general term in the series given in Eq. (8), which contains only coefficients b_n with n an odd integer, can be expressed in terms of b_1 by a similar analysis using the recurrence relation given by Eq. (5). We shall omit the details, and give the final result:

$$b_{2p+1} = \frac{2^p p!}{(2p+1)!} b_1, \qquad p = 1, 2, \ldots. \tag{10}$$

This completes Step 3.

We now insert Eqs. (9) and (10) into Eqs. (7) and (8). This gives

$$y_{\text{even}} = b_0 y_1(x) \qquad \text{and} \qquad y_{\text{odd}} = b_1 y_2(x)$$

where the functions $y_1(x)$ and $y_2(x)$ are defined by

$$y_1 = \sum_{p=0}^{\infty} \frac{1}{2^p p!} x^{2p} \qquad \text{and} \qquad y_2 = \sum_{p=0}^{\infty} \frac{2^p p!}{(2p+1)!} x^{2p+1}. \tag{11}$$

Since b_0 and b_1 are arbitrary constants, we observe by inserting these expressions for y_{even} and y_{odd} in Eq. (6) and then setting $b_0 = 1$ and $b_1 = 0$, that y_1 is a series solution of the DE. Similarly, y_2 is a series solution. The series given by y_1 is a power series representation of the function $e^{x^2/2}$. However, we are unable to determine the sum of the series for y_2 in terms of elementary functions.

We shall now demonstrate that

$$y = c_1 y_1(x) + c_2 y_2(x) \tag{12}$$

is a general solution of the DE. To show that Eq. (12) is a general solution we need only show that the Wronskian

$$W = y_1 y_2' - y_1' y_2$$

of y_1 and y_2 is not equal to zero for some value of x for which the solution is defined. We choose the value $x = 0$ because the series in Eq. (11) simplify at $x = 0$. By evaluating them at $x = 0$, we get $y_1(0) = 1$, and $y_2(0) = 0$. Thus we have $W(0) = y_2'(0)$. To find $y_2'(0)$ we differentiate the series for y_2 and then evaluate the result at $x = 0$. This gives $W(0) = 1 \neq 0$. Hence Eq. (12) is a general solution.

Exercises

1. Find a power series solution for the following DEs about the point $x = 0$. Find the radius of convergence of the series.
   (a) $y'' - y = 0$
   (b) $y'' - 2xy' - 4y = 0$
   (c) $(1 - x^2)y'' - xy' + y = 0$
   (d) $(1 + x^2)y'' + 2xy' + y = 0$
   (e) $y'' + xy' + y = 0$
   (f) $y'' - xy = 0$

   The solutions of Hermite's DE

$$y'' - 2xy' + 2Ny = 0,$$

where N is a constant, are used in quantum mechanics to study the spatial position of a moving particle that is undergoing simple harmonic motion in time. In quantum mechanics the exact position of a particle at a given time cannot be predicted, as it can in classical mechanics. It is possible to determine only the probability of the particle being at a given location at a given time. The unknown $y(x)$ in Hermite's DE is then related to the probability of finding the particle at the position x. The constant N is related to the energy of the particle.

2. Let $N = 1$ in Hermite's DE.
   (a) Show that $y = x$ is a solution.

(b) Use reduction of order (see Section 11.3) to show that

$$y_2 = x \int^x z^{-2} e^{z^2}\, dz$$

is a second linearly independent solution.

(c) Use the power series method to find a general solution. Compare your result with the solutions obtained in (a) and (b).

3. (a) Show by using the recurrence relation that if N is a nonnegative integer, then there is always a polynomial solution of Hermite's DE. (b) Verify that if N is not a nonnegative integer, then there are no polynomial solutions. (c) Show that for $N = 0, 1, 2$, and 3 the polynomials are 1, x, $1 - 2x^2$, and $x - (2/3)x^3$, respectively.

In Exercises 4 through 6 we study Legendre's equation

$$(1 - x^2)y'' - 2xy' + N(N + 1)y = 0, \qquad \text{for } -1 < x < 1,$$

where N is a fixed constant. It arises in problems such as the flow of an ideal fluid past a sphere, the determination of the electric field due to a charged sphere, and the determination of the temperature distribution in a sphere given its surface temperature.

4. Suppose that $N = 1$ in Legendre's equation. (a) Find a general solution of the DE by the power series method. Observe that one of the solutions is a polynomial. (b) Find the radius of convergence of the other solution.

5. (a) Generalize the results of Exercise 4 by showing that if N is a nonnegative integer, then there is always a polynomial solution. Show also that if N is not a nonnegative integer, then there are no polynomial solutions. (*Hint:* Study the recurrence relation.)

(b) Show that for $N = 0$, 1, 2, and 3, these polynomials are 1, x, $1 - 3x^2$, $x - (5/3)x^3$, and $1 - 10x^2 + (35/3)x^4$, respectively.

(c) When these polynomial solutions are multiplied by a suitable constant, then they are known as *Legendre polynomials*. The coefficient of the highest power in the Legendre polynomial $P_n(x)$ is always $(2n)!(n!)^{-2}2^{-n}$. Verify that

$$P_0(x) = 1, \qquad P_1(x) = x, \qquad P_2(x) = \frac{3}{2}x^2 - \frac{1}{2},$$

$$P_3(x) = \frac{5}{2}x^3 - \frac{3}{2}x, \qquad \text{and} \qquad P_4(x) = \frac{35}{8}x^4 - \frac{15}{8}x^2 + \frac{3}{4}.$$

(d) Show that the polynomials listed in (c) are given by

$$P_n(x) = \frac{1}{2^n n!} \frac{d^n}{dx^n}(x^2 - 1)^n.$$

This expression for $P_n(x)$ is in fact valid for all n. It is called *Rodrigue's formula.*

*6. By using a power series in powers of x, find a general solution of Legendre's DE for any N.

*7. Find a power series representation for a general solution in powers of $x + 1$ of the DE $x^2y'' + xy' + y = 0$. (*Hint:* Expand the coefficients x^2 and x in a Taylor series in powers of $(x + 1)$.)

*8. Find a general solution in powers of $x - 1$ of the DE $y'' - xy = 0$.

12.4 LIMITATIONS OF THE POWER SERIES METHOD

The following example shows that it is not always possible to obtain power series representations about $x = x_0$ for general solutions of DEs with variable coefficients. Thus if an IVP with initial conditions at x_0 is to be solved, the power series method may not give a general solution valid at x_0. At the end of this section we give sufficient conditions on the coefficients of the DE that insure the existence of power series representations.

EXAMPLE

Find a power series solution in powers of x of

$$x^2y'' + 4xy' + 2y = 0, \tag{1}$$

solution. We seek a power series solution in the form

$$y = \sum_{n=0}^{\infty} b_n x^n.$$

Substituting this series into the DE gives

$$x^2 \sum_{n=2}^{\infty} n(n-1)b_n x^{n-2} + 4x \sum_{n=1}^{\infty} nb_n x^{n-1} + 2 \sum_{n=0}^{\infty} b_n x^n = 0.$$

Then by introducing the new index m, rearranging the sums, and equating the coefficients of x^m, for $m = 0, 1, 2, \ldots$, to zero we obtain the recurrence relation

$$(m^2 + 3m + 2)b_m = 0, \qquad m = 0, 1, 2, \ldots.$$

Since $m^2 + 3m + 2 \neq 0$, for $m = 0, 1, 2, \ldots$, we conclude that

$$b_m = 0, \qquad m = 0, 1, 2, \ldots.$$

Thus all the coefficients b_m, for $m = 0, 1, 2, \ldots$, are zero. This shows that only the trivial solution of the DE, $y = 0$, has a power series representation in powers of x. Since the DE is a Cauchy-Euler equation, the reader can find the general solution and verify that it has no power series representation in powers of x.

Frequently it is possible to obtain one power series solution of a DE but not two linearly independent solutions, as we show in Exercises

2 and 3. It is the behavior of the coefficients $P(x)$ and $Q(x)$ in the DE

$$y'' + P(x)y' + Q(x)y = 0 \tag{2}$$

as $x \to x_0$ which determine the number of power series solutions in powers of $(x - x_0)$, as we shall now proceed to demonstrate.

For simplicity, we consider equations in which P and Q are rational functions. We recall that a *rational function is the ratio of two polynomials.* Clearly, in all the examples in this and the previous section $P(x)$ and $Q(x)$ are rational functions. A more general class of coefficients is discussed in Exercise 5.

To introduce the conditions on P and Q we study the behavior of the coefficients in these previous examples. In Example 1 of Section 12.3-1 we have $P(x) = -x$ and $Q(x) = -1$, and hence

$$\lim_{x \to 0} P(x) = 0 \qquad \text{and} \qquad \lim_{x \to 0} Q(x) = -1.$$

For the DE in Example 2 of Section 12.3-1 we have, $P(x) = -x/(1 - x^2)$ and $Q(x) = 4/(1 - x^2)$, and

$$\lim_{x \to 0} P(x) = 0 \qquad \text{and} \qquad \lim_{x \to 0} Q(x) = 4.$$

In these examples, where the general solution has a power series in powers of x, the limits as $x \to 0$ of P and Q are finite. In the Example of this section we see that as $x \to 0$, $P(x) \to \infty$ and $Q(x) \to \infty$, and there are no power series solutions. These examples suggest that the existence of finite limits of P and Q as $x \to 0$ plays an important role for the existence of two linearly independent power series solutions. Thus we give the following definition.

definition

If the coefficients $P(x)$ and $Q(x)$ are rational functions and if

$$\lim_{x \to x_0} P(x) \qquad \text{and} \qquad \lim_{x \to x_0} Q(x)$$

exist, then $x = x_0$ is an ordinary point of the DE

$$y'' + P(x)y' + Q(x)y = 0.$$

If at least one of these limits does not exist, then $x = x_0$ is a singular point.

As an example, consider the DE

$$y'' + \frac{x}{1 - x}\, y' + \frac{1}{x(1 - x)}\, y = 0.$$

Here $P(x) = x/(1 - x)$ and $Q(x) = 1/x(1 - x)$ are rational functions.

The point $x = 1$ is a singular point of this DE because P and Q do not have finite limits as $x \to 1$. The point $x = 0$ is also a singular point because Q does not have a finite limit as $x \to 0$. Since P and Q have finite limits for any $x \neq 0$ or 1, all other points are ordinary points for this DE.

We now state without proof sufficient conditions for a general solution of a second-order DE to have a power series representation.

theorem

If x_0 is an ordinary point of the DE

$$y'' + P(x)y' + Q(x)y = 0,$$

then there are two linearly independent power series solutions. These series converge in some interval $|x - x_0| < R$, where $R > 0$.

Exercises

1. Find all singular points of the following DEs.

   (a) $y'' + \frac{1}{x}y' + \frac{1}{x}y = 0$

   (b) $y'' - \frac{x}{(x+1)}y' + \frac{x-1}{x(x-2)}y = 0$

   (c) $x^2y'' - xy' + (1 - x^2)y = 0$ (d) $(x^2 - 9)y'' + xy' + y = 0$

2. Show that the Bessel equation of zero order

   $$x^2y'' + xy' + x^2y = 0$$

   does not have two linearly independent power series solutions in powers of x.

3. For the following DEs show that $x = 0$ is a singular point. Find all power series solutions about the point $x = 0$. Do these power series solutions provide a general solution?

   (a) $8xy'' + (x^3 - 8)y' + 2x^2y = 0$ (b) $xy'' + 2y' - 2xy = 0$

   (c) $y'' - \frac{1}{x}y' + \frac{1}{x^2(x+1)}y = 0$ (d) $6x^2y'' + y' + (1 - x^2)y = 0$

4. Solve the Cauchy-Euler equation given in Eq. (1) by the technique of Section 11.4. Show that the general solution can be expanded in a Taylor series in powers of $(x - x_0)$ for any $x_0 \neq 0$.

5. Power series solution have been discussed for the DE given in Eq. (2) where $P(x)$ and $Q(x)$ are rational functions. We shall now study a more general class of DEs. First we give the following definition.

definition

A function f is analytic at a point $x = x_0$ if its Taylor series about x_0 exists and converges to $f(x)$ for all x in an interval containing x_0.

We can show that $\sin x$, $\cos x$, e^x, and all polynomials are analytic, at every point $x = x_0$. Furthermore, a rational function is analytic at every point where the denominator is not zero. We extend the definition of an ordinary point given in the text. The point $x = x_0$ is an ordinary point of the DE given in Eq. (2) if $P(x)$ and $Q(x)$ are analytic at $x = x_0$. The theorem in the text is also valid when this extended definition of an ordinary point is used. Determine the singular points of the following DEs.

(a) $y'' - \dfrac{e^{-x}y}{x} = 0$ (b) $y'' + e^x y' + \sin xy = 0$

(c) $y'' + \dfrac{\sin x}{x} y = 0$

6. Find a general solution of the following DEs in powers of x.
 (a) $y'' - e^{-x}y = 0$. (*Hint:* First expand e^{-x} in a Taylor series at $x = 0$.)
 (b) $y'' - (\cos x)y = 0$

chapter 13

THE METHOD OF FROBENIUS

13.1 INTRODUCTION

We have seen in Chapter 12 that if $x = x_0$ is an ordinary point of

$$y'' + P(x)y' + Q(x)y = 0, \tag{1}$$

then the DE has two linearly independent power series solutions in powers of $(x - x_0)$. They converge in some interval about $x = x_0$. Furthermore, we observed from the Example in Section 12.4 that if x_0 is a singular point of the DE, then there may not be any linearly independent power series solutions in powers of $(x - x_0)$.

We will now generalize the power series method and obtain another infinite series representation of solutions that is valid arbitrarily near a singular point but not necessarily at the singular point itself. To motivate the analysis we consider the Example of Section 12.4, where we studied the DE

$$x^2y'' + 4xy' + 2y = 0, \qquad \text{for } x > 0. \tag{2}$$

We showed that there are no power series solutions about the singular point $x = 0$. However, the DE is a Cauchy-Euler equation, which was studied in Section 11.4. By using the techniques that were developed in that section, we obtain the general solution

$$y = \frac{c_1}{x} + \frac{c_2}{x^2}, \tag{3}$$

which is valid for $x > 0$. Furthermore, we recall that for the general Cauchy-Euler equation

$$x^2y'' + b_1xy' + b_2y = 0,$$

the solutions are of the form $y = x^kf(x)$, where k is a real number, and the specific form of the function $f(x)$ depends on the constants b_1 and b_2. This suggests that we modify the power series representation and seek infinite series solutions of DE (1) in the form

$$y = (x - x_0)^k \sum_{n=0}^{\infty} b_n(x - x_0)^n. \tag{4}$$

where the index k and the coefficients $b_0, b_1, \ldots$ are to be determined. Since k is not specified, there is no loss of generality in assuming that $b_0 \neq 0$ (see Exercise 3).

The modification given by Eq. (4) of the power series method was originally proposed by the German mathematician, Frobenius; consequently, it is called *the method of Frobenius.* A series in the form given by Eq. (4) is called a Frobenius series or a Frobenius solution about the point $x = x_0$. If k is a nonnegative integer, then Eq. (4) is equivalent to a power series, where the first term is $b_0(x - x_0)^k$. The factor x^k provides additional freedom in the class of solutions and hence in the type of DE that we can consider. Series involving negative or fractional powers and even complex powers are now possible.

Exercises

1. For each of the following DEs find a general solution by using techniques we studied in Chapter 2. Show that the solution can be written as a Frobenius series about $x = 0$ and find the value of k. Observe that the solutions do not have power series expansions about $x = 0$.
   (a) $xy' + y = 0$ (b) $xy' + (1 + x)y = 0$
   (c) $xy' + (\cos x)y = 0$ (d) $(1 - x)y' + \dfrac{1}{x}y = 0$

2. Show that if the power series

$$\sum_{n=0}^{\infty} b_nx^n$$

can be differentiated termwise, then the Frobenius series

$$\sum_{n=0}^{\infty} b_n x^{n+k}$$

can be differentiated termwise. (*Hint:* Use the product rule to differentiate

$$(x^k) \sum_{n=0}^{\infty} b_n x^n.)$$

3. Show that there is no loss of generality in assuming that $b_0 \neq 0$ in Eq. (4). (*Hint:* Show that if $b_0 = 0$ and $b_1 \neq 0$, then the series in Eq. (4) can be written as

$$(x - x_0)^{k+1} \sum_{m=0}^{\infty} b_{m+1}(x - x_0)^m$$

Replace $k + 1$ by k and b_{m+1} by b_m. Generalize to $b_n = 0$ for $n = 1, 2, \ldots, j$, $b_{j+1} \neq 0$.)

13.2 ELEMENTS OF THE METHOD

We illustrate the techniques and the difficulties in the method of Frobenius in the following examples.

EXAMPLE 1

Find an infinite series representation about $x_0 = 0$ of a general solution of

$$y'' + \frac{1}{2x} y' + \frac{1}{4x} y = 0, \qquad \text{for } x > 0.$$

solution. Since $P(x) = 1/2x$ and $Q(x) = 1/4x$, both P and Q are rational functions and they approach infinity as $x \longrightarrow 0$. Thus $x = 0$ is a singular point of the DE. We use the method of Frobenius and seek a solution in the form

$$y = x^k \sum_{n=0}^{\infty} b_n x^n = \sum_{n=0}^{\infty} b_n x^{n+k}, \qquad \text{with } b_0 \neq 0. \tag{1}$$

Differentiation of Eq. (1) yields

$$y' = \sum_{n=0}^{\infty} (n + k) b_n x^{n+k-1}$$

and

$$y'' = \sum_{n=0}^{\infty} (n + k)(n + k - 1) b_n x^{n+k-2}.$$

Substituting these series for y, y', and y'' into the DE, we get

$$4x \sum_{n=0}^{\infty} (n + k)(n + k - 1) b_n x^{n+k-2} + 2 \sum_{n=0}^{\infty} (n + k) b_n x^{n+k-1} + \sum_{n=0}^{\infty} b_n x^{n+k} = 0. \tag{2}$$

We now combine the three sums in Eq. (2) into a single sum by appropriate changes in the dummy indices of summation. We set $n = m + 1$, $n = m + 1$, and $n = m$ in the first, second, and third sums, respectively. Thus we get

$$\sum_{m=-1}^{\infty} 2(m + k + 1)(2m + 2k + 1)b_{m+1}x^{m+k} + \sum_{m=0}^{\infty} b_m x^{m+k} = 0. \tag{3}$$

In each of these sums the same powers of x occur, as we desired. If these sums are to be combined, they must extend over the same index ranges. The first sum starts at $m = -1$ but the last sum starts at $m = 0$. Thus we rewrite the first sum by removing the first term from under the summation sign. Then Eq. (3) can be written as

$$2k(2k - 1)b_0 x^{-1+k} + \sum_{m=0}^{\infty} [2(m + k + 1)(2m + 2k + 1)b_{m+1} + b_m]x^{m+k} = 0.$$

We equate the coefficients of each power of x to zero. Since $b_0 \neq 0$, we get

$$2k(2k - 1) = 0 \tag{4}$$

and

$$2(m + k + 1)(2m + 2k + 1)b_{m+1} + b_m = 0, \qquad m = 0, 1, \ldots. \tag{5}$$

Quadratic equation (4) is called the *indicial equation*. The roots of this equation are called the *indicial roots*. They are

$$k = 0 \qquad \text{and} \qquad k = \tfrac{1}{2}.$$

Corresponding to each of these roots, we get a recurrence relation from Eq. (5) for the coefficients $b_0, b_1, \ldots$.

First we consider $k = 0$. Then Eq. (5) gives

$$b_{m+1} = -\frac{1}{2(m + 1)(2m + 1)} b_m, \qquad m = 0, 1, \ldots. \tag{6}$$

From Eq. (1), the series corresponding to $k = 0$ is

$$y = \sum_{n=0}^{\infty} b_n x^n, \tag{7}$$

where the coefficients are given by Eq. (6). It is a power series solution and it could have been obtained directly by the power series method. It is easy to show, by using the ratio test and the recurrence relation given by (6), that the series given by Eq. (7) converges for all values of x. Furthermore, it can be shown that the general term in this series is

$$b_n = \frac{(-1)^n}{(2n)!} b_0.$$

This can be verified by mathematical induction. By inserting this expression for b_n in Eq. (7), and setting $b_0 = 1$ we get the solution

$$y_1 = \sum_{n=0}^{\infty} \frac{(-1)^n}{(2n)!} x^n. \tag{8}$$

The recurrence relation, see Eq. (5), for the other root $k = \frac{1}{2}$ of the indicial equation is

$$b_{m+1} = -\frac{1}{(2m+3)(2m+2)} b_m, \qquad m = 0, 1, \ldots. \tag{9}$$

Thus the series corresponding to $k = \frac{1}{2}$ is

$$y = x^{1/2} \sum_{n=0}^{\infty} b_n x^n, \tag{10}$$

where the coefficients b_n are given recursively by Eq. (9). By using the ratio test and the recurrence relation, it can be shown that the series in Eq. (10) converges for all values of x. However, the factor $x^{1/2}$ is not real valued for $x < 0$. Thus the solution given in Eq. (10) is real valued only for $x \geq 0$. Furthermore, we observe that Eq. (10) is not differentiable at $x = 0$ because the derivative of $x^{1/2}$ is infinite when $x = 0$. Thus y given by Eq. (10) is a solution of the DE for $x > 0$, and it is not a solution at the singular point $x = 0$. It could not have been obtained by the power series method.

The expression for b_n in terms of b_0 can be derived from the recurrence relation. It is

$$b_m = \frac{(-1)^m}{(2m+1)!} b_0.$$

When we use these coefficients in Eq. (10), with $b_0 = 1$ we obtain the solution

$$y_2 = x^{1/2} \sum_{n=0}^{\infty} \frac{(-1)^n}{(2n+1)!} x^n. \tag{11}$$

The solutions y_1 and y_2 given by Eqs. (8) and (11), respectively, are linearly independent for $x > 0$ (see Exercise 5). Thus

$$y = c_1 y_1 + c_2 y_2 = c_1 \sum_{n=0}^{\infty} \frac{(-1)^n}{(2n)!} x^n + c_2 x^{1/2} \sum_{n=0}^{\infty} \frac{(-1)^n}{(2n+1)!} x^n \tag{12}$$

is a general solution for $x > 0$.

In many problems it is necessary to determine a general solution for $x < 0$. In the above example we can do this by using the transformation $x = -t$ for $t > 0$ to change the independent variable from x to t. The transformed DE is then solved for $t > 0$ (which corresponds to $x < 0$) by the Frobenius method. The result of this computation shows that Eq. (12) is a general solution for $x < 0$ as well as for $x > 0$ if we replace $x^{1/2}$ in y_2 by $|x|^{1/2}$. The details of this computation are sketched in Exercise 3. It is usually shown in more advanced texts that solutions of a second-order linear DE obtained by the Frobenius method for $x > 0$ can be extended to $x < 0$ by replacing x^k by $|x|^k$. We shall assume that this is true in our subsequent work.

EXAMPLE 2

Find a series solution about $x = 0$ of

$$x^3y'' + y = 0, \qquad \text{for } x > 0.$$

solution. For this DE, $P(x) = 0$ and $Q(x) = 1/x^3$, which is infinite at $x = 0$. Thus $x = 0$ is a singular point of the DE and we seek a solution in the form of a Frobenius series. We substitute Eq. (1) and its derivatives into the DE. Then by introducing new dummy variables of summation and combining terms, we get

$$b_0x^k + \sum_{m=0}^{\infty} [(m + k)(m + k - 1)b_m + b_{m+1}]x^{m+k+1} = 0.$$

Equating the coefficients of each power of x to zero gives $b_0 = 0$ and the recurrence relation

$$b_{m+1} = -(m + k)(m + k - 1)b_m, \qquad m = 0, 1, 2, \ldots. \tag{13}$$

Since $b_0 = 0$, the assumption that $b_0 \neq 0$ in the series given by Eq. (1) is contradicted. Thus the Frobenius method fails to give a series solution of the DE. If we permit $b_0 = 0$, then the recurrence relation of Eq. (13) shows that $b_m = 0$ for $m = 1, 2, \ldots$. Then the Frobenius method yields only the trivial solution $y \equiv 0$.

We shall now give sufficient conditions on the coefficients $P(x)$ and $Q(x)$ for which the method of Frobenius always gives at least one solution. These conditions are related to the "rates" at which $P(x)$ and or $Q(x)$ approach infinity at a singular point. Thus, we introduce the following classifications of singular points. For simplicity we only consider coefficients that are rational functions. (See Exercise 10 for a more general treatment.)

definition

Let $P(x)$ and $Q(x)$ be rational functions. If $x = x_0$ is a singular point of the DE

$$y'' + P(x)y' + Q(x)y = 0$$

and if both

$$\lim_{x \to x_0} (x - x_0)P(x) \qquad \text{and} \qquad \lim_{x \to x_0} (x - x_0)^2Q(x)$$

exist, then $x = x_0$ is a regular singular point. Otherwise $x = x_0$ is an irregular singular point.

We illustrate the definition by classifying the singular points of several specific DEs. First we consider the DE in Example 1: $4xy'' + 2y' + y = 0$. The only singular point is $x_0 = 0$ because $P(x) = 1/2x$ and $Q(x) = 1/4x$. Since

$$\lim_{x \to 0} xP(x) = \frac{1}{2} \qquad \text{and} \qquad \lim_{x \to 0} x^2Q(x) = 0,$$

we conclude that $x = 0$ is a regular singular point of the DE. However, in Example 2 $x = 0$ is a singular point and, since $x^2Q(x) = x^{-1} \to \infty$ as $x \to 0$, it is an irregular singular point. In Example 1 we obtained Frobenius series representations of two linearly independent solutions of the DE with the series being expanded about the regular singular point. However, in Example 2 the method of Frobenius failed to give a series representation about the irregular singular point.

We now state, without proof, a theorem that guarantees that there is at least one solution in the form of a Frobenius series.

theorem

If $x = x_0$ is a regular singular point of the DE

$$y'' + P(x)y' + Q(x)y = 0$$

and the indical roots are real, then there is at least one solution

$$y = |(x - x_0)|^{k_2} \sum_{n=0}^{\infty} B_n(x - x_0)^n,$$

for x in the interval $0 < |(x - x_0)| < R$. Here k_2 is the larger indicial root and R is some positive number. Furthermore, if the indicial roots $k_2 > k_1$ are distinct and if they satisfy

$$k_2 - k_1 \neq N,$$

where N is any positive integer, then two linearly independent solutions in the interval $0 < |(x - x_0)| < R$ are

$$y_1 = |(x - x_0)|^{k_1} \sum_{n=0}^{\infty} A_n(x - x_0)^n \qquad \text{and} \qquad y_2 = |(x - x_0)|^{k_2} \sum_{n=0}^{\infty} B_n(x - x_0)^n.$$

In Example 1, $x_0 = 0$ is a regular singular point and $k_2 - k_1 = \frac{1}{2} \neq$ a positive integer. Two linearly independent series solutions were obtained by the method of Frobenius. We shall discuss the cases $k_2 = k_1$ and $k_2 - k_1 = N$ in the next section.

If the indicial roots are complex conjugates, then the factor $|(x - x_0)|^k$ in the Frobenius solution is $|(x - x_0)|$ raised to a complex number. This is a complex-valued function of the real variable x, which can be defined in terms of the complex-valued exponential function. This was done for the Euler equation in Exercise 5 of Section 11.4. It can be shown that the theorem is valid in this case. The solutions y_1 and y_2 are then complex-valued functions. To find real-valued solutions, appropriate linear combinations of y_1 and y_2 must be formed. However, we shall not pursue this case any further.

Exercises

1. For the following DEs, find the singular points and determine whether they are regular or irregular.
(a) $xy'' + y' + (x^2 - 1)y = 0$ (b) $x^2y'' - xy' + (\frac{1}{4} - x^2)y = 0$
(c) $(x - 1)^2y'' - 4(x - 1)y' + x^2y = 0$
(d) $x^2(x - 1)y'' + 9(x - 2)y' + y = 0$

2. Find a general solution of each of the following DEs, by using the Frobenius method about $x = 0$.
(a) $4xy'' + 2y' + y = 0$ (b) $2xy'' + (1 - 2x)y' - y = 0$
(c) $4xy'' + (6 - x)y' - y = 0$ (d) $2x^2y'' + xy' - (2x + 1)y = 0$

3. We observed in Example 1 that the general solution given in Eq. (12) is valid only for $x > 0$. In order to find a general solution for $x < 0$: (a) Use the change of independent variable $x = -t$ to transform the DE to

$$4t\frac{d^2y}{dt^2} + 2\frac{dy}{dt} - y = 0 \tag{i}$$

(b) Solve Eq. (i) for $t > 0$ by the Frobenius method to obtain the linearly independent solutions

$$y_1 = \sum_{n=0}^{\infty} \frac{t^n}{(2n)!} \quad \text{and} \quad y_2 = t^{1/2}\sum_{n=0}^{\infty} \frac{t^n}{(2n+1)!}. \tag{ii}$$

(c) Replace t by $-x$ in Eq. (ii) and conclude that Eq. (12) gives a general solution for $x < 0$ as well as for $x > 0$ if we replace $x^{1/2}$ in Eq. (12) by $|x|^{1/2}$.

4. For each of the following DEs, find a general solution valid for $x < 0$ as well as for $x > 0$ by proceeding as in Exercise 3.
(a) $2x^2y'' - xy' + (x - 5)y = 0$ (b) $2x^2y'' + 3xy' + 2xy = 0$

5. Show that y_1 given by Eq. (8) and y_2 given by Eq. (11) are two linearly independent solutions of the DE in Example 1. (*Hint:* Let $z = x^{1/2}$ in y_1 and y_2, and show that the resulting power series (in powers of z) are not proportional. Hence y_1 and y_2 are not proportional.) Observe that the technique of evaluating $W(x)$ at $x = 0$ that was used in the examples of Section 12.3-1 is not applicable here because $W(0)$ is not defined.

6. Show that $x = 1$ is a regular singular point of the DE $2(x - 1)^2y'' - (x - 1)y' + xy = 0$. Use Frobenius' method to find a general solution for $x < 1$ and $x > 1$. (*Hint:* Use the transformation $t = x - 1$.)

7. For each of the following DEs classify the point $x = 0$. Show that the method of Frobenius fails to yield a general solution about $x = 0$.
(a) $x^4y'' + y = 0, \quad x > 0$ (b) $x^3y'' + xy' + y = 0, \quad x > 0$

8. Generalize the results of Example 2 in the text by considering $x^Ny'' + y = 0$, where N is a positive integer. (a) Classify the singular point $x = 0$ for all values of N. Observe that the larger the value of N, the faster $|Q(x)| \to \infty$ as

$x \to 0$. (b) Show that for $N \geq 3$ there is no Frobenius solution. (c) Find all Frobenius solutions about $x = 0$ of the DE with $N = 1$. (d) Repeat (c) for $N = 2$.

*9. Generalize the results of Exercise 8 by considering

$$y'' + \frac{a}{x^p} y' + \frac{b}{x^q} = 0,$$

where p and q are positive integers, and a and b are constants.

*10. The method of Frobenius can be extended to the DE $y'' + P(x)y' + Q(x)y = 0$ when the functions $P(x)$ and $Q(x)$ are not rational functions of x. Then the following more general definition of regular singular point is needed. Refer to Exercise 5 of Section 12.4 for the definition of an analytic function.

definition

If either (or both) of the functions $P(x)$ and $Q(x)$ is not analytic at the point $x = x_0$, then x_0 is a singular point. If the functions $(x - x_0)P(x)$ and $(x - x_0)^2Q(x)$ are both analytic at the point $x = x_0$, then $x = x_0$ is called a regular singular point of the DE. Otherwise, $x = x_0$ is called an irregular singular point.

The theorem in the text is valid at a regular singular point as defined above. Classify the singular points of the following DEs and find the indicial roots.

(a) $2x^2y'' - (x \sin x)y' + y = 0$ (b) $2x^2y'' + 2xe^xy' - 3y = 0$
(c) $4x^2y'' - (\sin x)y' + (\cos x)y = 0$

13.3 SPECIAL CASES

The theorem in Section 13.2 guarantees that if x_0 is a regular singular point and if the indicial roots k_1 and k_2 are distinct and do not differ by an integer, then the method of Frobenius always gives series representations about x_0 of two linearly independent solutions of the DE

$$y'' + P(x)y' + Q(x)y = 0.$$

Clearly, if $k_2 = k_1$, the method cannot possibly yield two linearly independent solutions, since each indicial root provides only one solution of the DE. Then when one solution y_1 of the DE is obtained by the Frobenius method a second linearly independent solution y_2 can be found by reduction of order. In practice the computation of y_2 is complicated because the solution y_1 is given by an infinite series. It can be shown that y_2 has the form

$$y_2 = y_1 \ln |x - x_0| + |x - x_0|^k \sum_{n=0}^{\infty} B_n(x - x_0)^n$$

where k is the indicial root, x_0 is the regular singular point, and $B_0, B_1, \ldots,$ are constants.[1]

The occurrence of a "log term" in the second solution could have been anticipated by considering the Cauchy-Euler equation

$$x^2y'' + b_1xy' + b_2y = 0, \tag{1}$$

where b_1 and b_2 are constants (see Section 11.4). The reader can verify that $x_0 = 0$ is a regular singular point and by using the Frobenius method that the indicial equation is

$$k^2 + (b_1 - 1)k + b_2 = 0.$$

If $(b_1 - 1)^2 - 4b_2 = 0$, the indicial root is repeated and the Frobenius method yields only the solution

$$y_1 = x^{(1-b_1)/2}.$$

A second linearly independent solution was obtained in Section 11.4 by reduction of order. It is

$$y_2 = y_1 \ln x.$$

The difficulty that arises when $k_2 - k_1 =$ integer is not apparent from our previous work. For the Cauchy-Euler DE given in Eq. (1), the indicial roots can differ by an integer but the Frobenius method gives series representations about $x_0 = 0$ of two linearly independent solutions. However, as we show in Exercise 4, this is not true in general. Then a second solution is obtained by reduction of order. It is difficult to tell without examining the recurrence relation for each specific problem whether we will get one or two solutions by the Frobenius method when $k_2 - k_1 =$ integer.

This discussion is summarized in the following theorem.

theorem

Let $x = x_0$ be a regular singular point of the DE

$$y'' + P(x)y' + Q(x)y = 0,$$

and let k_1 and k_2 be the indicial roots with $k_2 \geq k_1$.

I. If $k_1 = k_2 = k$, then two linearly independent solutions for $|(x - x_0)| > 0$ are given in the form

$$y_1(x) = |(x - x_0)|^k \sum_{n=0}^{\infty} A_n(x - x_0)^n$$

[1] It can be shown that the term $\ln |x - x_0|$ cannot be represented by a Frobenius series about $x = x_0$ and thus y_2 cannot possibly be obtained from the Frobenius method.

and

$$y_2(x) = y_1 \ln |(x - x_0)| + |(x - x_0)|^k \sum_{n=0}^{\infty} B_n(x - x_0)^n.$$

II. If $k_2 - k_1 =$ a positive integer, then two linearly independent solutions are given by

$$y_1 = |(x - x_0)|^{k_2} \sum_{n=0}^{\infty} A_n(x - x_0)^n$$

and

$$y_2 = Cy_1 \ln |(x - x_0)| + |(x - x_0)|^{k_1} \sum_{n=0}^{\infty} B_n(x - x_0)^n.$$

The constant C may be zero. Then both solutions are found by the Frobenius method. The power series in parts I and II converge for all $0 < |(x - x_0)| < R$ for some $R > 0$.

As we have observed the computation of y_2 by reduction of order may be complicated. An alternate, and usually preferable, procedure is to substitute the forms for y_2 given in parts I or II of the theorem into the DE and then solve for the constants B_n.

Exercises

1. Show that the method of Frobenius yields two linearly independent solutions for the DE $x^2y'' + xy' + (x^2 - \frac{1}{4})y = 0$. The indicial roots are $\pm\frac{1}{2}$ and thus differ by an integer. (*Hint:* Use the indicial root $k = -\frac{1}{2}$ in the recurrence relation and show that a general solution is obtained. Show that the indicial root $k = \frac{1}{2}$ yields no new solutions.)

2. (a) Find all Frobenius series solutions about $x = 0$ of the DE $x^2y'' + (3x + x^3)y' + (1 + 3x^2)y = 0$ which are valid for $x > 0$.
   (b) Show that your answer in (a) is a constant multiplied by $y_1 = x^{-1}e^{-x^2/2}$.
   (c) Use the solution y_1 and reduction of order to obtain a second linearly independent solution

$$y_2 = y_1(x) \int^x z^{-1}e^{(z^2/2)}\, dz.$$

   Expand the integrand in a series in powers of z and integrate term by term to show that y_2 has the form given in the theorem with $C \neq 0$.

3. For each of the following DEs find a general solution for $x > 0$ in the form of a Frobenius series about $x = 0$.

(a) $xy'' - (2x + 1)y' + (x + 1)y = 0$ (b) $xy'' + 2y' + xy = 0$
(c) $x^2y'' - 2x^2y' + 2(2x^2 - 1)y = 0$ (d) $3x^2y'' + x(9 - x)y' - xy = 0$

4. For each of the following DEs show that the indicial roots are either equal to or differ by an integer, and that only one Frobenius series solution about $x = 0$ exists. Find it and then find the second solution by using the method of reduction of order. (See Exercise 2.)
(a) $x(1 - x)y'' - 3xy' - y = 0$ (b) $xy'' + y' - (x + 1)y = 0$
(c) $x^2y'' + 2(2x - x^2)y' + 2(1 - x)y = 0$
(d) $xy'' + y' + xy = 0$

chapter 14

NONLINEAR DIFFERENTIAL EQUATIONS

14.1 INTRODUCTION

Our study of differential equations has been limited to solving linear DEs. However, as we have seen in several of the applications, the predictions obtained by using linear equations are at variance with experimental and observational evidence. For example, in Section 2.4-1 we studied the growth and decay of an isolated population of a single organism by using the IVP

$$y' = ky, \qquad y(0) = a_0 \tag{1}$$

for the population $y(t)$ at the time t. Here the constant $k = m - n$, where m and n are the birth and death rates respectively, and a_0 is the prescribed initial population size. The solution of this IVP is $y = a_0e^{kt}$. Thus for example, if $k > 0$—that is, if the birth rate exceeds the death rate—then $y(t) \to \infty$ as $t \to \infty$ and the population grows to infinity exponentially. This is an unsatisfactory result because observations of many populations suggest that they approach a limiting value or they oscillate with bounded amplitude.

The assumption that the birth and death rates are constants is not realistic for large populations. Because as the size of the population increases, food, space, and other limited resources become scarce. Then we would expect the birth rate to decrease and the death rate to increase. Thus both rates should be functions of the population size y.

A widely used model that includes population-dependent birth and death rates is obtained by assuming that

$$m = b_1 - b_2 y \qquad \text{and} \qquad n = b_3 + b_4 y, \tag{2}$$

where $b_1, \ldots, b_4$ are specified positive constants. We choose positive values for b_2 and b_4 so that the birth rate m decreases and the death rate n increases as the population size y increases. Thus the coefficient k in Eq. (1) is given by $k = m - n = (b_1 - b_3) - (b_2 + b_4)y$. By defining the constants α and A by

$$\alpha = b_2 + b_4 \qquad \text{and} \qquad A = \frac{b_1 - b_3}{b_2 + b_4},$$

we can rewrite k as, $k = \alpha(A - y)$. Then by inserting this in Eq. (1) we get the new IVP for population dynamics as

$$y' = \alpha(A - y)y, \qquad y(0) = a_0. \tag{3}$$

The DE in Eq. (3) is called the logistic equation. It was derived during the first half of the nineteenth century by the Belgian demographer and mathematician, P. Verhulst. Verhulst, and later the American biologists and demographers R. Pearl and L. Reed, used this DE to predict the population growth of various countries, including the United States. It is a fundamental equation in demography and in the mathematical theory of ecology. It has been applied in mathematical studies of the spread of rumors and other mathematical problems in psychology and sociology. We shall solve it in Section 14.5-1. The logistic equation is a first-order nonlinear DE because it contains the term $-y^2$.

A more general first-order DE is given by

$$y' = f(x, y) \tag{4}$$

where $f(x, y)$ is a specified function of the two variables x and y. If $f(x, y) = -q(x)y + r(x)$, then Eq. (4) is the first-order linear DE. For any other choice of f the DE is nonlinear. In particular, if $f(x, y) = \alpha(A - y)y$, then Eq. (4) is the logistic equation.

In this chapter we shall consider several special classes of functions $f(x, y)$ for which it is possible to solve the IVP for the DE given in Eq. (4). The function $f(x, y)$ will always be real-valued and we shall study only real-valued solutions. For example, in the next section we shall study the

DE

$$y' = \frac{G(x)}{H(y)}. \qquad (5)$$

That is, $f(x, y)$ is the quotient of a function of x only and a function of y only. Thus the dependence of f on x and y is "separated" and any DE of the form given in Eq. (5) is called a *separable equation.* We observe that the logistic equation given by Eq. (3) is of this form with $G(x) = 1$ and $H(y) = 1/[\alpha(A - y)y]$. The linear homogeneous DE $y' = -q(x)y$ is also of this form, with $G(x) = -q(x)$ and $H(y) = 1/y$.

Exercises

1. Determine which of the following DEs are separable.
   (a) $(yx^2 + xy)y' + 2y^2x = 0$
   (b) $(x^2 + xy)y' + 2x = 0$
   (c) $y' + ye^{x+2y} = 0$
   (d) $y' + 3x^2y = x^2$
   (e) $(\sin x)y' + x \cos y = 0$
   (f) $y' + xye^{x-y} = 0$
   (g) $y' + 3x^2y = x$
   (h) $[\cos (x + y)]y' + \cos y = \cos x$
2. For what choices of the functions $r(x)$ and $q(x)$ is the first-order linear DE $y' + q(x)y = r(x)$ separable?
3. (a) Repeat Exercise 2 for the Bernoulli equation $y' + q(x)y = r(x)y^n$, where n is a constant.
   (b) Show that the change of dependent variable $u = y^{1-n}$, $n \neq 0, 1$ reduces the Bernoulli equation to a linear DE for $u(x)$.
4. (a) Consider the DE $y' + a(x)y + b(x)y^2 = r(x)$. This is called a Ricatti equation. It occurs in many problems in control theory and in other applications of DEs. Determine all choices of $r(x)$, $a(x)$, and $b(x)$ for which the DE is separable.
   (b) Let $y = y_1$ be a solution of the Ricatti equation. Show that the change of variables $y = y_1 + 1/u(x)$ reduces the Ricatti equation to a linear DE for $u(x)$.
5. Assume that due to seasonal factors the constants b_1 and b_3 in Eq. (2) vary in some specified way with time. Will the resulting DE be separable?
6. Generalizations of the logistic equation can be derived by assuming that the birth and death rates depend on higher powers of y. For example, $n = b_1 - b_2y + b_3y^2$ and $m = b_4 + b_5y + b_6y^2$. Here $b_1, b_2, \ldots, b_6$ are constants. Using these values, derive a DE for the population size y. Is the DE separable?
7. Derive the logistic equation by assuming that the relative rate of growth of a population at time t, y'/y, is proportional to the difference between the population size at time t and a constant L. The constant L can be interpreted as the largest population that can be supported by the environment in question. It is equal to the constant A in Eq. (3).

8. (a) Consider the logistic DE given by Eq. (3). Show that if $y > A$, then $y' < 0$, and conclude that the population cannot increase without bound. (b) Interpret the result of (a) in terms of the interpretation given to L in Exercise 7.

14.2 SEPARABLE EQUATIONS

We shall study the separable DE which was introduced in Section 14.1

$$y' = \frac{G(x)}{H(y)}. \tag{1}$$

First we recall our method for solving the linear DE,

$$y' = -q(x)y. \tag{2}$$

We divide both sides of Eq. (2)[1] by y or, equivalently, we multiply both sides by $H(y) = 1/y$ and obtain

$$\frac{y'}{y} = -q(x). \tag{3}$$

We solved Eq. (3) by integrating both sides of the equation. The indefinite integral of y'/y was explicitly evaluated because $y'/y = (\ln |y|)'$ is an exact derivative. However, this indefinite integral can also be evaluated by transforming the dummy variable of integration. This gives

$$\int^x \frac{y'}{y}\,dz = \int^x \frac{1}{y}\frac{dy}{dz}\,dz = \int^{y(x)} \frac{dy_1}{y_1} = \ln |y| + c.$$

We have used y_1 as the new integration variable and c is a constant of integration.

Since this technique for solving Eq. (2) does not use the linearity of the DE, it can be applied to other (nonlinear) separable DEs, as we illustrate in the following example.

EXAMPLE 1

Solve the DE

$$y' = -\frac{x}{y}, \qquad \text{for } y \neq 0.$$

solution. The DE is separable because it can be written in the form of Eq. (1) with $G(x) = -x$ and $H(y) = y$. We multiply both sides of the DE by y and

[1] We have observed already that Eq. (2) is a special case of Eq. (1) and hence it is a separable DE.

express it as $yy' = -x$. By integrating both sides of this equation we get

$$\int^x yy'\, dz = -\int^x z\, dz + c_1, \tag{4}$$

where c_1 is the arbitrary integration constant. Since

$$\int^x yy'\, dz = \int^x y \frac{dy}{dz}\, dz = \int^y y_1\, dy_1 = \frac{y^2}{2}$$

and[2]

$$\int^x z\, dz = \frac{x^2}{2}.$$

Eq. (4) can be expressed as

$$y^2 = -x^2 + c, \tag{5}$$

where $c = 2c_1$. Therefore, if the DE has a solution, it must satisfy Eq. (5).

In order for the solutions to be real, we must have $y^2 \geq 0$. Hence c must be positive and x must be restricted to the interval $x^2 \leq c$, or $|x| \leq \sqrt{c}$. Then the DE has the two solution candidates

$$y_1(x) = (c - x^2)^{1/2} \quad \text{and} \quad y_2(x) = -(c - x^2)^{1/2}, \tag{6}$$

which are obtained by solving Eq. (5) for y. It is easy to verify by direct substitution that y_1 and y_2 are indeed solutions of the DE for $|x| < \sqrt{c}$. Graphs of these solutions for representative values of the constant c are sketched in Figure 14.1. Furthermore, we observe that y_1' and y_2' are not defined at $x = \pm\sqrt{c}$, and thus y_1 and y_2 are not solutions at these points.

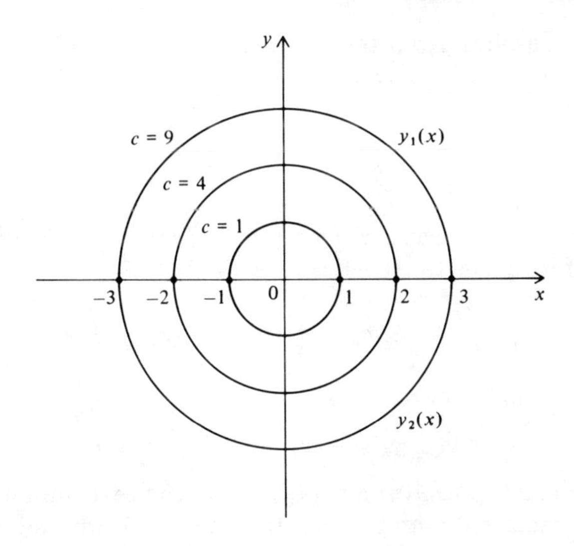

Figure 14.1

[2] Alternately, we can evaluate the indefinite integral of yy' by observing that $yy' = (y^2/2)'$ is an exact derivative.

EXAMPLE 2

Solve the IVP

$$y' = -\frac{x}{y}, \qquad y(0) = a_0,$$

where a_0 is a specified constant.

solution. To solve the IVP we use the solutions of the DE that were determined in Example 1; see Eq. (6). We select c so that the initial condition $y(0) = a_0$ is satisfied. If $a_0 > 0$, we must select the positive solution y_1. By substituting $x = 0$ and $y_1 = a_0$ in Eq. (6) we get $a_0 = \sqrt{c}$. Therefore $c = a_0{}^2$, and a solution of the IVP is

$$y = \sqrt{a_0{}^2 - x^2}.$$

Similarly, if $a_0 < 0$, we use y_2 and get a solution of the IVP as

$$y = -\sqrt{a_0^2 - x^2}.$$

If $a_0 = 0$, then $c = a_0 = 0$ and the functions y_1 and y_2 given in Eq. (6) are not solutions of the IVP.

We now return to solving DE (1). We multiply Eq. (1) by $H(y)$ and integrate both sides of the resulting equation to get

$$\int^x H(y) \frac{dy}{dz}\, dz = \int^x G(z)\, dz + c,$$

where c is an arbitrary constant of integration. Then by transforming the variable of integration from z to y_1 in the integral on the left of this equation we obtain

$$\int^y H(y_1)\, dy_1 - \int^x G(z)\, dz = c. \tag{7}$$

Thus we have shown that if y is a solution of DE (1), then it must satisfy Eq. (7).[3] If we define the functions $h(y)$ and $g(x)$ by

$$h(y) = \int^y H(y_1)\, dy_1 \qquad \text{and} \qquad g(x) = \int^x G(z)\, dz, \tag{8}$$

then Eq. (7) can be rewritten as

$$\Phi(x, y) = h(y) - g(x) - c = 0. \tag{9}$$

The indefinite integrals $h(y)$ and $g(x)$ are, of course, known only to within additive constants of integration. In Example 1 we had $h(y) = y^2/2$, $g(x) = -x^2/2$, and thus $\Phi(x, y) = y^2/2 + x^2/2 - c_1$.

[3] We assume that $H(y)$ and $G(x)$ are continuous. Then the integrals in Eq. (7) exist.

To find a solution of the DE explicitly, we must solve Eq. (9) for y in terms of x and c. If we can find such a solution $y(x)$ and, in addition, if $y'(x)$ exists, then $y(x)$ is a solution of the DE. A proof of this result is discussed in Exercise (5). However, we wish to emphasize that every solution of Eq. (9) is not necessarily differentiable and hence not necessarily a solution of the DE given in Eq. (1).

When it is possible to solve Eq. (9) for y, and y' exists, the result is called an *explicit solution* of the DE. If the transcendental equation, Eq. (9), has a differentiable solution but we do not determine it explicitly, then $\Phi(x, y) = 0$ is called an *implicit solution* of the DE. For the DE in Example 1 we have $\Phi(x, y) = y^2 + x^2 - c = 0$ (see Eq. (5)). The explicit solutions given in Eq. (6) were determined by solving the equation $\Phi(x, y) = 0$ for $y(x)$.

It is not always obvious how to solve Eq. (9) for y as a function of x. For example, the reader can verify that for the DE

$$(y + \sin y)y' - \frac{1}{x} = 0,$$

$\Phi(x, y) = y^2/2 - \cos y - \ln |x| = 0$. It is not possible to explicitly solve $\Phi(x, y) = 0$ for y as a function of x to determine explicit solutions $y(x)$ of this DE. Indeed, some equations have no solutions, whereas others, for example, see Eq. (5), have more than one solution. Furthermore, if there is a solution, it may not be possible to obtain the explicit representation in terms of elementary functions. The problem of how and when an equation $\Phi(x, y) = 0$ can be solved for y is a fundamental problem of mathematics.

Exercises

1. Solve each of the following DEs and find explicit solutions where possible.

(a) $y' - \dfrac{x^3}{y} = 0$ (b) $y' + \dfrac{x^2}{y^3} = 0$

(c) $e^{-x}y' - \csc y = 0$ (d) $2(y + xy)y' + (1 + y^2) = 0$

(e) $yy' + \sqrt{4 - y^2} = 0$ (f) $(e^y \sin y)y' + x = 0$

(g) $xy' + x^3(1 + y^2) = 0$ (h) $(y^3 + 1)y' + x^2 = 0$

2. Sketch representative solution curves for the following DEs.

(a) $y' - \dfrac{x^2}{y^2} = 0$ (b) $yy' + \sqrt{16 - y^2} = 0$

(c) $y' + y^2 = 0$ (d) $2e^y y' + x = 0$

3. Solve the following IVPs. Find a solution in explicit form and determine the interval in which the solution is valid.
(a) $(y + 1)y' - (6x^2 + 2) = 0, \quad y(0) = 1$
(b) $y' - 2x \sin y = 0, \quad y(0) = \pi/2$
(c) $y' = \dfrac{1 + x}{1 + y}, \quad y(0) = 3$
(d) $(x^2 + 1)(\cos y)y' + 4x = 0, \quad y(1) = \pi/4$
4. Find two solutions of the IVP $y' = y^{1/3}, \quad y(0) = 0$.
5. Use the chain rule to prove that if Eq. (9) has a differentiable solution $y(x)$, then $y(x)$ is a solution of the DE given in Eq. (1).
6. (a) Find an implicit solution to the IVP $y' = e^{y^2}$, $y(0) = 0$. (b) Use the error function, which is defined by

$$\operatorname{erf}(z) = \frac{2}{\sqrt{\pi}} \int_0^z e^{-u^2}\, du$$

to express x as a function of y. We observe that this definite integral cannot be explicitly evaluated. However, because of its importance in probability theory and heat conduction problems, extensive tables for erf (z) exist.

In the next three exercises we study a special class of first-order DEs that can be reduced to separable equations by a transformation of the dependent variable y.

7. If $f(x, y)$ depends on x and y only through the ratio y/x (or x/y), then we call the DE $y' = f(x, y)$ homogeneous. In this case $f(x, y) = F(y/x)$. It is not always easy to recognize a homogeneous DE by inspection. A simple test to determine whether f is of this form is derived by letting $y = vx$.
(a) Show that if $f(x, xv) = f(1, v)$, then the DE is homogeneous.
(b) Show that the change of dependent variable $y = xv$ transforms any homogeneous DE $y' = F(y/x)$ into the separable DE

$$v' = \frac{F(v) - v}{x}$$

8. Test each of the following DEs for homogeneity, and solve those that are homogeneous.

(a) $y' = \dfrac{x - 2y}{2x + y}$ (b) $y' = \dfrac{3y^2}{xy + x^2}$

(c) $y' = \dfrac{y}{x}\left(\ln \dfrac{y}{x} + 1\right)$

*9. Consider the DE

$$y' = \frac{ax + by + A}{cx + dy + B},$$

where a, b, c, d, A, and B are prescribed constants.

(a) Show that if $ad - bc \neq 0$, then this DE can always be reduced to a homogeneous DE by the change of variables $x = X + h$ and $Y = Y + k$, where h and k are solutions of the equations

$$ah + bk + A = 0, \qquad ch + dk + B = 0$$

(b) If $ad - bc = 0$, show that either of the changes of variables

$$v = ax + by + A, \qquad \text{or} \qquad v = cx + dy + B$$

reduces the DE to a separable DE.

14.3 EXACT EQUATIONS

In Section 2.3, we solved the linear DE

$$y' + q(x)y = 0$$

by using an integrating factor. That is, we multiplied the DE by the integrating factor $e^{Q(x)}$, where $Q(x)$ is any indefinite integral of $q(x)$. Then the resulting equation was written as $(ye^Q)' = 0$. Since the left side of this equation is an exact derivative of the function $U(x, y) = ye^{Q(x)}$, the solution was found easily by a single integration. Thus the DE was reduced to the form

$$[U(x, y(x))]' = 0$$

for a function U of x and a solution $y(x)$.

We now wish to study nonlinear DEs, $y' = f(x, y)$, that can be solved by a similar method. In the following analysis it is convenient to express f as $f = -M(x, y)/N(x, y)$. This can always be done by setting $N(x, y) = 1$ and $f(x, y) = -M(x, y)$. However, other choices of M and N are frequently more convenient. Thus we shall consider the DE in the form

$$N(x, y)y' + M(x, y) = 0. \tag{1}$$

In particular, we shall study the special class of DEs for which the left side of Eq. (1) is an exact derivative. Then the DE can be expressed as

$$N(x, y)y' + M(x, y) = [U(x, y(x))]' = 0. \tag{2}$$

More general classes of DEs are discussed in Exercises 10–13.

If there exists a function $U(x, y)$ such that the DE can be written in the form of Eq. (2), then it is called an *exact* DE. It follows from Eq. (2) that if an exact DE has a solution $y(x)$, then

$$\Phi(x, y) = U(x, y) - c = 0$$

is an implicit solution, where c is an arbitrary constant.

An easily applied test to determine whether the DE given in Eq. (1) is exact is obtained by first using the chain rule of differentiation to calculate U'. This gives

$$[U(x, y(x))]' = \frac{\partial U}{\partial x} + \frac{\partial U}{\partial y} y' = U_x + U_y y', \tag{3}$$

where $\partial U/\partial x = U_x$ and $\partial U/\partial y = U_y$ are the partial derivatives of U. Then by equating the left side of Eq. (2) and the right side of Eq. (3), we conclude that the resulting equation is satisfied if

$$M = U_x \qquad \text{and} \qquad N = U_y. \tag{4}$$

If, in addition, M_y and N_x are continuous functions, then by differentiating the first equation in Eq. (4) with respect to y and the second equation with respect to x and using $U_{xy} = U_{yx}$, we get[4]

$$M_y = N_x. \tag{5}$$

As we shall see in Theorem 2 at the end of this section, if the coefficients M and N satisfy Eq. (5), then the DE is exact and conversely if they do not satisfy Eq. (5), then the DE is not exact. Thus Eq. (5) is an easily applied test for exactness. The application of this exactness test is illustrated in the following examples.

EXAMPLE 1

Show that the DE

$$(x + y^2)y' + y = 0 \tag{6}$$

is exact. Obtain an implicit solution.

solution. Here we have $M_y = 1$ and $N_x = 1$, and hence $M_y = N_x$. This shows, by using Eq. (5), that Eq. (6) is an exact DE. We shall now solve the DE by expressing it as $[U(x, y)]' = 0$.

First we observe the differentiation formulas,

$$(xy)' = xy' + y \qquad \text{and} \qquad \left(\frac{y^3}{3}\right)' = y^2 y'.$$

By regrouping the terms in Eq. (6) into groups, each of which is an exact derivative, and using these formulas, we obtain

$$\begin{aligned}(x + y^2)y' + y &= (xy' + y) + y^2 y' \\ &= (xy)' + \left(\frac{y^3}{3}\right)' = \left[xy + \frac{y^3}{3}\right]' = 0,\end{aligned}$$

[4] The condition $U_{xy} = U_{yx}$, which implies that the order of differentiation with respect to x and y is immaterial, is valid if U_{xy} and U_{yx} are continuous functions. Functions $U(x, y)$ for which $U_{xy} \neq U_{yx}$ are rarely encountered in studying DEs.

which also shows that the DE is exact. Therefore, $U = xy + y^3/3$, and we have the implicit solution $xy + y^3/3 - c = 0$.

We solved the exact DE in Example 1 by appropriately grouping terms and using differentiation formulas to recognize that $Ny' + M = 0$ could be written as $[U(x, y)]' = 0$. For many exact DEs it is not obvious how to do this. A systematic procedure for constructing the function $U(x, y)$ for exact DEs is illustrated in the following example.

EXAMPLE 2

Solve the DE

$$(x + y)y' + (y + 3x) = 0.$$

solution. The DE is exact because $M_y = N_x = 1$. It is easy to construct an implicit solution of this exact DE by using the formulas, $xy' + y = (xy)'$, $yy' = (y^2/2)'$, and $3x = (3x^2/2)'$. However, we shall not proceed in this manner. Instead, we shall determine the function U by using the relations in Eq. (4). Thus we have

$$U_x = M = y + 3x, \tag{7}$$

and

$$U_y = N = x + y. \tag{8}$$

These are two equations that U must satisfy. We integrate both sides of Eq. (7) with respect to x, keeping y fixed. This gives

$$U(x, y) = yx + \frac{3x^2}{2} + A(y), \tag{9}$$

where $A(y)$ is an arbitrary *function* of integration. Here A depends on y because U was integrated with respect to the first variable x only, with y considered as a fixed parameter. To verify that Eq. (9) is correct, we need only take its partial derivative with respect to x.

Since U must also satisfy Eq. (8), we determine $A(y)$ by differentiating Eq. (9) with respect to y and then inserting the result in the left side of Eq. (8). This gives

$$U_y = x + A'(y) = x + y$$

where $A'(y) = dA/dy$. This equation is equivalent to $A' = y$, and hence

$$A(y) = y^2/2 + c.$$

The constant of integration, c, is not a function of x because A is independent of x. By inserting this expression for A in Eq. (9) we get

$$U(x, y) = yx + \frac{3x^2}{2} + \frac{y^2}{2} + c. \tag{10}$$

Since the DE is exact, $U(x, y) = 0$ is an implicit solution.

This problem can also be solved by first integrating Eq. (8) with respect to y and then using Eq. (7) to determine the resulting arbitrary function of x (see Exercise 8).

The technique used in Example 2 can be applied to construct $U(x, y)$ for an arbitrary exact DE. Specifically we can prove (see Exercise 9):

theorem 1

The exact DE

$$N(x, y)y' + M(x, y) = 0,$$

where M and N are continuous functions of x and y has the implicit solution

$$U(x, y) = \int^x M(z, y)\, dz + \int^y \left[N(x, s) - \frac{\partial}{\partial s}\int^x M(z, s)\, dz\right] ds = \text{constant}.$$

We summarize our discussions in the following theorem.

theorem 2

Let M, N, M_y and N_x be continuous functions of x and y in the rectangular region, $a \le x \le b$, $c \le y \le d$. If

$$M_y = N_x, \tag{11}$$

then $Ny' + M = 0$ is an exact DE. Conversely if $Ny' + M = 0$ is an exact DE, then Eq. (11) holds.

A proof of this theorem is given in *An Introduction to Ordinary Differential Equations*, by E. A. Coddington, Englewood Cliffs, N.J.: Prentice Hall, 1961.

Exercises

1. Test each of the following DEs for exactness and solve those that are exact.
   (a) $(2y - x)y' - (y - 2x) = 0$ (b) $(2y - x)y' + (y - 2x) = 0$
   (c) $(x \cos y + y^2)y' - \sin y = 0$ (d) $(x \cos y + y^2)y' - x \sin y = 0$
   (e) $(y^2 + 2x)y' + 2y = 0$ (f) $(4xy^3 + 3x^2)y' + (y^4 + 6xy) = 0$
   (g) $(x^2y - x)y' + (y^2x - y) = 0$ (h) $(e^y + x)y' + (3e^x + y) = 0$

2. Solve each of the following DEs and sketch some typical solutions.
   (a) $(2y - x)y' - y = 0$ (b) $2xy' + (y^2 - 2x) = 0$

3. Solve the following IVPs.
   (a) $(y + 4x)y' + (4y + 2x + 3) = 0, \quad y(1) = 1$
   (b) $(x^3 + x^2 + y)y' + (3x^2y + 2xy) = 0, \quad y(0) = 0$
   (c) $(\sin y - x^2)y' + (x^2 - 2yx) = 0, \quad y(0) = 1$

4. Show that the DE

$$(ay + bx)y' + (cy + dx) = 0, \qquad a,\ b,\ c,\ d \text{ constants}$$

is exact if $b = c$. Solve the DE in this case.

5. Find all values of the constants a and b so that the following DEs are exact
(a) $(ax^2 + bxy)y' + (6x^2 + ay^2) = 0$ (b) $(x^a y^2 + bx)y' + by + x^3 y^b = 0$

6. Find all possible choices of $q(x)$ and $r(x)$ for which the linear DE $y' + q(x)y = r(x)$ is exact.

7. For what choices of the functions G, H, P, and Q is the DE $P(y)G(x)y' + Q(x)H(y) = 0$ exact?

8. Solve the DE of Example 2 by first integrating both sides of Eq. (8) with respect to y and then requiring that the resulting expression for U satisfy Eq. (7).

*9. Prove Theorem 1. (*Hint:* Proceed as in Example 2.)

10. *A function $F(x, y)$ is an integrating factor of the DE in* Eq. (1), *if there is a function $U(x, y(x))$ such that,*

$$F(x, y)[N(x, y)y' + M(x, y)] = [U(x, y(x))]' = 0. \tag{i}$$

Show that each of the functions given in the right-hand column is an integrating factor and solve the DE.

(a) $x^2y' + y^2 = 0$ $\qquad (xy)^{-2}$
(b) $xy' - y + 2xy^2 = 0$ $\qquad y^{-2}$
(c) $2x^2yy' + xy^2 = x^4$ $\qquad x^{-1}$
(d) $e^y y' + (1 + x)e^{-y} = 0$ $\qquad e^y$

11. Show that any constant times and integrating factor is an integrating factor.

12. Show that x^2, x^5y, and $y^{-2/3}$ are integrating factors for the DE $xy' + 3y = 0$.

13. By using the chain rule of differentiation on $[U(x, y)]'$ in Eq. (i) show that the resulting equation is satisfied if

$$FM = U_x \quad \text{and} \quad FN = U_y. \tag{ii}$$

This equation is the analogue of Eq. (4). By taking $\partial/\partial y$ of the first equation in Eq. (ii) and $\partial/\partial x$ of the second and then assuming that U_{xy} and U_{yx} are continuous so that $U_{yx} = U_{xy}$, show that

$$MF_y - NF_x + (M_y - N_x)F = 0.$$

Since M and N are known functions of x and y, the unknown integrating factor satisfies a differential equation that involves partial derivatives of F. This is called a partial differential equation. In general it is difficult to find solutions to this equation and thus integrating factors are known for only certain nonlinear first-order DEs.

14.4 GENERAL REMARKS ON FIRST-ORDER DIFFERENTIAL EQUATIONS

In this section we shall discuss certain general features of the first-order nonlinear DE $y' = f(x, y)$. We discuss the geometric interpretation of the corresponding IVP in Section 14.4-1. In Section 14.4-2 we indicate some of the difficulties that arise in the study of nonlinear differential equations, which do not occur for linear DEs.

14.4-1 Geometric interpretation and graphical solutions

In the previous sections of this chapter we have shown that it is possible to find an explicit solution of the nonlinear DE

$$y' = f(x, y) \tag{1}$$

only for special classes of functions, $f(x, y)$. It is impossible to determine the solutions of this equation for other, more extensive, classes for which Eq. (1) is known to possess a solution (see Section 14.4-2). Consequently, a variety of approximate and graphical methods have been developed to solve Eq. (1). Some of these approximate methods are discussed in Chapter 15.

For many applications it is not essential to know an exact solution or even an accurate approximation. Only qualitative features of the solution are required. Then, graphical or curve-sketching procedures may provide sufficient information. Such techniques are widely used in studying the nonlinear oscillations of electrical, mechanical, and biological systems. We shall discuss a simple graphical procedure for constructing solutions. To describe this procedure, we shall first consider a geometrical interpretation for the solutions of DE (1).

If $y(x)$ is a solution of the DE, then the graph of the solution is a curve in the xy-plane that is called an *integral curve*; see Figure 14.2. At any point (x_1, y_1) of this curve, the corresponding slope of the curve, $y'(x_1)$, is obtained by evaluating DE (1) at that point, that is, $y'(x_1) = f(x_1, y_1)$. For example, it is easily seen that $y = 1/(1 + x^2)$ is a solution of the DE $y' = -2xy^2$. The slope of the corresponding integral curve at $x = x_1$ is given by

$$y'(x_1) = \frac{-2x_1}{(1 + x_1^2)^2}.$$

This is equal to

$$f(x_1, y(x_1)) = -2x_1y^2(x_1) = \frac{-2x_1}{(1 + x_1^2)^2}.$$

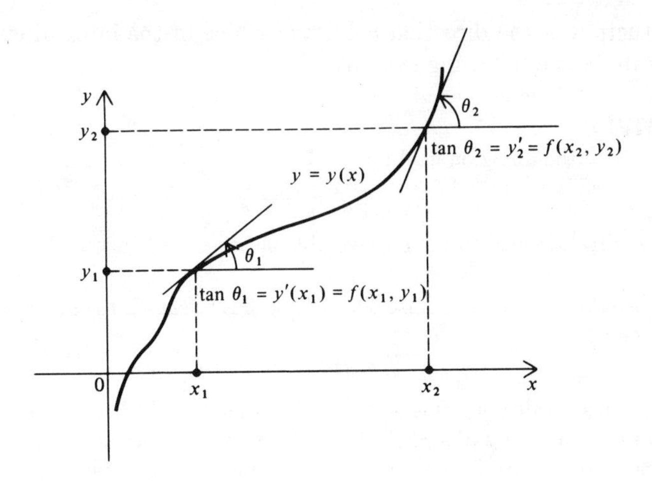

Figure 14.2

In general, the integral curves are not known explicitly. Nevertheless, their tangent lines are easily determined because $f(x, y)$ is the slope of the tangent line to the integral curve[5] that passes through the point (x, y). This is illustrated in Figure 14.2. Thus the DE assigns to every point in the xy-plane for which $f(x, y)$ is defined, a direction or tangent line. This tangent line must coincide with the tangent line to the integral curve passing through the point. The collection of directions is called the *direction* or *tangent field* of DE (1).

The foregoing discussion suggests the following procedure for sketching the integral curves of a DE. We draw small segments of the tangent lines (the direction field) of the DE at a sufficiently dense number of points in the xy-plane. Since the integral curves are tangent to these segments, the features of the integral curves are, in many cases, immediately obvious from the direction field, as we shall show in the Example. The construction of the direction field is simplified by using the level curves of $f(x, y)$. The level curves are determined by the equation

$$f(x, y) = c,$$

which defines a family of curves in the xy-plane that depend on the parameter c. For each admissible value of c we obtain one member of the family. These level curves are also called the *isoclines* of the differential equation. Since $y' = f(x, y) = c$ on an isocline, the slopes of all the integral curves that intersect a given isocline are equal. This facilitates the graphical

[5] We assume that only one integral curve passes through each point (x, y). As we shall see in Section 14.4-2, this is equivalent to assuming that the IVP for the DE of Eq. 1 has a unique solution.

construction of the direction field, and hence of the integral curves, as we illustrate in the following example.

EXAMPLE

Draw the direction field for the DE

$$y' = x + y.$$

Use the direction field to sketch several typical integral curves.

solution. We use the isoclines of the DE to sketch the direction field. They are given by

$$f(x, y) = x + y = c.$$

For each fixed value of c this equation defines a straight line in the xy-plane. As c varies we get a family of straight lines. They are isoclines of the differential equation. Several of these lines are drawn in Figure 14.3. In addition, we have drawn

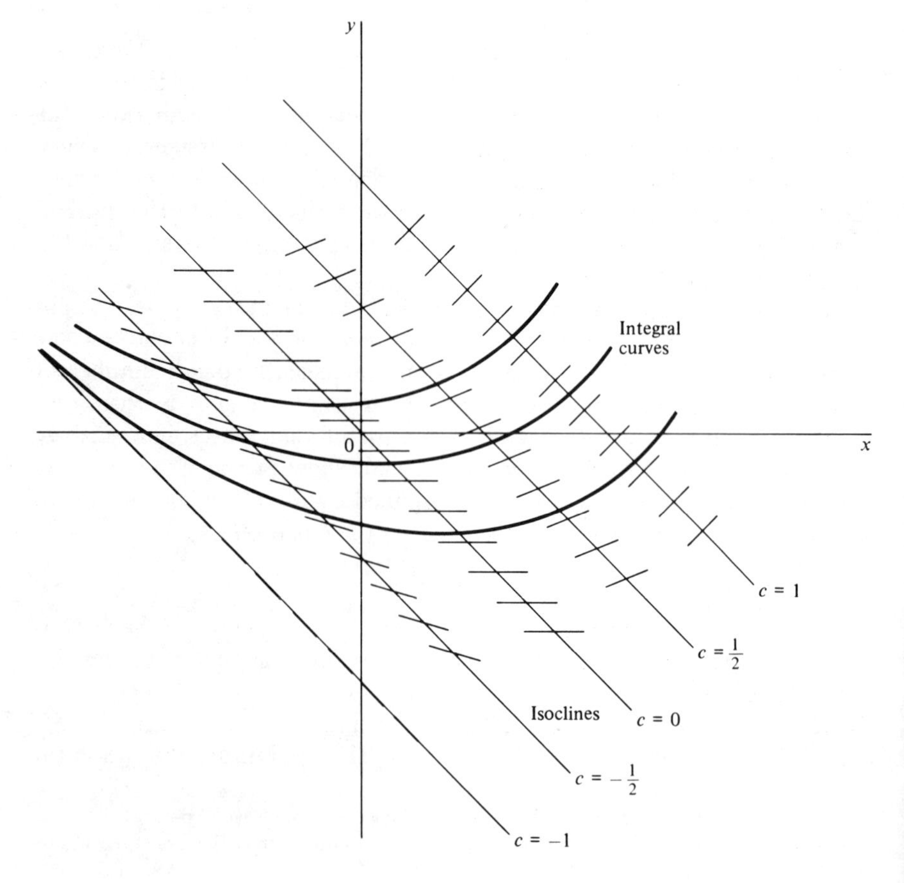

Figure 14.3

line segments of the direction field on these isoclines. For example, $c = 0$ corresponds to the isocline $x + y = 0$. The corresponding slope of any integral curve which intersects this isocline is zero. Thus the line segments in the direction field are horizontal along the line $y = -x$. Similarly, along the isocline $x + y = 1$, corresponding to $c = 1$, the slope of the integral curves must be equal to 1. Then the integral curves make an angle of 45° with the horizontal. After the direction field is sketched for sufficiently many isoclines, the integral curves are easily drawn through the direction field. We have sketched several integral curves in Figure 14.3.

Exercises

1. For each of the following DEs sketch the direction field using the method of isoclines. Use the direction field in each case to sketch typical integral curves.
   (a) $y' - y = 0$ (b) $y' + y = 0$
   (c) $y' + y^2 = 0$ (d) $y' = x^2 + y^2$
   (e) $y' + 2e^y = 0$ (f) $y' + (1 - y)(2 - y) = 0$
   (g) $y' - \sin y = 0$ (h) $y' + y(1 - y)(2 + y) = 0$

   There are many DEs, such as those that arise in studying nonlinear oscillations, where it is possible for more than one integral curve or even no integral curve to pass through a given point. Such situations can arise for the DE $y' = f(x, y)$, when $f(x, y)$ is not defined at the point in question. More specifically, if $f(x, y)$ is the quotient of two continuous functions $f = g(x, y)/h(x, y)$, and if $g(x_0, y_0) = h(x_0, y_0) = 0$, then (x_0, y_0) is called a *critical point* of the DE. The following exercises illustrate some of the possibilities.

2. Consider the DE $yy' + x = 0$. (a) Show that $(0, 0)$ is a critical point. (b) observe that the integral curves have vertical tangents along the x axis (except at the critical point). (c) Find the isoclines. Use them to sketch the direction field. (d) Sketch the integral curves. (e) Compare your sketch with the exact solutions. (f) Show that no integral curve passes through the critical point.

3. Consider the DE $xy' - y = 0$. (a) Show that $(0, 0)$ is a critical point. (b) Find the isoclines. Observe that all the isoclines intersect at the critical point, and hence conclude that the direction field is not defined by the DE at the origin. (c) Show that the isoclines are also integral curves, and hence all the integral curves intersect at the critical point.

4. Find the critical points, and sketch the isoclines and integral curves for the following DEs.

   (a) $y' = \dfrac{2y - x}{x}$ (b) $y' = \dfrac{x + y}{y}$

   *(c) $y' = \dfrac{\sin x}{y}$ *(d) $y' = \dfrac{x + y}{x - y}$

14.4-2 Existence of solutions

We shall now illustrate some of the fundamental difficulties that occur in solving nonlinear differential equations that are not encountered in studying linear differential equations.

In Chapter 2 we showed that if $q(x)$ and $r(x)$ are continuous functions, then the first-order linear DE

$$y' + q(x)y = r(x) \tag{1}$$

always has a solution, and every solution is contained in the general solution

$$y = ce^{-Q(x)} + e^{-Q(x)} \int^x e^{Q(s)} r(s)\, ds, \tag{2}$$

where $Q(x)$ is any indefinite integral of $q(x)$ and c is an arbitrary constant. That is, each solution of Eq. (1) is obtained from Eq. (2) by choosing a specific value of c. Furthermore, we saw that the IVP for Eq. (1) has a unique solution. We shall now present several examples that illustrate that these properties may not be shared by nonlinear DEs. We shall be concerned primarily with nonlinear DEs

$$y' = f(x, y) \tag{3}$$

and the corresponding IVP

$$y' = f(x, y), \qquad y(x_0) = a_0. \tag{4}$$

If a nonlinear first-order DE has a set of solutions depending on an arbitrary constant c, (a one parameter family of solutions), then every solution is not necessarily obtained from this set by choosing a special value of c, as it is for the linear DE. Solutions which are not contained in any one parameter family of solutions are called singular solutions. To illustrate this we consider the DE

$$y' = -\frac{x}{2} + \frac{1}{2}(4y + x^2)^{1/2}. \tag{5}$$

It has the one parameter family of solutions

$$y = cx + c^2 \tag{6}$$

for all values of the constant parameter c, as is easily verified by substituting Eq. (6) into the DE.[6] However, the DE also has the solution $y = -x^2/4$. This solution cannot be obtained from Eq. (6) for any choice of c (see Figure 14.4), or from any other one parameter family of solutions. It is a singular solution.

[6] This DE is discussed further in Exercise 1.

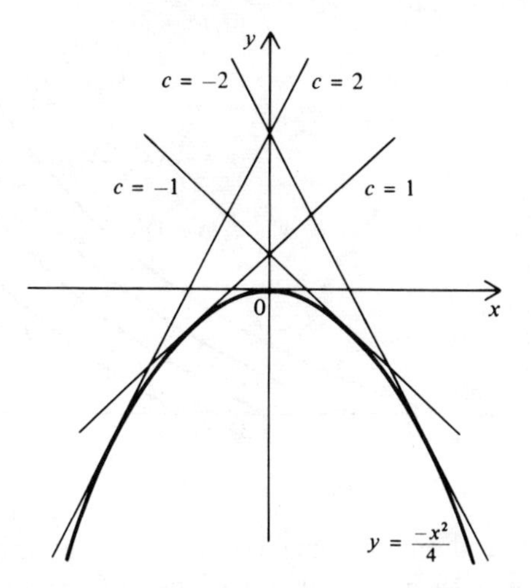

Figure 14.4

The IVP for the linear DE given by Eq. (1) has only one solution. We now show that there are nonlinear DEs for which the IVP has more than one solution. As the reader can verify, the IVP

$$y' = y^{2/3}, \qquad y(0) = 0 \tag{7}$$

has the two solutions

$$y_1 = \frac{1}{27}\,x^3 \qquad \text{and} \qquad y_2 = 0,$$

which are valid for all values of x. Furthermore, we observe that

$$y = \frac{1}{27}\,(x + c)^3 \tag{8}$$

is a solution of the DE for all values of the constant c and for all values of x. The solution y_1 is a special case of Eq. (8) when $c = 0$. We can construct other solutions of the IVP by patching together the solution $y = 0$ and a solution given by Eq. (8) at any point $x = x_1 > 0$. In order for the patched solutions to be equal at $x = x_1$, we must choose $c = x_1$ in Eq. (8). Thus we have an infinite number of solutions of the IVP given by Eq. (7) of the form

$$y = \begin{cases} 0, & \text{for } 0 \le x \le x_1 \\ \dfrac{1}{27}\,(x - x_1)^3, & \text{for } x > x_1, \end{cases}$$

where $x_1 \geq 0$ is any constant (see Figure 14.5).

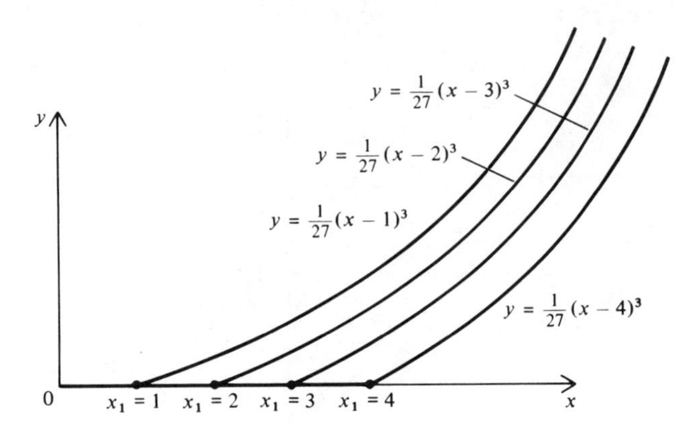

Figure 14.5

The preceding discussion suggests some of the inherent difficulties that are associated with nonlinear equations and are absent from linear equations. Thus there has been a concentrated effort by mathematicians for more than 100 years to determine conditions on $f(x, y)$ for which DE (3) has a solution and for which IVP (4) has only one solution. This has resulted in many theorems, which are called existence and uniqueness theorems. They are usually studied in more advanced courses in differential equations. As an illustration, we now state one of these theorems without proof.[7]

theorem

If in the DE $y' = f(x, y)$ the functions f and f_y are continuous in a region R of the xy-plane, then in some interval containing the point x_0 there is one and only one solution $y(x)$ of the DE for which $y(x_0) = a_0$, where $x = x_0$ and $y = a_0$ is any point in the region R.

We note that this theorem does not indicate the size of the interval containing the point x_0 for which there is a unique solution. This requires additional analysis.

Finally, we recall from Examples 1 and 2 of Section 14.2 that solutions of nonlinear DEs and their IVPs may exist only on a finite interval. Furthermore, for IVPs the length of the interval may depend on the given initial value. This does not occur for the linear equation given in Eq. (1), where $q(x)$ and $r(x)$ are continuous functions for all values of x. Then there is a unique solution for all values of x.

[7] Proofs of this and related theorems are given in "Lectures on Ordinary Differential Equations" by W. Hurewicz, Cambridge, Mass.: MIT Press, 1958.

Exercises

1. (a) Solve DE (5) by introducing the new dependent variable $z = 4y + x^2$. (b) Verify that for each c the curves $y = cx + c^2$ are tangent to the parabola $y = -x^2/4$. Thus the singular solution is tangent at each point to a curve of the one parameter family. It is called an envelope of the one parameter family.

2. (a) Use Exercise 1 to solve the DE

$$y' = 2x + 2\sqrt{x^2 - y}.$$

and thus show that $y = cx - c^2/4$ is a solution for every value of c. (b) Show that $y = x^2$ is a singular solution. (c) Verify that for each fixed c the curve $y = cx - c^2/4$ is tangent to the parabola $y = x^2$, and find the point of tangency. (d) Sketch the solution $y = cx - c^2/4$ for $c = 0, \pm\frac{1}{2}, \pm 1, \pm\frac{3}{2}$, and ± 2. On the same diagram sketch $y = x^2$.

3. (a) Verify that $y_1 = 0$ for all x and $y_2 = (\frac{3}{4}x)^{4/3}$ are solutions of the IVP

$$y' = y^{1/4}, \qquad y(0) = 0.$$

(b) Solve the separable DE in (a) and determine a family of solutions depending on an arbitrary constant c. (c) Patch the solutions found in (b) with the solution y_1 of the IVP at $x = x_1$ to construct an infinite number of solutions of the IVP. (d) Sketch the solutions for $x_1 = 1/2$, 1, and 3/2.

4. Analyze the IVP $y' = (y - 2)^{1/2}$, $y(1) = 2$ along the lines of Exercise 3.

5. Consider the IVP $y' = y^P$, $x > 0$ and $y(0) = 0$. (a) Show that if $0 < P < 1$, then there are an infinite number of solutions of the IVP. (b) For all other nonnegative values of P show that $y = 0$ for all x is the only solution. (c) Discuss the case $P < 0$. (d) For what values of P does $\partial f/\partial y$ satisfy the requirements of the theorem in the text?

6. Find a solution of $y' = y^2$, $y(x_0) = a_0$, where x_0 and a_0 are prescribed constants. For what values of x is the solution valid? Observe that the interval containing a_0 on which the solution is valid depends on the initial value.

7. Find the largest interval containing $x = 0$ for which the solutions of the following IVPs are valid.
   (a) $y' = y^3$, $\quad y(0) = a_0$ $\qquad$ (b) $y' = e^y$, $\quad y(1) = a_0$
   (c) $y' = \sin y$, $\quad y(0) = a_0$ $\qquad$ (d) $y' = y^{4/3}$, $\quad y(0) = a_0$

8. We have seen that the IVP $y' = y^{2/3}$, $y(0) = 0$ has more than one solution. Show that this IVP does not invalidate the theorem in the text.

14.5 APPLICATIONS OF FIRST-ORDER NONLINEAR DIFFERENTIAL EQUATIONS

In the introduction to this chapter we have indicated the importance of the logistic DE in mathematical studies of population dynamics. In Section 14.5-1 we shall solve the logistic equation and discuss the remarkable results it yields concerning the growth of actual populations. In Section 14.5-2 we shall show how the logistic equation and other nonlinear DEs arise in mathematical studies in the behavioral and biological sciences.

The original motivation for the study of nonlinear DEs was the development of the science of mechanics and the discovery of Newton's laws of motion. In Section 14.5-4 we present some simple mechanics problems that lead to first-order nonlinear DEs.

14.5-1 Population dynamics: the logistic equation

We now study in detail the logistic equation

$$y' = \alpha(A - y)y \qquad \text{for } y > 0 \tag{1}$$

where α and A are constants. In the exercises we discuss several other DEs that have been used in population studies. Equation (1) is a separable DE. To solve it, we divide both sides of Eq. (1) by $(A - y)y$ and then take the indefinite integral of the resulting equation. Then as in Section 14.2 we obtain

$$\int^y \frac{dy_1}{(A - y_1)y_1} = \alpha \int^t dz + c_1 = \alpha t + c_1. \tag{2}$$

The integral on the left side of Eq. (2) is evaluated by using the partial fraction expansion

$$\frac{1}{(A - y)y} = \frac{1}{A}\left[\frac{1}{A - y} + \frac{1}{y}\right].$$

Then by substituting this expansion into the integral on the left side of Eq. (2) and evaluating the resulting integrals, we get

$$\ln\left|\frac{A - y}{y}\right| = -A\alpha t + c_2,$$

where $c_2 = -Ac_1$, or, equivalently,

$$\left|\frac{A - y}{y}\right| = e^{-A\alpha t + c_2} = e^{c_2}e^{-A\alpha t} = ce^{-A\alpha t}, \tag{3}$$

where c is an arbitrary constant. It is positive because $e^{c_2} > 0$ for all values of c_2. If $(A - y)/y > 0$, then we can remove the absolute-value signs in Eq.

(3) and solve for $y(t)$. Similarly if $(A - y)/y < 0$, then $|(A - y)/y| = -(A - y)/y$ and Eq. (3) can be solved for y. The solutions in both cases can be combined into the single formula

$$y(t) = \frac{A}{1 + ce^{-A\alpha t}}, \tag{4}$$

if we allow the arbitrary constant c to take on either positive or negative values. Thus Eq. (4) is an explicit solution of the DE.

To solve the IVP for Eq. (1), we require that Eq. (4) satisfy the initial condition $y(0) = a_0$. This yields $c = (A/a_0) - 1$. Substituting this in Eq. (4), we have

$$y(t) = \frac{A}{1 + (A/a_0 - 1)e^{-A\alpha t}} \tag{5}$$

as a solution of the IVP.

We shall now discuss some of the qualitative features of this solution when $\alpha > 0$ and $A > 0$. We recall from Section 14.1 that in the population dynamics problem

$$\alpha(A - y) = \text{birth rate} - \text{death rate}. \tag{6}$$

Thus for small values of the population, $\alpha(A - y) > 0$ and the birth rate exceeds the death rate. Since $e^{-A\alpha t} \to 0$ as $t \to \infty$ when $A\alpha > 0$, we conclude from Eq. (5) that

$$\lim_{t\to\infty} y(t) = \frac{A}{1 + (A/a_0 - 1)\lim_{t\to\infty} e^{-A\alpha t}} = A.$$

A sketch of the solution in Eq. (5) is given in Figure 14.6 for two different initial conditions. They are called logistic curves. Since the population $y(t)$ approaches A as $t \to \infty$ for any choice of the initial population size a_0, the quantity A is called the limiting population that the system can support. If $a_0 > A$, then the population decreases for all time; however, if $a_0 < A$, then $y(t)$ increases for all time.

To further interpret A, we observe from DE (1) that when $y = A$, $y' = 0$. That is, the population is not changing. Thus $y = A$ corresponds to an equilibrium state of the population. Furthermore, when $y = A$ the birth and death rates are equal, as we see from Eq. (6).

The linear DE population model that we studied in Section 2.4.1 gave an unbounded population as $t \to \infty$ when the birth rate exceeded the death rate. The logistic DE, which accounts for decreases in resources and increases in competition as the population increases, gives the more realistic result that the population approaches a saturation or limiting population A as $t \to \infty$.

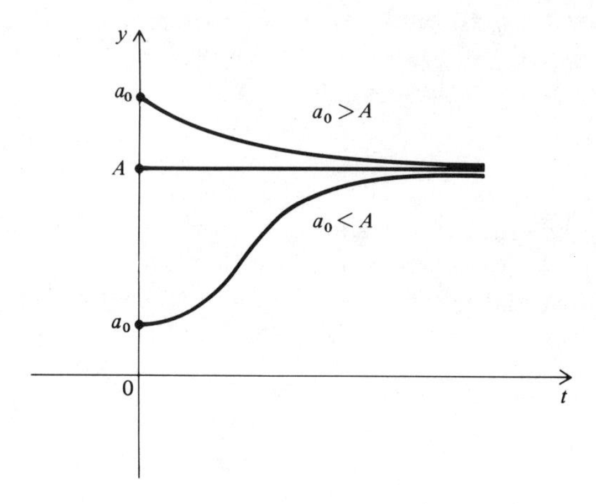

Figure 14.6

The solution given by Eq. (5) of the logistic IVP contains three constants: a_0, A, and α. Their specific values must be determined from quantitative observations of the populations. Typically, the population $y(t)$ is measured at three times t_1, t_2, and t_3 as y_1, y_2, and y_3, respectively. Then a_0, α, and A are determined so that at t_1, t_2, and t_3 the solution given by Eq. (5) equals these values. That is,

$$y(t_i) = y_i = \frac{A}{1 + (A/a_0 - 1)e^{-A\alpha t_i}}, \qquad i = 1, 2, 3. \tag{7}$$

Then Eq. (7) gives three transcendental equations to determine the three constants. As an example of this we shall consider the population of the United States. Approximate census figures are listed in Table 14.1.

TABLE 14.1

t (*year*)	$y(t)$†	t	$y(t)$†
1790	3.9	1890	62.9
1800	5.3	1900	76.2
1810	7.2	1910	92.0
1820	9.6	1920	106.0
1830	13.0	1930	123.0
1840	17.1	1940	132.3
1850	23.2	1950	151.7
1860	31.4	1960	180.0
1870	38.6	1970	205.4
1880	50.2		

† Population in millions.

The objective is to predict the subsequent population of the United States from the early census figures. We choose the reference year ($t = 0$) as 1800 and take t_1, t_2, and t_3 as $t_1 = 0$, $t_2 = 40$, and $t_3 = 80$. They correspond to the years 1800, 1840, and 1880. The corresponding values of y_1, y_2, and y_3 are given in the table. We substitute these values in Eq. (7). The solution of the resulting equations is lengthy and is discussed in Exercise 4. It gives the following values:

$$A = 270.57 \times 10^6, \; (A/a_0 - 1) = 50.051, \text{ and } A\alpha = .03042. \tag{8}$$

Thus the corresponding solution, from Eq. (5) is

$$y(t) = \frac{270.57 \times 10^6}{1 + 50.051e^{-.03042t}}. \tag{9}$$

A graph of this solution is given in Figure 14.7. The census figures are shown by the dots in the figure. The logistic curve gives an accurate prediction of the population except for the period of the Great Depression, 1930–1940, and the post-World War II period. During the depression the U.S. birth rate dropped significantly, and during the postwar "baby boom" of the 1950s the birth rate increased significantly. More sophisticated theories than the logistic equation are required to describe the effects of these random economic and political events on the birth and death rates.

Equation (9) predicts that as $t \to \infty$ the population of the United States will approach the limiting value of $A = 270.57$ million people as compared to the 1970 census figure of 205.4 million. This represents a 31 percent increase in the population.

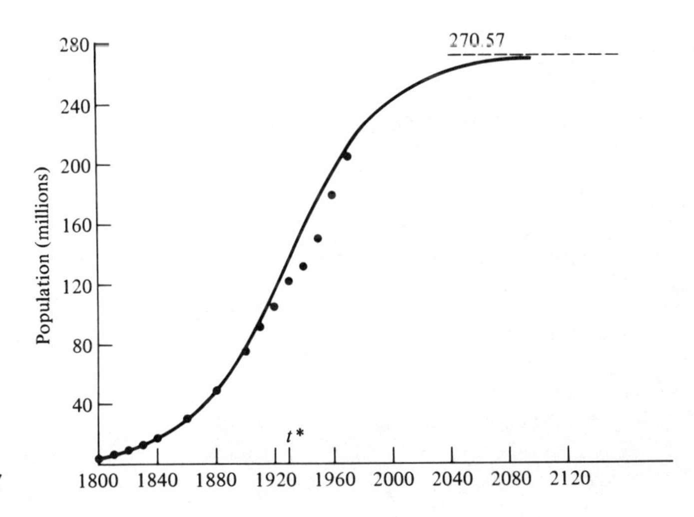

Figure 14.7

For small values of t the logistic curve has positive curvatures, and for large values of t the curvature is negative. The change from positive to negative curvature occurs at the inflection point $t = t^*$, where t^* is the root of the equation $y''(t^*) = 0$.

The reader should show (see Exercise 3) that if $a_0 < A$, then

$$t^* = \frac{1}{\alpha A} \ln \left(\frac{A}{a_0} - 1 \right). \tag{10}$$

Using the values of α, a_0, A given in Eq. (8), we get

$$t^* = \ln \frac{50.051}{.03042} \approx 129.$$

This corresponds to the year 1929 (1800 + 129). Thus, there was a transition in 1929 in the rate of change of population in the United States from an increasing to a decreasing rate. However, the census figures suggest that this transition actually occurred during the 1950s.

The solution given in Eq. (9) was obtained by using the data for the years 1800, 1840, and 1880. If the data from three other years are used to determine the constants a_0, A, and α, slightly different solutions are obtained (see the exercises). In demography, statistical methods that utilize all the available data are used to determine these constants.

The logistic equation does not give a good description of the growth of the world's population. Many attempts have been made to mathematically model the population of the human race. In Exercise 8 we describe one of these models and its striking conclusion.

Exercises

1. Solve each of the following IVPs and sketch the solution.
   (a) $y' = 2y(6 - y)$, $\quad y(0) = 3$ $\qquad$ (b) $y' = y(1 - y)$, $\quad y(0) = \frac{1}{4}$

2. Show that if $A < 0$, $\alpha > 0$, and $a_0 > 0$, then the solution of the logistic equation approaches zero as $t \to \infty$. Observe that $A < 0$ implies that the death rate is greater than the birth rate.

3. (a) Show that if $0 < a_0 < A$, then the logistic curve given by Eq. (5) has a point of inflection at the value $t = t^*$ given in Eq. (10). Show that $y = A/2$ at $t = t^*$. Observe that $y'' > 0$ for $t < t^*$ and $y'' < 0$ for $t > t^*$. Since y'' is the rate of change of y', we say that the population is accelerating for $t < t^*$ and decelerating for $t > t^*$.
   (b) If $a_0 > A$, show that Eq. (5) has no inflection point. Hence the population is always decelerating since $y'' < 0$ for all $t > 0$.

*4. In this exercise we discuss the solution of the three equations given in Eq. (7). Show that if $t_1 = 0$ and $t_3 = 2t_2$, they can be written as

$$y_1 = \frac{A}{B}, \qquad y_2 = \frac{A}{1 + (B - 1)e^{-\beta t_2}}, \qquad \text{and} \qquad y_3 = \frac{A}{1 + (B - 1)e^{-\beta t_3}}, \tag{i}$$

where $B = A/a_0$ and $\beta = A\alpha$. These are now three equations for the three constants A, B, β. From the first expression in Eq. (i), we have

$$A = By_1. \tag{ii}$$

By substituting this result into the last two equations in Eq. (i) and then solving this for the exponentials, show that

$$e^{-\beta t_2} = \frac{By_1/y_2 - 1}{B - 1} = K_2 \qquad \text{and} \qquad e^{-\beta t_3} = \frac{By_1/y_3 - 1}{B - 1} = K_3. \tag{iii}$$

Then by taking logarithms of these two equations and using $t_3 = 2t_2$, show that B must satisfy the quadratic equation

$$\left[\left(\frac{y_1}{y_2}\right)^2 - \frac{y_1}{y_3}\right]B^2 + \left[-\frac{2y_1}{y_2} + \frac{y_1}{y_3} + 1\right]B = 0. \tag{iv}$$

Determine B by solving Eq. (iv). Then A is determined from Eq. (ii). Use Eq. (iii) to show that $\beta = -(\ln K_2)/t_2 = -(\ln K_3)/t_3$.

*5. Verify the values given in Equation (8) by using the results of Exercise 4 and the following census data (from Table 14.1): $t_1 = 0$, $y_1 = 5.3$; $t_2 = 40$, $y_2 = 17.1$; $t_3 = 80$, $y_3 = 50.2$.

*6. In 1844, Verhulst modeled various populations, including that of the United States of America, by using the logistic law. He used census data from 1790 to 1840 and then predicted the population of the United States in 1940. (a) Use the census data from 1790, 1810, and 1830 to determine the constants in the logistic curve. (*Hint:* Use Exercise 4 with $t = 0$ for 1790, $t = 1$ for 1810, and $t = 2$ for 1830. Thus t is measured in generations.) (b) What does your solution predict for the 1940 population? Compare this result with the actual figure.

*7. In addition to human populations the logistic equation has been used as a model for growth of bacterial and fruit fly populations. The following data on the growth of a bacterial colony were taken by H. G. Thornton.[8] The size of the colony is obtained by measuring its area.

Age (days)	0	1	2	3	4	5
Area (cm^2)	.24	2.78	13.53	36.30	47.50	49.40

[8] See *Elements of Mathematical Biology* by A. J. Lotka, New York, N.Y.: Dover, 1956.

Using $t_1 = 0$, $t_2 = 1$, and $t_3 = 2$ and the results of Exercise 4, determine the area at $t = 3$ and 4 days, and compare your results with the experimental values. In the case of bacterial colonies one of the factors that limits growth is the inability of the colony to rid itself of its own waste products, which actually inhibit further growth.

The following exercise discusses a model proposed by Foerster, Mora and Amiot[9] to describe the population of the world. They observed that death rates are declining much faster than birth rates because of the rapid advances in modern medical science. Since the difference between the birth and death rates is increasing, they assumed that y'/y = birth rate − death rate $= ky^n$, where k and n are positive constants to be determined.

8. (a) Solve the IVP $y' = ky^{n+1}$, $y(0) = a_0$, where n is a positive constant. (b) Show that there is always a time t_1 such that, as $t \to t_1$, $y(t) \to \infty$. (c) Foerster, *et al.* fitted the world human population to the solution found in (a) and found that t_1 = A.D. 2026! Thus the world's population will become unbounded in less than 60 years. Discuss the deficiencies of the model that leads to the result of an unbounded population. (*Hint:* Consider what happens to birth rate − death rate as $t \to t_1$. Is this possible?)

9. Another model for population dynamics that produces a limiting population size is the Gompertz equation. It was proposed in 1825 to study mortality rates, and has since been used to study growth problems. It is given by $y' = -k \ln (y/L)$, where L and k are positive constants. (a) Observe that if $y > L$, then $y' < 0$ and y will decrease. (b) Solve this DE.

14.5-2 Applications in the behavioral and biological sciences

In this section we discuss the application of DEs to the study of contagion phenomena and chemical and biochemical kinetics. We begin with a mathematical model that is applicable to the spread of rumors, fads, or diseases.

Spread of rumors and diseases

We shall now derive a simple mathematical model to describe how a rumor spreads through a community. We define the quantities A, $y(t)$, and $x(t)$ by:

A = total population of the community (a constant);
$y(t)$ = the number of people at time t who have heard the rumor, and are available to spread it;
$x(t)$ = the number of people at time t who have not heard the rumor, but could possibly hear it.

[9] *Science*, vol. 12, no. 3, pp. 1291–1295 (1960).

We assume that each person in the community is either in category x or category y. This implies that $x + y = A$. This assumption is not valid, for example, if some people refuse to spread the rumor.

The rate of spread of the rumor at time t is $y'(t)$. It depends on the number of contacts between people in categories x and y. It is customary to assume that the number of contacts is a function of the product xy. This assumption can be justified by appealing to elementary probability theory. A simple model is obtained if we assume that y' is proportional to xy. This gives $y' = \alpha xy$, where α is the proportionality constant. Since $x = A - y$, we can rewrite this equation as

$$y' = \alpha(A - y)y. \tag{1}$$

This is precisely the logistic equation. The solution of this DE, which satisfies the initial condition $y(0) = a_0$ = number of people who have heard the rumor at $t = 0$, is given by (see Section 14.5-1)

$$y(t) = \frac{A}{1 + (A/a_0 - 1)e^{-A\alpha t}}.$$

To determine the largest number of people who have heard the rumor we observe that $a_0 < A$. Thus $y(t)$ is monotonically increasing and $\lim y(t) \to A$ as $t \to \infty$. The interpretation of this result is that all the people will eventually hear the rumor.

DE (1) is also a simple model for the spread of an infectious disease, if we interpret y as the number of infected people (infectives). Then x is the number of susceptible people (susceptibles) who have not been infected. Since $y(t) \to A$ as $t \to \infty$ for any value of a_0 (the initial number of infectives), this model predicts that all susceptible people in the community will eventually catch the disease. This is not a realistic result for many diseases, since it does not include the possibility of only a small number of the susceptibles catching the disease. Consequently, more sophisticated models of the spread of infectious disease involving systems of nonlinear equations have been developed. These models allow for the removal of infectives from the population by quarantine, death, recovery, subsequent immunity, and so forth. A simple model to account for recovery in an epidemic is discussed in Exercise 4.

Chemical and biochemical kinetics

The law of mass action for a single reacting chemical was given in Section 2.4-2 in the form

$$y' = F(y), \tag{2}$$

where $y(t)$ is the concentration of the given chemical and F is a known function. The reaction is said to be of the first order if $F = -ky$, where k is the rate constant. First-order reactions were studied in Section 2.4-2. However, many chemical and biochemical reactions involving only one reacting chemical are not first-order reactions. An example of this is the decomposition of hydrogen iodide into hydrogen and iodine, which can be written as

$$2HI \rightarrow H_2 + I_2,$$

Experimental data indicates that this reaction obeys DE(2) with $F(y) = -ky^2$, where k is a positive constant. It is called a second-order reaction. Another example of a second-order reaction is the reassociation of separated DNA (deoxyribonucleic acid) strands. In this case $y(t)$ is the concentration of separated DNA. The data obtained by studying this reassociation has been used to obtain information about the genetic code (see Exercise 10).

The IVP for a second-order reaction involving one reacting chemical is thus given by

$$y' = -ky^2, \qquad y(0) = a_0, \tag{3}$$

where a_0 is the initial concentration. Since $k > 0$, the minus sign implies that the concentration decreases as the reaction proceeds. The DE in Eq. (3) is separable. Thus the solution of the IVP is easily found to be

$$y = \frac{a_0}{1 + a_0kt}. \tag{4}$$

The solution approaches zero as $t \rightarrow \infty$. The time T at which the concentration has been reduced to half of its initial value (half-life) is given by $T = 1/a_0k$. In contrast to the first-order reaction, where the half-life is $(\ln 2)/k$, the half-life depends on the initial concentration a_0.

Enzyme catalyzed reactions

The rate at which a chemical reaction proceeds can be affected by the products of the reaction or by other chemicals that emerge virtually unchanged at the end of the reaction. These chemicals are called catalysts. For example, high-octane gasoline is obtained by means of a catalyzed reaction, which changes the straight chain structure of certain hydrocarbons in gasoline. Many of the biochemical reactions that take place in man and other living organisms are catalyzed by proteins called enzymes. If the end product of a chemical reaction enhances its progress, then the reaction is said to be autocatalytic. For example the enzyme pepsin which aids in the digestion of food is produced from the protein pepsinogen in an autocatalytic reaction. We discuss autocatalytic reactions in Exercise 11.

We shall now consider the Michaelis-Menten theory for enzyme catalyzed reactions. It is a valid model for many biochemical reactions. In this theory, the reaction involves an enzyme concentration $E(t)$ and a substrate concentration $S(t)$. They react to form an intermediate product called the enzyme-substrate complex, denoted by ES. This complex then decomposes to yield the enzyme E and a new product P. The reaction can be written as $E + S \rightarrow ES \rightarrow E + P$. Thus at the end of the reaction the enzyme emerges unchanged. The total effect of the reaction has been to convert the substrate S into the product P. The Michaelis-Menten IVP,[10] which describes the rate of change of the substrate, is

$$S' = -\frac{VS}{K+S}, \qquad S(0) = a_0. \tag{5}$$

Here V and K are positive constants and a_0 is the prescribed initial value of the substrate concentration S. The constant K involves the reaction rates of the reactions. It is called the Michaelis constant. The constant V depends on the enzyme concentration. As we shall see, it gives an upper bound on the rate at which the reaction can proceed for a fixed enzyme concentration.

Useful information concerning the reaction is obtained by studying the reaction rate $-S'$, which is given by DE (5) as $VS/(K + S)$. It is sketched in Figure 14.8 as a function of S for fixed values of K and V. If S is large compared to K, then we can expand the reaction rate in a Taylor series in S^{-1} to get

$$-S' = V\left[\frac{S}{K+S}\right] = V\left[\frac{1}{1+K/S}\right] = V\left[1 - \frac{K}{S} + \left(\frac{K}{S}\right)^2 + \cdots\right] \approx V.$$

Thus for large S the reaction rate is approximately constant. In particular, it is independent of the substrate concentration S. We see from the figure that V is an upper bound for the reaction rate. If S is small compared to K, then we can expand $S/(K + S)$ in a Taylor's series in S to obtain

$$V\left[\frac{S}{K+S}\right] = \frac{VS}{K} + \cdots.$$

Equation (5) is then reduced to $S' = -VS/K$. This describes a first-order reaction, since the right side of the DE is linear in S. Thus the Michaelis-Menten reaction rate is constant for large substrate concentrations and linear for small concentrations.

[10] See *Enzyme Kinetics* by K. M. Plowman, New York, N.Y.: McGraw-Hill, 1972.

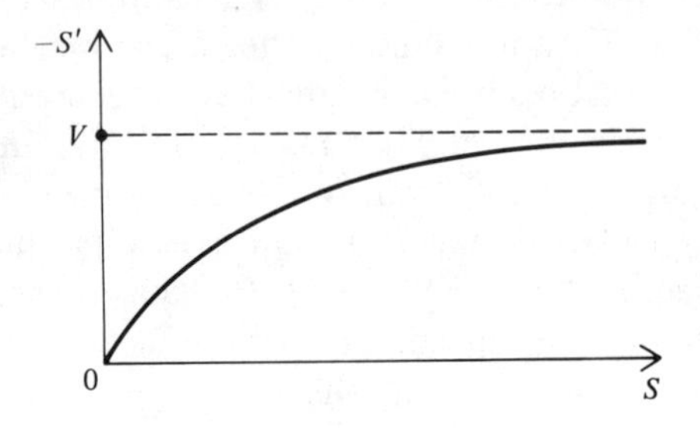

Figure 14.8

To find the actual concentration at any time, we must solve the IVP given by Eq. (5). The DE in Eq. (5) is separable. Therefore by using the techniques of Section 14.2, we can show that an implicit solution of the IVP is

$$K \ln\left(\frac{S}{a_0}\right) + S = a_0 - tV. \tag{6}$$

The constants K and V in Eq. (6) must be determined from experimental data.

Exercises

1. (a) If initially only one person in a population of 1000 has a disease, then find the time at which half the population is infected. Assume that DE (1) describes the spread of the disease, and that α is prescribed. (b) Find the time at which 500 people are infected if the constant α in (a) is doubled.

2. Repeat Exercise 1, assuming that initially 100 people are infectious.

3. Assume that DE (1) describes the spread of a given disease. (a) Find the time at which the number of infectives is $(A - a_0)/2$. The epidemic has run half its course at this time. (b) Suppose that α is doubled. How does this effect your answer to (a)?

4. We can generalize the model discussed in the text by assuming that at any time t a number of people recover from the disease and become susceptible again. This is typical of the common cold, for example. If we assume that the rate of recovery is proportional to the number of infectives, then DE (1) becomes

$$y' = \alpha(A - y)y - \beta y,$$

where $\beta > 0$ is the constant of proportionality. (a) Show that if $A - \beta/\alpha < 0$, then the rate of recovery is rapid, and for any initial number of infectives the number of infectives approaches zero as $t \to \infty$. That is, the epidemic eventually disappears. (b) Show that if $A - \beta/\alpha > 0$, then the number of infectives approaches a nonzero value as $t \to \infty$. Does the number of infectives approach the total population as $t \to \infty$?

5. Consider the epidemic model given in Exercise 4, with a total population of 1000, the rate of recovery $\beta = 1$, and $\alpha = 1$. (a) Find the limiting epidemic size. (b) Find the time for the initial number of infectives a_0 to double. (c) Find the time at which the epidemic has reached half its final size.

6. A chemical dissociates in a second-order reaction into two products. If the concentration decreases by 50 percent in 10 minutes and the initial concentration is 10 moles per liter, find the time at which 90 percent of the chemical has disassociated.

7. The following two measurements of concentration y are made during the dissociation of a chemical in a second-order reaction:

$$t = 5, \qquad y = 20 \text{ moles per liter,}$$

$$t = 20, \qquad y = 10 \text{ moles per liter.}$$

Determine the reaction rate k and the initial concentration of the reactant.

8. To determine whether a given reaction is of the second order, experimental data are used. The concentration y of the reacting chemical is measured at various values of the time t, and a graph of $1/y$ vs. t is plotted. Show, using Eq. (4), that if the reaction is second order, then this graph should be a straight line. What is the slope of this line?

9. We saw that the half-life T for the second-order reaction given in Eq. (4) is dependent on the initial concentration. (a) Show that the time for the concentration to reduce from $a_0/2$ to $a_0/4$ is greater than T. In general, the time required for successive halving of the concentration increases. Hence, the reaction slows down as the concentration decreases. (b) Discuss the difference between the result of (a) and the analogous results for a first-order reaction.

10. DNA is the complex chemical in the cell nuclei, which is responsible for the translation and transcription of genetic information. All organisms from viruses to man have their own type of DNA. DNA is in the form of two very long molecular strands, which are wound helically around each other and attached through cross-links. If DNA is heated, these links break and the double strands dissociate, and single strands result. When the separated strands are allowed to cool, then they re-form double strands. This process of reassociation is found experimentally to involve two different second-order chemical reactions: a very rapid reassociation for some fraction of the DNA and a much slower reassociation for the remaining DNA. Both reassociations obey the DE $y' = -ky^2$ where $y(t)$ is the concentration of the dissociated fractions and k is the same for both reactions. (a) Show that if a_1 and a_2 are the initial concentrations of each fraction of DNA, then $a_1 = (T_2/T_1)a_2$ where T_1 and T_2 are the half-lives for the two reactions. (b) A typical example is calf DNA, where it is found that $T_2/T_1 = 100{,}000$. Conclude that this implies that one fraction of the DNA is 100,000 times more concentrated than the remaining portion. Using other biological knowledge, it can be concluded that the basic genetic information lies in the more concentrated segment and it is repeated in hundreds

of thousands of copies. It is believed that this great number of copies may account for the great reliability of genetic transmittal and the slowness of evolution.

11. In autocatalytic reactions the rate of reaction is increased by the presence of the substance being produced. Suppose that in an autocatalytic reaction a chemical A is being converted into a chemical P with y and z as the concentrations of A and P, respectively. We assume that the rate of change of A (or P) is proportional to the product yz. Show that $y(t)$ satisfies the IVP $y' = -ky(\beta + \alpha - y)$, $y(0) = \alpha$, given that, $z(0) = \beta$, and that $\alpha - y(t) = z(t) - \beta$. What is the final concentration of A? of P?

12. (a) Given that the initial substrate concentration $S(0) = .5$ moles per liter and that $S(1) = .2$ and $S(2) = .1$, determine the constants V and k in the Michaelis-Menten DE given in Eq. (5). (b) Find the half-life of the substrate S.

13. Find the half-life for the substrate S in the Michaelis-Menten reaction.

14.5-3 Orthogonal trajectories

The equation $y = g(x)$ is the explicit representation of a curve in the xy-plane. If we permit g to depend on a parameter c, then

$$y = g(x; c) \tag{1}$$

is an equation for a one-parameter family of curves in the xy-plane. That is, for each value c, Eq. (1) represents a curve. As c varies, we get a family of curves, one for each value of c. For example, for each value of c, $y = cx$ is a straight line of slope c passing through the origin; see Figure 14.9. As c varies, we get a family of straight lines.

More generally, for each value of the parameter c,

$$F(x, y; c) = 0 \tag{2}$$

is the implicit representation of a curve in the xy-plane. As c varies, we get a family of curves. For example,

$$F = x^2 + y^2 - c^2 = 0 \tag{3}$$

is an equation for a family of concentric circles with center at the origin and radius c; see Figure 14.9.

An orthogonal trajectory of a given family of curves such as Eq. (1) or (2) is a curve that intersects each curve of the family at right angles. For example, the straight line $y = c_1x$ of slope c_1 that passes through the origin intersects each of the curves of the family of concentric circles given by Eq. (3) at right angles. (Why?) More precisely, the straight lines are perpendicular to the tangent lines to the circles at the points of intersection. Since these lines are orthogonal trajectories for any value of c_1, they also form a one-parameter family of curves; see Figure 14.9. Conversely, the circles given by Eq. (3) are orthogonal trajectories of the family of straight lines $y = c_1x$.

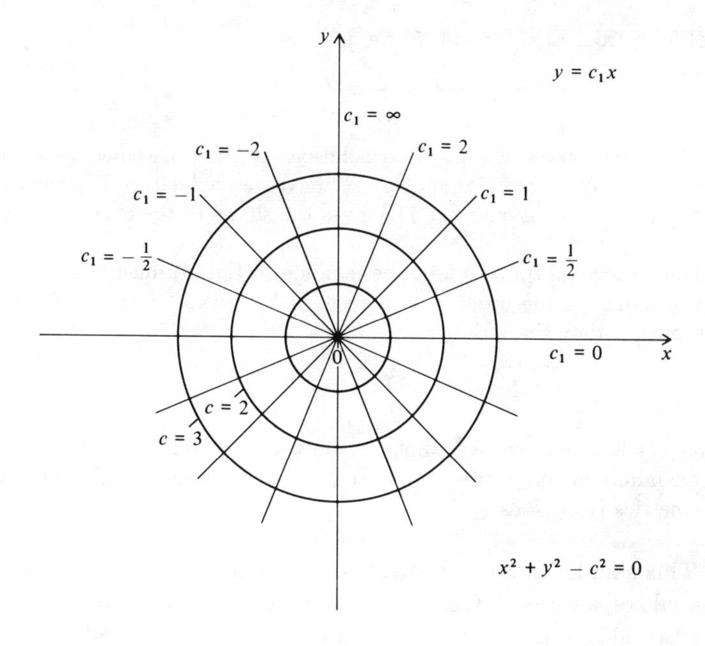

Figure 14.9

Families of curves that are mutually perpendicular are called orthogonal trajectories of each other. An important application of orthogonal trajectories occurs in studying the propagation of waves, such as light waves, sound waves, or electromagnetic waves. It is known that the light rays are perpendicular to the wavefronts. For two-dimensional waves, the light rays are frequently a one-parameter family of straight lines and the wavefronts are a one-parameter family of perpendicular curves. Thus the rays and the wavefronts are orthogonal trajectories of each other.

The orthogonal trajectory problem is: Given a family of curves (for example, the rays), determine a family of orthogonal trajectories (the wavefronts). In the next example we illustrate a general procedure for determining orthogonal trajectories.

EXAMPLE

Determine a family of orthogonal trajectories of the family of straight lines

$$y = c_1x. \tag{4}$$

solution. We first find the slopes of the curves in the given family by differentiating Eq. (4). This gives

$$\frac{dy}{dx} = c_1. \tag{5}$$

Eliminating c_1 from Eqs. (4) and (5) we find that

$$\frac{dy}{dx} = \frac{y}{x}. \tag{6}$$

We note that the curves of Eq. (4) are solutions of DE (6). Hence, by eliminating the parameter c_1 from Eqs. (4) and (5), we have derived a DE that has the given family of curves as solutions. The DE gives the slopes of the curves at the points (x, y).

The orthogonal trajectories corresponding to Eq. (4) must have slopes at the point (x, y) equal to the negative reciprocal of Eq. (6). Thus the orthogonal trajectories must satisfy the DE

$$\frac{dx}{dx} = -\frac{x}{y}. \tag{7}$$

Equation (7) is separable. An implicit solution is $y^2 + x^2 = c$, which is a one-parameter family of concentric circles. They are orthogonal trajectories of Eq. (4), as is geometrically obvious.

This simple example illustrates a general technique for determining orthogonal trajectories. Given a one-parameter family of curves in the form give by Eq. (1) we attempt to eliminate c_1 from Eq. (1) and its derivative $y' = g'(x; c_1)$. If this is possible, this gives a DE in the form $y' = f(x, y)$ for the slope at each point (x, y) of the curves given by Eq. (1). The orthogonal trajectories must then satisfy the DE

$$y' = -\frac{1}{f(x, y)}.$$

Exercises

1. In each of the following families of trajectories find the family of orthogonal trajectories. Sketch typical members of each family on the same graph.
   (a) $x - 2y = c$ (b) $y = cx^2$
   (c) $y = c/x$ (d) $y = ce^{-x}$
   (e) $y = cx^3$ (f) $y = e^{cx}$
2. If a one-parameter family of curves is given in the implicit form $F(x, y; c) = 0$, then implicit differentiation can be used to determine the orthogonal trajectories. For example, $F(x, y; c) = y^2 + x^2 = c$ can be treated by differentiating with respect to x, treating y as a function of x. This yields $2yy' + 2x = 0$. Thus $y' = -x/y$ is the DE that the given family satisfies. Use this to determine the orthogonal trajectories.
3. Determine the orthogonal trajectories for the following equations. Sketch the given family and these trajectories on the same graph.
   (a) $y^2 = cx$ (b) $y^2 = x^2 - c$
   (c) $x^2 + 4y^2 = c$ (d) $x^2 + y^2 = cx$

14.5-4 Rectilinear motion of a particle

In elementary mechanics, the motion of a particle or of a rigid body is described by Newton's second law. If the particle moves in a straight line (rectilinear motion) and $y(t)$ is its displacement along the line, then Newton's law is

$$F = my'' = mv' \tag{1}$$

where m is the mass and $v(t) = y'(t)$ is the velocity of the particle. In general, the force F is a function of t, y, and v.

In this section we consider the motion of a falling body through a resisting medium. Examples of this are the return of a space capsule through the earth's atmosphere or the sinking of a heavy macromolecule, such as a protein, through a viscous biological fluid. When a body moves through a fluid, such as air, its motion is resisted by the fluid. The force of resistance R usually depends on the velocity of the particle and not on its position. For small velocities, R is approximately proportional to the velocity; that is, $R = kv$, where k is the proportionality constant. The motion of a particle for which $R = kv$ was studied in Section 3.3-3.

The precise relationship between R and v for larger values of v is not completely known. For intermediate velocities (below the speed of sound), experimental results suggest that $R = kv^2$, where k is a positive constant that depends on the shape of the body and the properties of the fluid.

We shall now study the motion of a particle of mass m falling through the air, where the resistance force is given by $R = kv^2$. We choose the downward displacement $y(t)$ measured from the initial altitude as positive. Thus we have $y(0) = 0$. Then, by referring to Figure 14.10, we find that the resultant force acting on the particle at time t is

$$F = mg - kv^2. \tag{2}$$

Here mg is the weight of the particle, which we assume to be constant.

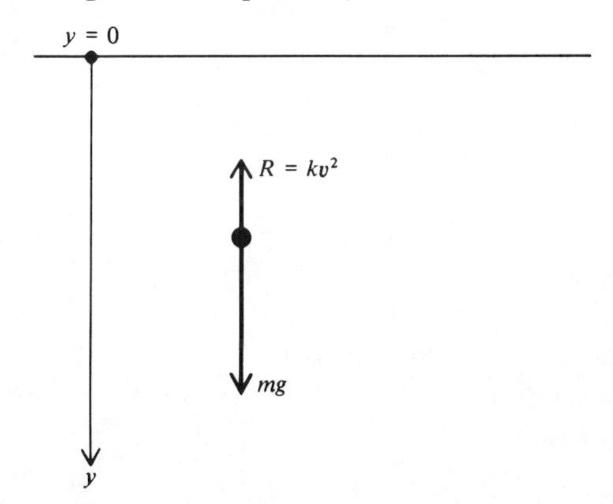

Figure 14.10

The minus sign occurs in Eq. (2) because in opposing the motion the resisting force acts upward. If the particle were rising instead of falling, then the resisting force would act downward.

The equation of motion of the particle is obtained by substituting Eq. (2) into Eq. (1). This gives the nonlinear DE

$$v' = g - \frac{k}{m} v^2. \tag{3}$$

If the initial downward velocity at $t = 0$ of the particle is given as $v_0 \geq 0$, then the IVP for the falling particle is to determine a solution of Eq. (3) that satisfies the initial condition,

$$v(0) = v_0. \tag{3}$$

When the IVP is solved for $v(t)$, the position of the particle is determined by solving the IVP $y' = v$, $y(0) = 0$.

Equation (3) is separable. Hence an implicit solution is given by

$$\int^v \frac{dv_1}{g - (k/m)v_1^2} = \int^t dt_1 = t + c_1, \tag{5}$$

where c_1 is an arbitrary constant and v_1 is a dummy integration variable. The integral on the left side of Eq. (5) is easily evaluated by using a partial fraction expansion. This gives

$$\ln \left| \frac{g + \alpha v}{g - \alpha v} \right| = 2\alpha(t + c_1), \qquad \alpha = \left(\frac{gk}{m}\right)^{1/2}.$$

We solve this equation for v and use the initial condition in Eq. (4) to evaluate c_1. This gives the solution of the IVP as,

$$v = \frac{g}{\alpha}\left[\frac{c - e^{-2\alpha t}}{c + e^{-2\alpha t}}\right], \qquad c = \frac{1 + \alpha v_0/g}{1 - \alpha v_0/g}. \tag{6}$$

We now interpret the solution given by Eq. (6). Since $\alpha > 0$ and $e^{-2\alpha t} \to 0$ as $t \to \infty$, we have

$$\lim_{t \to \infty} v = \frac{g}{\alpha} = \left(\frac{gm}{k}\right)^{1/2}.$$

That is, as $t \to \infty$, the solution approaches the limiting value $V = g/\alpha = (gm/k)^{1/2}$. The quantity V, which is called the terminal velocity, is independent of the initial velocity v_0. Thus as the particle falls, its velocity approaches V. If $v_0 > V$, then the velocity of the particle decreases monotonically to V and conversely if $v_0 < V$, then $v(t)$ increases monotonically to V. (See Exercise 6.)

In Section 3.3-3 we studied the falling particle where the resistance was proportional to v, that is $R = k_1 v$. Then the limiting velocity is gm/k_1. Thus different resistance laws result in substantially different terminal velocities.

The position of the particle $y(t)$ is determined by integrating the DE $y' = v$, where v is given by Eq. (6); see Exercise 5.

Exercises

1. (a) A skydiver jumps from an airplane and immediately opens her chute. It is observed experimentally that the coefficient of resistance k depends on the cross-sectional area of the parachute and is approximately 1 for a parachute that is 30 ft in diameter. Use Eqs. (4) and (5) to determine the terminal velocity of the skydiver. The skydiver weighs 128 lbs. (b) Suppose that instead of immediately opening her chute the skydiver falls freely. In this case the value of k is approximately .1. Determine the terminal velocity and compare with the result of (a).

2. An object weighing 1 pound is projected downward with an initial velocity of 1500 ft/sec from a height of 8 mi above the earth. The resistance force is $R = 10^{-4}v^2$, where v is the velocity. Determine the velocity of the object as a function of time.

3. (a) If the external force F on a particle is a function of only the velocity v, then Newton's second law $F(v) = mv'$ is a first-order DE for the velocity $v(t)$. Show that if the initial velocity is $v(0) = v_0$, then an implicit solution for the velocity is given by

$$t = m \int_{v_0}^{v} \frac{dv_1}{F(v_1)}.$$

(b) Use the result of (a) to determine $v(t)$ if $F(v) = v^n$, where n is a constant.

4. (a) Consider the motion of a particle that has an initial velocity v_0, resisted by a force $R = kv^n$, where k and n are constants. Assuming that this is the only force acting, determine the largest value of n for which the resisting force causes the body to stop in a finite period of time. (b) Find an expression for the stopping time. (c) Use the fact that $y' = v$ to find the largest value of n so that the total distance traveled by the particle is finite.

5. (a) Integrate Eq. (6) in the special case $v_0 = 0$ and use the initial condition $y = 0$ at $t = 0$ to obtain the displacement

$$y(t) = \frac{V^2}{g} \ln\left(\frac{e^{\alpha t} + e^{-\alpha t}}{2}\right).$$

(b) Suppose that $v_0 \neq 0$ in Eq. (6). Integrate Eq. (6) and use the initial condition $y = 0$ at $t = 0$ to obtain

$$y(t) = \frac{V^2}{g} \ln \left\{ \frac{e^{\alpha t} + e^{-\alpha t}}{2} + \frac{v_0}{V} \left[\frac{e^{\alpha t} - e^{-\alpha t}}{2} \right] \right\} \tag{i}$$

6. Show that if $v_0 \neq V$, then the velocity $v(t)$ of a falling particle is either monotonically increasing or monotonically decreasing to V as $t \rightarrow \infty$, in two ways. First use the exact solution given in Eq. (6). Then sketch the direction field directly from the DE (3) and draw the same conclusion.

chapter 15

NUMERICAL METHODS

15.1 INTRODUCTION

We have seen in previous chapters that in principle it is always possible to solve the initial value problem explicitly for a single linear differential equation with constant coefficients and for a system of such equations. However, when the order n of the DE or of the system is large, then there are practical difficulties involved in obtaining this solution: The roots of an nth-degree characteristic equation must be determined, and a system of n simultaneous linear algebraic equations for the n constants in a general solution must be solved to satisfy the initial conditions. If n is sufficiently large, then it is usually impossible to determine the characteristic roots and the constants explicitly. Furthermore, we have seen that explicit solutions can be obtained only for very special second-order linear equations with variable coefficients and first-order nonlinear equations. Thus, in general, it is not possible to solve IVPs (and BVPs) explicitly. Consequently, mathematicians and scientists have developed methods to construct approximate solutions of differential equations.

Before describing some of these methods, we indicate how the need for approximate methods arises naturally in a typical application. In Sections 14.1 and 14.5.1 we discussed the growth of an isolated population according to the logistic equation

$$y' = \alpha(A - y)y. \tag{1}$$

However, the population is not isolated in many studies of population dynamics; members are added or subtracted by immigration and emigration (see Section 3.3-2). For these problems we must modify the logistic equation by adding an immigration term $B(t, y)$ to the right side of Eq. (1). The form of B depends upon the specific problem that is under consideration and usually cannot be determined precisely. We assume for illustrative purposes that B is of the form

$$B(t, y) = \beta e^{-ky} \tag{2}$$

where $\beta > 0$ and $k > 0$ are constants. B depends on the size y of the population. Since B is small for large values of y, we see that there is less immigration when the population is large than when it is small. This decrease in immigration for a human population may be caused, for example, by the imposition of quotas, or by overcrowding of the region and a resulting deterioration of the favorable conditions that had attracted immigrants. Thus we consider the modified IVP

$$y' = \alpha(A - y)y + \beta e^{-ky}, \qquad y(0) = a_0, \tag{3}$$

for population dynamics with immigration.

The DE is separable, but it is difficult to evaluate the integrals necessary to solve it by the methods used for such equations. Thus we shall develop methods for approximating the solution of the IVP given by Eq. (3) and, more generally, for the IVP

$$y' = f(x, y), \qquad y(x_0) = a_0, \tag{4}$$

where $f(x, y)$ is a specified function and x_0 and a_0 are specified numbers. These methods can be extended to obtain approximate solutions of systems of first-order DEs and single DEs of higher order.

In this chapter we shall briefly describe a class of techniques for obtaining approximate solutions. They are called difference methods.[1] These methods are based on approximating all derivatives in the DE by algebraic expressions. The impetus for the development of these methods in the nineteenth and early twentieth centuries came principally from problems in astronomy. In particular, astronomers had to solve complicated systems of ordinary differential equations in order to compute the orbits

[1] The reason for the name, difference methods, will become apparent later in the section.

of planets. Approximate methods were invented because these equations could not be solved exactly.

Difference methods are particularly well suited for high-speed electronic digital computers. Calculations that can now be routinely completed in minutes or even seconds on modern computers formerly required months or years when they were performed by hand or mechanical desk calculators. There is a continuing effort to devise new and improved approximate methods for solving DEs. Most computing centers have in their computing systems, programmed subroutines of these methods.

For simplicity of presentation we shall apply some of these methods to the specific IVP

$$y' = y + 1, \qquad y(0) = 0. \tag{5}$$

We can then easily evaluate and compare the accuracy of the methods because the exact solution of this problem is

$$y = e^x - 1. \tag{6}$$

The methods are applicable to the more general IVP given by Eq. (4). They can also be generalized to higher-order DEs and to systems of DEs.

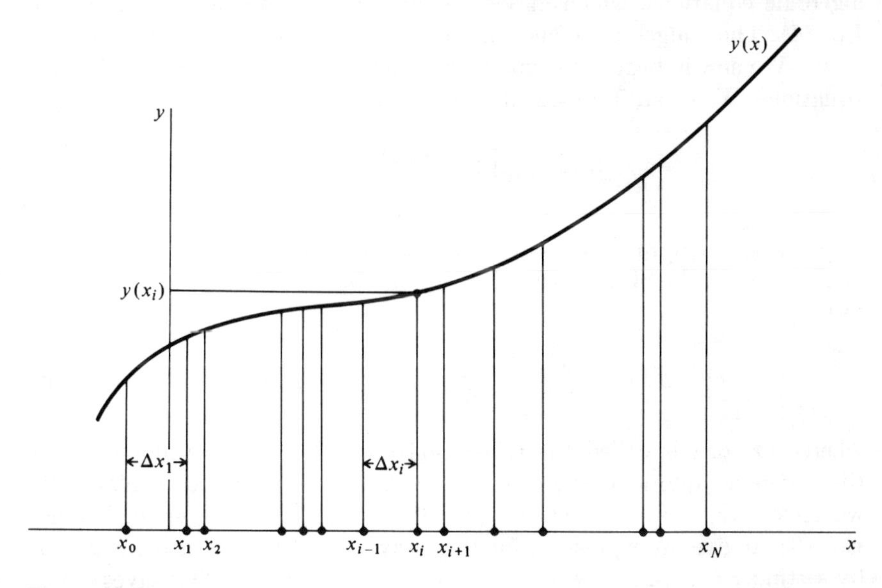

Figure 15.1

The first step in obtaining a numerical or approximate solution by a difference method is to partition the interval $[x_0, x_N]$ on which the solution is desired into a finite number of subintervals by the points $x_0 < x_1 < x_2 < \cdots < x_N$; see Figure 15.1. The points are called the *mesh points* or the *grid points*. The spacings between the points are

$$\Delta x_i = x_i - x_{i-1}, \qquad i = 1, 2, \ldots, N.$$

They are called the mesh spacings. In many applications the mesh points are spaced uniformly. This means that

$$\Delta x_i = h = \text{constant}, \qquad \text{for } i = 1, 2, \ldots, N.$$

Then the mesh points are given by

$$x_i = x_0 + ih, \qquad \text{for } i = 1, 2, \ldots, N. \tag{7}$$

For simplicity, we shall assume throughout this chapter that the mesh points are spaced uniformly, although the methods we shall develop also apply to variable mesh spacing.

At each mesh point x_i we seek a number u_i, which is an approximation to the value of the solution $y_i = y(x_i)$ at the point x_i. The set of numbers $\{u_i\} = u_0, u_1, u_2, \ldots, u_N$ is called a *mesh function* since $\{u_i\}$ is only defined for the mesh points. The mesh function is a *numerical* or approximate *solution of the IVP*. The numbers $\{u_i\}$ are determined from a set of algebraic equations, which in some sense approximate the IVP given by Eq. (4). These algebraic equations are called the difference equations.

We now indicate an elementary idea that is used to derive difference equations. We recall from calculus that $y'(x)$ is defined by

$$y'(x) = \lim_{\Delta x \to 0} \left[\frac{y(x + \Delta x) - y(x)}{\Delta x} \right]. \tag{8}$$

The bracketed term in Eq. (8) is called the difference quotient of $y(x)$ at the point x. We observe that Eq. (8) can be rewritten in the equivalent form

$$y'(x) = \frac{y(x + \Delta x) - y(x)}{\Delta x} + e(x, \Delta x) \tag{9}$$

where $e(x, \Delta x)$ is called the truncation error. It gives the amount that the difference quotient at x deviates from the derivative at x. Naturally we must have $e(x, \Delta x) \to 0$ as $\Delta x \to 0$ for Eq. (8) to be valid. We use Eq. (9) to give an expression for the derivative at the mesh point $x = x_i$ by setting $x = x_i$, $\Delta x = h$, and $x + \Delta x = x_i + h = x_{i+1}$. This gives

$$y'(x_i) = \frac{y_{i+1} - y_i}{h} + e(x_i, h) \tag{10}$$

because $y(x_i) = y_i$ and $y(x_i + h) = y(x_{i+1}) = y_{i+1}$. A difference approximation to $y'(x_i)$ is obtained by neglecting the error in Eq. (10). Then we get

$$y'(x_i) \approx \frac{y_{i+1} - y_i}{h}. \tag{11}$$

The right side of Eq. (11) is now an approximation to the derivative. The geometric interpretation of this approximation is indicated in Figure 15.2. Since $e(x_i, h) \to 0$ as $h \to 0$, the approximation should improve as $h \to 0$.

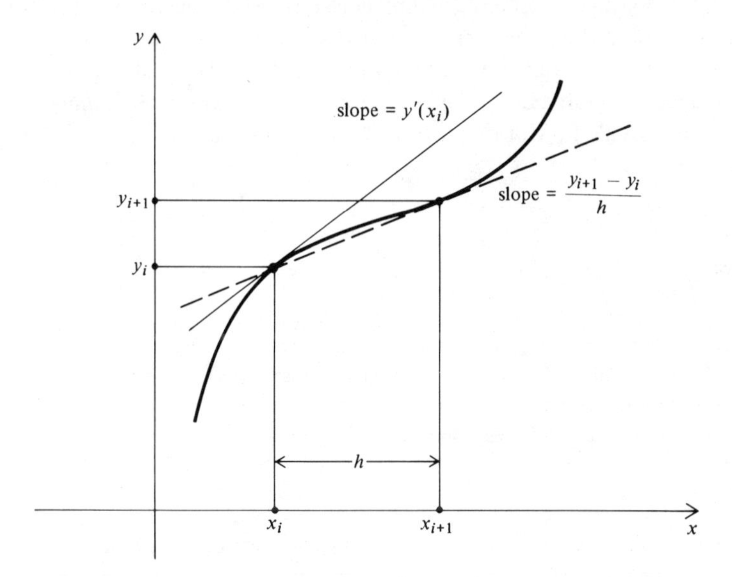

Figure 15.2. The slope $y'(x_i)$ of the tangent line to the curve at the point x_i is approximated by the slope of the dashed line (secant line) connecting the points y_i and y_{i+1}.

We use the approximation in Eq. (11) for y' to obtain a difference approximation to the DE, $y' = f(x, y)$ at the mesh point x_i. This gives

$$\frac{u_{i+1} - u_i}{h} = f(x_i, u_i), \qquad i = 0, 1, 2, \ldots, N - 1. \tag{12}$$

We have substituted the approximations u_i for y_i and u_{i+1} for y_{i+1} in Eq. (12) because Eq. (12) is an approximation to the DE. That is, we have neglected the error introduced by replacing the derivative in the DE by the difference quotient. By rearranging terms we reduce Eq. (12) to

$$u_{i+1} = u_i + f(x_i, u_i)h, \qquad i = 0, 1, \ldots, N - 1. \tag{13}$$

For obvious reasons, the algebraic equations given in Eq. (13) are called difference equations. In particular, they are called the Euler difference equations.

In the next section we shall derive the Euler difference equations by another technique. This technique will be generalized to obtain more accurate difference approximations to the DE, in succeeding sections. In general, the difference equations should be such that as the mesh spacing $h \to 0$, that is, as the number of mesh points increase indefinitely, their solution approaches the exact solution. Furthermore, the difference equations should be "easily solvable" for the mesh function $\{u_i\}$.

For detailed mathematical discussions of the accuracies of difference methods and the development of more sophisticated methods, the reader is referred to specialized texts on numerical methods and analysis such as: E. Isaacson and H. B. Keller, *Analysis of Numerical Methods*, Wiley, New York, 1966; P. Henrici, *Elements of Numerical Analysis*, Wiley, New York, 1964.

Exercises

1. Place a uniform mesh with mesh width $h = 1/4$ on the interval $[0, 1]$. For each of the following functions compare the actual value of the derivative with the approximate value, given by the difference quotient, at each point of the mesh. Why must the mesh point $x = 1$ be excluded?
   (a) x (b) $x^2 + 1$
   (c) $\sin x$ (d) e^x

2. (a) Consider $f(x) = x^2 + 1$ on the interval $[0, 1]$. Compare the actual derivative with the difference quotient approximation using a mesh with uniform spacing $h = .1$. (b) Compare your result with that of Exercise 1(b), for the point $x = 1/2$. Which approximation is more accurate?

3. The Euler difference approximation is sometimes called a forward difference, since we have $y'(x_i) \approx (u_{i+1} - u_i)/h$ and thus the derivative at x_i depends on the approximate value of y at the mesh point immediately to the right of x_i. A *backward difference* approximation is $y'(x_i) \approx (u_i - u_{i-1})/h$. (a) Sketch a diagram that illustrates this approximation. (b) Using this backward difference approximation find approximate values for the derivative of x and $x^2 + 1$ on the interval $[0, 1]$ with equally spaced meshes of spacing $h = .25$ and $h = .1$.

4. Use the Taylor series

$$f(x) = f(b) + f'(b)(x - b) + \frac{f''(b)}{2!}(x - b)^2 + \cdots$$

   to derive a series expression for the truncation error $e(x, h)$ in Eq. (10). (*Hint:* Let $b = x_i$ and $x = x_{i+1}$ in the Taylor expansion.)

5. Repeat Exercise 4 for the backward difference truncation error.

15.2 EULER'S METHOD

Euler's equations and many other difference equations can be derived by the following technique. The DE $y' = f(x, y)$ is first integrated from any mesh point x_i to the neighboring mesh point $x_{i+1} = x_i + h$. This gives

$$\int_{x_i}^{x_{i+1}} y'(x)\, dx = \int_{x_i}^{x_{i+1}} f(x, y(x))\, dx.$$

By using

$$\int_{x_i}^{x_{i+1}} y'\, dx = y(x_{i+1}) - y(x_i),$$

in this equation, we get

$$y(x_{i+1}) = y(x_i) + \int_{x_i}^{x_{i+1}} f(x, y(x))\, dx. \tag{1}$$

The difference equations are then obtained by numerically approximating the integral in Eq. (1). A numerical scheme for approximating an integral is called a quadrature formula.

There are a variety of quadrature formulas of increasing accuracy and complexity. The simplest quadrature formula, which we shall now describe, is the *rectangle rule*. The integral

$$I = \int_{x_i}^{x_{i+1}} F(x)\, dx, \tag{2}$$

where $F(x)$ is a prescribed function, can be interpreted as the area under the curve $y = F(x)$ between the points x_i and x_{i+1}.

We approximate this area, see Figure 15.3, by the area of the rectangle whose height is the ordinate of the left endpoint, $F(x_i)$, and whose base is $h = x_{i+1} - x_i$. This is the hatched area in Figure 15.3. Thus

$$I \approx F(x_i)h. \tag{3}$$

The difference between I and $F(x_i)h$ is called the error. It is equal to the area of the shaded region in the figure. If $F(x)$ does not vary too rapidly in the interval (x_i, x_{i+1}) and h is small, then this error is small. We apply the approximation in Eq. (3) to the integral in Eq. (1) with $F(x) = f(x, y(x))$. This gives

$$\int_{x_i}^{x_{i+1}} f(x, y(x))\, dx \approx f(x_i, y(x_i))h.$$

We substitute this approximation in Eq. (1). Then, by using u_i and u_{i+1}

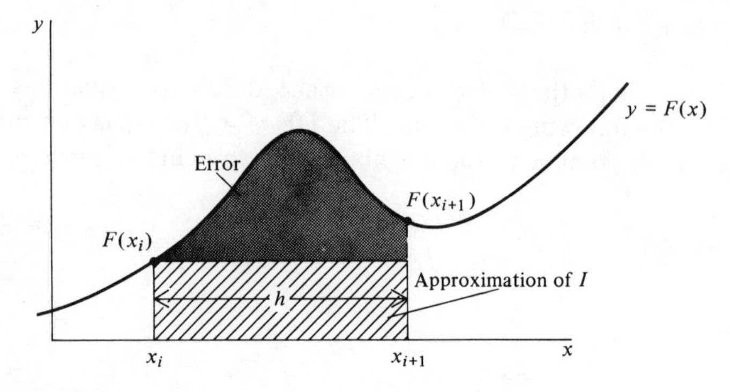

Figure 15.3

in place of the exact values $y(x_i)$ and $y(x_{i+1})$, we get

$$u_{i+1} = u_i + f(x_i, u_i)h, \qquad i = 0, 1, \ldots, N - 1. \tag{4}$$

The initial condition yields

$$u_0 = y(x_0) = a_0. \tag{5}$$

The algebraic equations given in Eqs. (4) and (5) are precisely the Euler difference equations.

The numbers $u_1, u_2, \ldots, u_N$ are determined successively by setting $i = 0, 1, \ldots, N - 1$ in Eq. (4) and using Eq. (5). Thus we have by setting $i = 0$ in Eq. (4),

$$u_1 = u_0 + f(x_0, u_0)h. \tag{6}$$

Since $u_0 = a_0$ is a specified number and the function $f(x, y)$ is specified, we can evaluate $f(x_0, u_0)$ and, consequently, u_1 from Eq. (6). Similarly, by setting $i = 1$ in Eq. (4), we get $u_2 = u_1 + f(x_1, u_1)h$. Thus u_2 is evaluated from this equation because u_1 is already given by Eq. (6), and therefore we can compute $f(x_1, u_1)$. The remaining u_i, $i = 3, 4, \ldots, N - 1$, are successively computed in this fashion. An effective way of illustrating the sequential nature of such computations is by means of a flow chart. The flow chart in Figure 15.4 summarizes Euler's method.

We can interpret Euler's method geometrically by recalling that the DE $y' = f(x, y)$ defines, for each point in the xy-plane, the slope y' of the tangent line to the solution curve $y(x)$. Thus at the point $(x_0, y(x_0)) = (x_0, a_0)$, the DE gives the slope as $y'(x_0) = f(x_0, a_0)$. Hence we can construct the tangent line at x_0 (see Figure 15.5). We move along the tangent line to obtain the approximation u_1 of $y(x_1)$. Thus

$$y(x_1) \approx u_1 = u_0 + y'(x_0)(x_1 - x_0) = u_0 + f(x_0, u_0)h.$$

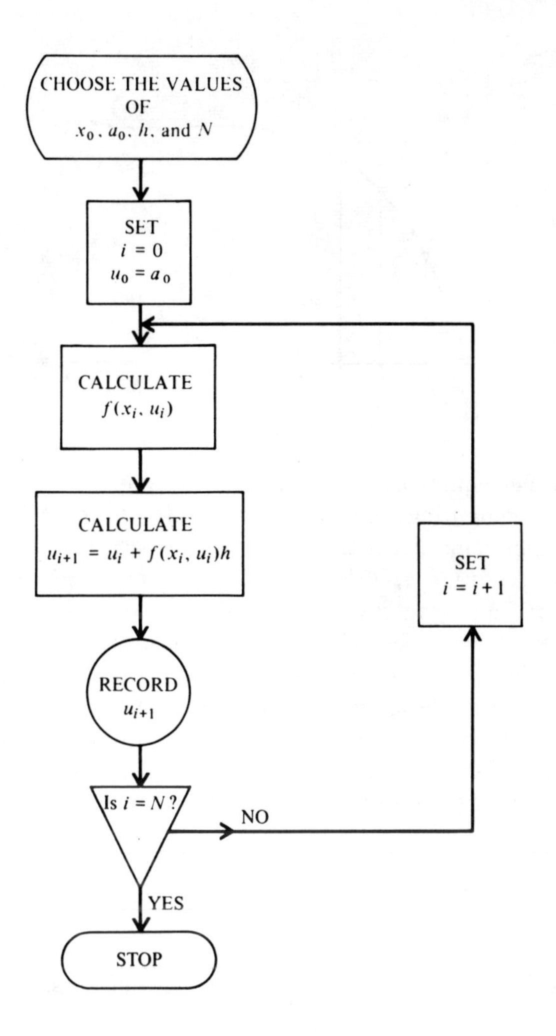

Figure 15.4

We compute an approximate value u_2 of $y(x_2)$ by evaluating the slope $y'(x_1) = f(x_1, y_1) \approx f(x_1, u_1)$ at $x = x_1$ and then moving along the line through (x_1, u_1) with this slope. This gives

$$u_2 = u_1 + y'(x_1)(x_2 - x_1) = u_1 + f(x_1, u_1)h.$$

In general, we obtain

$$u_{i+1} = u_i + f(x_i, u_i)h,$$

which is the Euler difference equation.

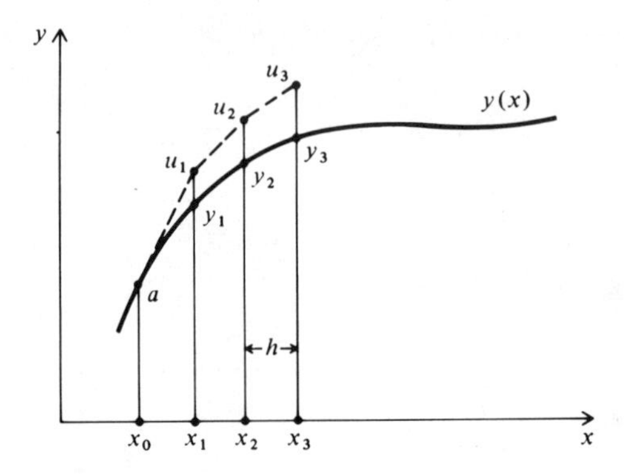

Figure 15.5

The Euler equations given by Eqs. (4) and (5) provide a simple procedure for computing approximations u_i to the exact solution. The error in the approximation is the difference between the exact solution at $x = x_i$ and the solution u_i of Eq. (4). It seems clear from the method of approximating the integral given in Eq. (3) that as the mesh width $h \to 0$, the solution u_i of the Euler difference equations, Eq. (4), must approach the exact solution. This convergence of the approximate to the exact solution is rigorously established in textbooks on numerical analysis.

Most of the computational work that is required to determine each u_i is expended in evaluating the numbers $f(x_i, u_i)$. If $f(x, y)$ is a complicated function, then this evaluation is time-consuming.

EXAMPLE 1

Obtain approximations to the solution of the IVP

$$y' - y = 1, \qquad y(0) = 0 \tag{7}$$

on the interval $0 \leq x \leq 1$ by using Euler's difference equations.

solution. We observe that for this IVP, $f(x, y) = y + 1$. Since $x_0 = 0$, the mesh points are defined by $x_i = ih$, $i = 0, 1, \ldots, N$, where h is the mesh width. The Euler difference equations are

$$u_{i+1} = u_i + f(x_i, u_i)h = u_i + (u_i + 1)h \tag{8}$$

where we have used the fact that $f(x_i, u_i) = u_i + 1$.

We can now evaluate the quantities $u_1, u_2, \ldots, u_N$ by successively setting $i = 0, 1, \ldots, N - 1$ in Eq. (8). To actually perform these calculations we must first prescribe a numerical value for the mesh width h. Furthermore, we must decide on the number of decimal places to be used in the computations. We first choose $h = .5$. Thus the mesh consists of the mesh points $x_0 = 0$, $x_1 = x_0 + h = .5$, $x_2 = x_1 + h = .5 + .5 = 1.0$. Furthermore, we stipulate that in the computations we shall round all numbers to two decimal places—that is, two digits to the right of the decimal point. For example the number 2.134 is rounded to 2.13 and the number 2.135 is rounded to 2.14.

The errors that occur due to rounding of decimals are called *roundoff errors.* These errors play an important role in numerical computations. We shall illustrate the effects of decreasing the mesh width and increasing the number of decimal places that are carried in the computations in several examples in this chapter.

Thus with $h = .5$, we have from Eq. (8),

$$u_{i+1} = u_i + .5(u_i + 1) = 1.5u_i + .5. \tag{9}$$

From the initial condition $y(0) = 0$, we obtain $u_0 = 0$. Then by substituting this value in Eq. (9) with $i = 0$, we obtain $u_1 = 1.5u_0 + .5 = .5$. Setting $i = 1$ in Eq. (9) and using $u_1 = .5$, we get

$$u_2 = 1.5u_1 + .5 = 1.5(.5) + .5 = .75 + .5 = 1.25.$$

To assess the accuracy of the numerical solution $\{u_0, u_1, u_2\}$ we evaluate the exact solution $y = e^x - 1$ at the mesh points x_0, x_1, and x_2 by using tables for e^x. To avoid the introduction of additional errors due to rounding in the evaluation of the exact solution $y(x_i)$, the values of e^x were rounded to three decimal places instead of two. The *relative error* E_i of the approximation at the mesh point x_i is defined by

$$E_i = \frac{y(x_i) - u_i}{y(x_i)} \times 100. \tag{10}$$

The results are presented in Table 15.1. The error at x_1 is already greater than 20 percent. Errors of this large magnitude are usually unacceptable in applications. We observe that the magnitude of E_i increases as x_i increases. This is typical of many problems, since errors tend to accumulate as $| x - x_0 |$ increases.

TABLE 15.1 Numerical Solution Using Euler's Method with $h = .5$

i	x_i	u_i	$y(x_i)$	E_i *(percent)*
0	0.0	0.0	0.0	0.0
1	0.5	0.5	0.649	22.96
2	1.0	1.25	1.718	27.24

Furthermore, we observe that in obtaining the values of u_i it was unnecessary to round any numbers. That is, the numerical solutions in this example are unaffected by roundoff errors.

EXAMPLE 2

Use Euler's method to solve the IVP of Example 1 numerically with a mesh width of $h = .25$. Round all decimals to two places.

solution. Since $h = .25$, we have the mesh points $x_0 = 0$, $x_1 = x_0 + h = .25$, $x_2 = .5$, $x_3 = .75$, and $x_4 = 1$. The Euler difference equations, given by Eq. (4), are

$$u_{i+1} = u_i + (u_i + 1)(.25) = 1.25u_i + .25.$$

Then, proceeding as we did in the previous example, by successively setting $i = 0, 1, 2, 3$ in this equation we obtain the numerical solution $\{u_0, u_1, u_2, u_3, u_4\}$. We have, for example, $u_0 = 0$, $u_1 = .25$, and $u_2 = (1.25 \times .25) + .25 = .31 + .25 = .56$, where we have rounded $1.25 \times .25 = .3125$ to two places (.31). The results are summarized in Table 15.2.

TABLE 15.2 Numerical Solution Using Euler's Method with $h = .25$ and Two-Decimal-Place Accuracy

i	x_i	u_i	$y(x_i)$	E_i	E_i (*with* $h = .5$ *mesh*)
0	0.0	0.0	0.0	0.0	0.0
1	0.25	0.25	0.284	11.97	—
2	0.50	0.56	0.649	13.71	22.96
3	0.75	0.95	1.117	14.95	—
4	1.0	1.44	1.718	16.18	27.24

By comparing the last two columns in Table 15.2 we observe that a substantial decrease in the relative error was achieved by decreasing the mesh width to one-half its previous value. However, E_i is still large. A further decrease in E_i can be achieved by reducing the mesh width. For example, with a mesh width of $h = .1$ and retaining two-decimal-place accuracy, we get for $x = 1$ ($x_{10} = 1$), the approximation $u_{10} = 1.61$. The error for this result is 6.29 percent. With the same mesh width ($h = .1$) but rounding to five decimal places, we get $u_{10} = 1.59375$. This gives a 7.25 percent error.[2] Thus by retaining greater arithmetic precision, we have increased the error! However, by using a mesh width of $h = .05$, we find that the numerical solution at $x = 1$ is $u = 1.65$ when we retain two decimal places and $u = 1.655$ when we retain three decimal places. The errors are respectively 3.96 percent and 3.67 percent. Thus when $h = .05$, greater arithmetic precision leads to a smaller error in the numerical solution, as we would expect. In fact, a careful mathematical analysis of the Euler method including roundoff errors shows that the numerical solutions converge to the exact solution as $h \to 0$. This result and the anomalous decrease in accuracy with an increase in arithmetic precision that was obtained with the $h = .1$ mesh suggest that the mesh $h = .1$ is too coarse.

Exercises

1. Determine the exact solution of the IVP $y' + 3y = x^2$, $y(0) = 1$. Obtain numerical solutions of the IVP on the interval $0 \leq x \leq 1$ by using Euler's method.

[2] In computing this error we rounded the exact value to six decimal places.

(a) Use a mesh width of $h = .5$ and two-decimal-place accuracy. Determine the error at each mesh point by using the exact solution.
(b) Use a mesh width of $h = .25$ and two-decimal-place accuracy. Determine the error by using the exact solution.
(c) Use a mesh width of $h = .1$ and two-decimal-place accuracy and determine the error.

2. Repeat the calculations of Exercise 1 using three-decimal-place accuracy. Compare with the results of Exercise 1.

3. Repeat the steps in Exercises 1 and 2 for the IVP $y' + y = \sin x$, $y(0) = 0$.

4. Repeat the steps in Exercises 1 and 2 for the logistic IVP $y' + 2y(y - 1) = 0$, $y(0) = \frac{1}{2}$.

5. Prove the convergence of the numerical solution obtained by Euler's method to the exact solution for the IVP $y' - y = 0$, $y(0) = 1$. (*Hint:* By using mathematical induction show that Euler's difference equation for this IVP leads to

$$u_i = (1 + h)^i, \qquad i = 1, 2, \ldots.$$

Since $x_i = ih$, we have

$$u_i = (1 + h)^{x_i/h} = [(1 + h)^{1/h}]^{x_i}.$$

Then use the fact that $(1 + h)^{1/h} \to e$ as $h \to 0$ to prove that as $h \to 0$, $u_i \to e^{x_i}$.)

6. Write a digital computer program to solve the IVP $y' = f(x, y)$, $y(x_0) = a$ by Euler's method, where $f(x, y)$ is any function and x_0 and a are any given numbers. Try the program on the IVPs given in Exercises 1–4.

7. Approximate the integral given by Eq. (2) by taking the height of the rectangle as the ordinate of the right endpoint. Use this result to derive another set of difference equations.

8. Another way to derive Euler's equations and other difference equations for the IVP $y' = f(x, y)$, $y(x_0) = a_0$ is to consider the Taylor series about the point b,

$$y(x) = y(b) + y'(b)(x - b) + \frac{y''(b)}{2!}(x - b)^2 + \cdots.$$

Let $b = x_i$ and $x = x_{i+1}$, where x_i and x_{i+1} are adjacent mesh points and derive the formula

$$y(x_{i+1}) = y(x_i) + f(x_i, y(x_i))(x_{i+1} - x_i) + \frac{y''(x_i)}{2}(x_{i+1} - x_i)^2 + \cdots$$

Retain two terms in this formula to obtain the Euler difference equations.

9. If we retain the third term in the Taylor series in Exercise 8, then we should

expect a more accurate approximation. Use the fact that

$$y''(x_i) = \frac{\partial f}{\partial x}(x_i, y(x_i)) + \frac{\partial f}{\partial y}(x_i, y(x_i))\frac{dy(x_i)}{dx}$$

to derive difference equations corresponding to this three-term Taylor approximation.

10. Use the result of Exercise 9 to solve the IVP $y' - y = 1$, $y(0) = 0$, with $h = .5$ and $h = .25$, and compare your results with those of Euler's method (Tables 15.1 and 15.2). Retain two decimal places and compute the relative errors.

15.3 DISCUSSION

The computations in the examples in the preceding section show that as the mesh width h is decreased, the error in the numerical solutions decreases. It was possible to determine the error because the exact solution to the IVP is known explicitly. Thus we can determine the mesh widths which give approximations within a desired accuracy. However, in almost all applications of numerical methods the explicit solution is not known, which is, of course, the purpose in using a numerical method. Thus we require criteria to select mesh widths that will give sufficiently accurate numerical approximations. The selection of appropriate mesh widths usually is determined by practical limitations such as the size of the computer, and time and accuracy requirements, rather than a precise mathematical analysis. We shall now describe a procedure that is commonly used in actual computations.

First a mesh width $h = h_1$ is selected that is believed to give an accurate numerical solution. This choice is based to a large extent on experience and trial and error. Then the computations are repeated with a second and more refined mesh width $h = h_2 < h_1$. It is customary to choose $h_2 = h_1/2$, although this is not essential. If the numerical approximations with the refined ($h = h_2$) mesh differ from the approximations with $h = h_1$ by more than an amount preassigned by the investigator, then a third computation using a more refined mesh $h = h_3 < h_2$ is performed. The differences in the numerical solutions with the h_2 and h_3 meshes are determined. This refinement of the mesh width is continued until the investigator decides that the differences in the numerical solutions for two successive mesh widths are less than a preassigned small number.[3]

[3] If the numerical solutions converge to the exact solution as $h \to 0$, then the differences in the numerical solutions for successively smaller mesh widths converge to zero as $h \to 0$. Thus the convergence of the differences in the numerical solutions is a necessary condition for convergence to the exact solution. However, we wish to emphasize that it is not necessarily a sufficient condition.

For example, let us consider the change in the numerical solutions of the IVP studied in the examples in the previous section as h is decreased. We recall that we used the mesh widths $h_1 = .5$, $h_2 = .25$, $h_3 = .1$, and $h_4 = .05$. We denote the corresponding approximations by $u(h_1)$, $u(h_2)$, $u(h_3)$, and $u(h_4)$. In Table 15.3 we summarize the results for the point $x = 1$. The relative percent differences in the approximations, at a fixed value of x and as the mesh width is refined, are denoted by

$$R_j = \frac{u(h_{j+1}) - u(h_j)}{u(h_{j+1})} \times 100, \qquad j = 1, 2, 3. \tag{1}$$

TABLE 15.3

j	h_j	$u(h_j)$	R_j	Error (percent)
1	0.5	1.25		27.24
			$\frac{1.44 - 1.25}{1.44} = 13.19\%$	
2	0.25	1.44		16.18
			$\frac{1.61 - 1.44}{1.61} = 10.56\%$	
3	0.1	1.61		6.29
			$\frac{1.65 - 1.61}{1.65} = 2.42\%$	
4	0.05	1.65		3.96

The last column in Table 15.3 gives the error. Thus, for example, if we had decided in advance of the computation that we should select a mesh width h such that at $x = 1$ the relative differences R_j satisfy $| R_j | < 3$ percent, then the mesh width $h = h_4 = .05$ would be acceptable. As we observe from Table 15.3, this results in a relative error of 3.96 percent in the approximation at $x = 1$.

Exercises

1. Obtain numerical solutions of the IVP $y' + 3y = x^2$, $y(0) = 1$ on the interval $0 \leq x \leq 1$ by Euler's method, for a sequence of decreasing values of the mesh width h. Determine appropriate values of h by the condition that the maximum value over the mesh of the relative differences as defined by Eq. (1) is (a) less than 10 percent and (b) less than 5 percent. For the values of h deter-

mined in (a) and (b), find the maximum error in the numerical solution. Use three decimal place accuracy in the calculations.

2. Solve the IVP $y' + \sin y = 1$, $y(0) = \frac{1}{2}$ on the interval $0 \le x \le 2$ by Euler's method. Determine an appropriate mesh value of h by solving the IVP for a sequence of values of h until the maximum value over the mesh points of the relative differences is less than 3 percent.

3. Repeat Exercise 2 for the IVP $y' + \sin y = 1 + \sin x$, $y(0) = \frac{1}{2}$. The IVPs given in Exercises 2 and 3 occur in mathematical studies of the Josephson tunnel junction. A Josephson junction consists of two thin superconducting strips separated by a thin dielectric. Current is passed down each of the two superconducting strips. When the current is increased (or the temperture of the superconductors is lowered), a critical value is reached such that a current passes through the dielectric from one superconductor to another. This is called a tunneling current. Josephson received the 1973 Nobel prize in physics for discovering this phenomenon. The tunneling current is proportional to $\sin y$. The IVPs in Exercises 2 and 3 are believed to mathematically describe the behavior of a "point" Josephson junction. The inhomogeneous terms 1 and $\sin x$ represent exterior current sources due, for example, to weak magnetic fields. Josephson junctions are sensitive detectors of electromagnetic radiation. They are used to detect weak magnetic radiation that is received from distant stars and galaxies.

15.4 EXTRAPOLATION OF THE MESH WIDTH

Since a final mesh width is usually selected by solving the IVP for several mesh sizes, approximations u_i are available for at least two mesh widths. We shall now show how to combine these available approximations to obtain a more accurate approximation without further reducing the mesh width. This technique, which was invented by L. Richardson,[4] is called mesh extrapolation.

To explain Richardson's procedure, we shall use $u(x, h)$ to denote a numerical solution of the IVP

$$y' = f(x, y), \qquad y(x_0) = a, \tag{1}$$

[4] Lewis Richardson was a famous British meteorologist. He made many fundamental contributions to meteorology and mathematics. He was the first scientist to seriously use difference methods to obtain numerical solutions of the differential equations of dynamic meteorology. Many of his ideas are now used routinely in numerical weather forecasting. However Richardson is most famous for his book, "Arms and Insecurity." Richardson's service as an ambulance driver in France during World War I exposed him to the horrors and deprivations of war. In his book he tried to demonstrate mathematically the futility of war as a means of resolving human conflict.

at the point x of the mesh, where h is the mesh width. We assume that $u(x, h)$ converges to the exact solution $y(x)$ as $h \to 0$. In addition, we assume that $u(x, h)$ can be expressed as a power series in h,

$$u(x, h) = y(x) + a(x)h + R(x, h), \tag{2}$$

where $R(x, h) \equiv b(x)h^2 + \cdots$ is the remainder. It is small when h is small. The leading term in the power series is the exact solution $y(x)$ because we have assumed that u converges to y as $h \to 0$. The coefficients $a, b, \ldots$ are functions of x since, for a given h, the difference between the exact solution $y(x)$ and the numerical solution $u(x, h)$ will depend on the point x at which the difference is taken, as well as on the mesh width h.

We assume that numerical solutions $u(x, h_1)$ and $u(x, h_2)$ corresponding to the mesh widths $h_1 > h_2$ have been evaluated by Euler's method. For simplicity, we select these meshes so that every point x in the mesh with $h = h_1$ is also in the mesh with $h = h_2$. For example, this is the case if $h_2 = h_1/2$. By setting $h = h_1$ and $h = h_2$ in Eq. (2), we get

$$u(x, h_1) = y(x) + a(x)h_1 + R(x, h_1) \tag{3}$$

and

$$u(x, h_2) = y(x) + a(x)h_2 + R(x, h_2). \tag{4}$$

We solve Eqs. (3) and (4) for $y(x)$ by eliminating the terms $a(x)h_1$ and $a(x)h_2$. This gives

$$y(x) = \frac{h_1u(x, h_2) - h_2u(x, h_1)}{h_1 - h_2} + \frac{h_2R(x, h_1) - h_1R(x, h_2)}{h_1 - h_2} \tag{5}$$

Now we neglect the terms in Eq. (5) containing the remainders, since the remainders are assumed to be small. Then $y(x)$ in Eq. (5) is no longer the solution; instead, it is an approximation to the solution, which is called the Richardson approximation $y_R(x)$. Thus we have for the Richardson approximation

$$y_R(x) = \frac{h_1u(x, h_2) - h_2u(x, h_1)}{h_1 - h_2} \tag{6}$$

In particular, if $h_2 = h_1/2$, we have from Eq. (6)

$$y_R(x) = 2u(x, h_2) - u(x, h_1). \tag{7}$$

EXAMPLE 1

Euler's method was used in Examples 1 and 2 of Section 15.2 to obtain numerical approximations of the IVP $y' - y = 1$, $y(0) = 0$, by using mesh widths $h_1 = .5$, $h_2 = .25$, $h_3 = .1$, and $h_4 = .05$. The relevant results for $u(x, h)$ at $x = .5$ and $x = 1.0$ are summarized in Table 15.4.

TABLE 15.4

h	$u(.5, h)$	*Error (percent)*	$u(1.0, h)$	*Error (percent)*
0.5	0.5	22.96	1.25	27.24
0.25	0.56	13.71	1.44	16.18
0.1	—	—	1.61	6.29
0.05	—	—	1.65	3.96

Use the mesh widths $h_1 = .5$ and $h_2 = .25$ to obtain a Richardson approximation at the points $x = .5$, $x = 1.0$.

solution. Since $h_2 = h_1/2$, we use Eq. (7) for the Richardson approximation. The computations necessary to determine y_R are summarized in Table 15.5.

TABLE 15.5

x	0.5	1.0
$u(x, .5)$	0.5	1.25
$u(x, .25)$	0.56	1.44
$y_R(x)$	0.62	1.63
Percent error	4.47%	5.12%
$= \dfrac{y(x) - y_R(x)}{y(x)} \times 100$		

We see from Table 15.4 that in solving the IVP with a mesh width of $h = .1$ we obtain an error at $x = 1$ of 6.29 percent. This error is greater than the error of the Richardson approximation (5.12 percent) using the relatively coarse meshes of $h_1 = .5$ and $h_2 = .25$.

EXAMPLE 2

Find the Richardson approximation for the solution of the IVP of Example 1 at $x = 1$ by using the mesh widths $h_1 = .1$ and $h_2 = .05$.

solution. Since $h_2 = h_1/2$, we find by using the results given in Table 15.4 that the Richardson approximation at $x = 1$ is

$$y_R(1) = 2u(1, h_2) - u(1, h_1) = 2 \times 1.65 - 1.61 = 1.69.$$

Since the exact solution at $x = 1$ to three decimal places is $y(1) = 1.718$, the error at $x = 1$ is 1.63 percent. The error in the numerical solution using the mesh $h = .05$ was 3.96 percent. Thus substantially greater accuracy is achieved, with less computation, by using the Richardson approximation than by obtaining a third numerical solution $u(1, h_3)$ with $h_3 = h_2/2$. This would also be the case at the other mesh points of the mesh $h = .1$.

An analysis of the improvement in accuracy that is obtained by using Richardson's method is discussed in Exercise 3.

Exercises

1. (a) Use the mesh extrapolation procedure to find Richardson approximations for the solution of the IVP of Exercise 1 of Section 15.2 using the mesh widths of $h = .5$ and $h = .25$ and the numerical solutions obtained by Euler's method in that exercise. (b) Determine the error in the Richardson approximation by using the exact solution. Compare the Richardson approximation with the numerical solution obtained in Exercise 1 of Section 15.2 with a mesh width of $h = .1$.

2. Repeat Exercise 1 for the IVPs and numerical solutions obtained by Euler's method in (a) Exercise 3 of Section 15.2. (b) Exercise 4 of Section 15.2.

3. The Richardson approximation given in Eq. (6) is obtained without computing a new numerical solution with finer meshes $h < h_2$. To show why it gives a more accurate approximation to the exact solution than $u(x, h_2)$ we reexamine Eq. (2). We define the error $E(x, h)$ at the point x by

$$E(x, h) = u(x, h) - y(x) = a(x)h + b(x)h^2 + \cdots. \tag{i}$$

Thus if $a(x) \neq 0$, the decrease in the error in Euler's method is essentially proportional to h as $h \to 0$. The error $E_R(x)$ in the Richardson approximation is defined by

$$E_R(x) = y_R(x) - y(x).$$

Use the expression given in Eq. (6) for y_R and the definition of E given in Eq. (i) to get

$$E_R(x) = \frac{h_1}{h_1 - h_2} E(x, h_2) - \frac{h_2}{h_1 - h_2} E(x, h_1). \tag{ii}$$

Finally, insert the expressions in Eq. (i) for $E(x, h_1)$ and $E(x, h_2)$ into Eq. (ii) and obtain $E_R = -bh_1h_2 + \cdots$. For example, if $h_2 = h_1/2$, then $E_R = -(b/2)h_1^2 + \cdots$. Thus the decrease in the error in the Richardson approximation is proportional essentially to h^2 as $h \to 0$. Then if $a(x) \neq 0$ in Eq. (i), a smaller error is achieved with the Richardson approximation than with Euler's method, and this is done with relatively little extra computation.

4. (a) Derive an expression for the Richardson two-mesh error for the case $b(x) = 0$ for all x.
   (b) What power of h is the error proportional to if $h_2 = h_1/2$?

15.5 THE MODIFIED EULER METHOD

The accuracy of numerical solutions obtained by Euler's method is improved by using finer mesh widths, as we showed in the example in Section 15.2. However, the number of computations that are required to evaluate the approximations at a fixed value of x increase as the mesh width decreases. Therefore it is natural to seek other methods that will give greater accuracy than Euler's method for the same size of the mesh width. However, as we shall see, greater accuracy usually requires more computations per mesh point. Thus, in choosing a numerical method we must consider a trade-off between the accuracy of the method, and hence the computational time per mesh point, and the size of the mesh width.

To indicate one procedure for constructing more accurate numerical methods we again integrate the DE $y' = f(x, y)$ from the mesh point x_i to the mesh point x_{i+1}. This gives

$$y(x_{i+1}) = y(x_i) + \int_{x_i}^{x_{i+1}} f(x, y(x))\, dx. \tag{1}$$

In deriving Euler's method we used the rectangle rule to approximate the integral in Eq. (1). In effect, we approximated the area under the curve $y = F(x) = f(x, y(x))$ from x_i to x_{i+1} by the area of a rectangle (see Figure 15.3). We now approximate the area under the curve by a trapezoid. The resulting approximation of the integral

$$I = \int_{x_i}^{x_{i+1}} F(x)\, dx$$

is called the trapezoidal rule.

The trapezoidal rule is obtained by first connecting the points $(x_i, F(x_i))$ and $(x_{i+1}, F(x_{i+1}))$ by a straight line, as shown in Figure 15.6. Then I is approximated by the trapezoidal region, which is shown cross-hatched in the figure. The area of the trapezoid A_T is the sum of the area of the rectangle of height $F(x_i)$ and base $(x_{i+1} - x_i)$ and the area of the triangle with the same base, and altitude $F(x_{i+1}) - F(x_i)$. Thus we have

$$A_T = F(x_i)(x_{i+1} - x_i) + \tfrac{1}{2}(x_{i+1} - x_i)[F(x_{i+1}) - F(x_i)]$$

$$= \frac{[F(x_i) + F(x_{i+1})]}{2} h.$$

This shows that A_T is the product of the average of the endpoint heights of the trapezoid and the length of the base. The trapezoidal rule then gives $I \approx A_T$. If we apply this approximation to the integral in Eq. (1) with $F(x) = f(x, y(x))$ and replace $y(x_i)$ and $y(x_{i+1})$ by their approximations

u_i and u_{i+1}, we get

$$u_{i+1} = u_i + \frac{[f(x_i, u_i) + f(x_{i+1}, u_{i+1})]}{2} h, \qquad i = 1, 2, \ldots, N - 1. \quad (2)$$

This equation is to be used to evaluate u_{i+1} when u_i is known. It is difficult to evaluate u_{i+1} from Eq. (2) because the unknown u_{i+1} appears in one of the arguments of f on the right side. For example, if $f(x, y) = \sin y$, then Eq. (2) becomes

$$u_{i+1} = u_i + \frac{[\sin u_i + \sin u_{i+1}]}{2} h$$

and it is hard to solve for u_{i+1} explicitly.

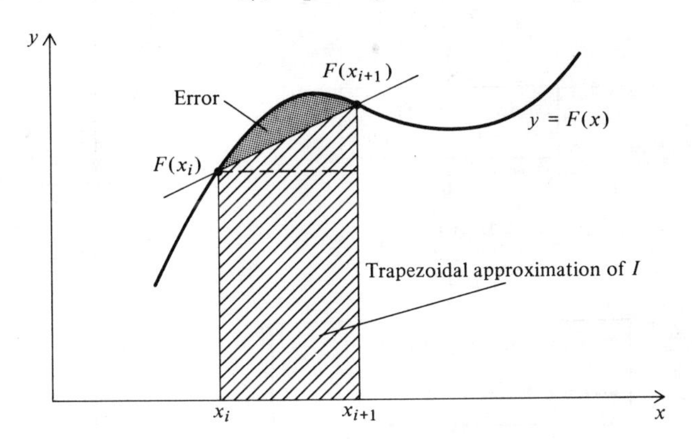

Figure 15.6

In the modified Euler method this difficulty is circumvented by replacing u_{i+1} in the right side of Eq. (2) by the quantity $\tilde{u}_{i+1}$, which is obtained by the Euler difference formula

$$\tilde{u}_{i+1} = u_i + f(x_i, u_i)h. \quad (3)$$

Then the *modified Euler difference equation* is given by

$$u_{i+1} = u_i + \frac{f(x_i, u_i) + f(x_{i+1}, \tilde{u}_{i+1})}{2} h \quad (4)$$

where $\tilde{u}_{i+1}$ is defined in Eq. (3). A flowchart for the modified Euler method is given in Figure 15.7. Thus to evaluate u_{i+1} when u_i is known, we must compute $f(x_i, u_i)$ as in the Euler method, and in addition compute the quantity $f(x_{i+1}, \tilde{u}_{i+1})$. This extra computation per mesh point must be compensated for by greater accuracy if the modified Euler method is to be an improvement. We shall demonstrate this in the following example.

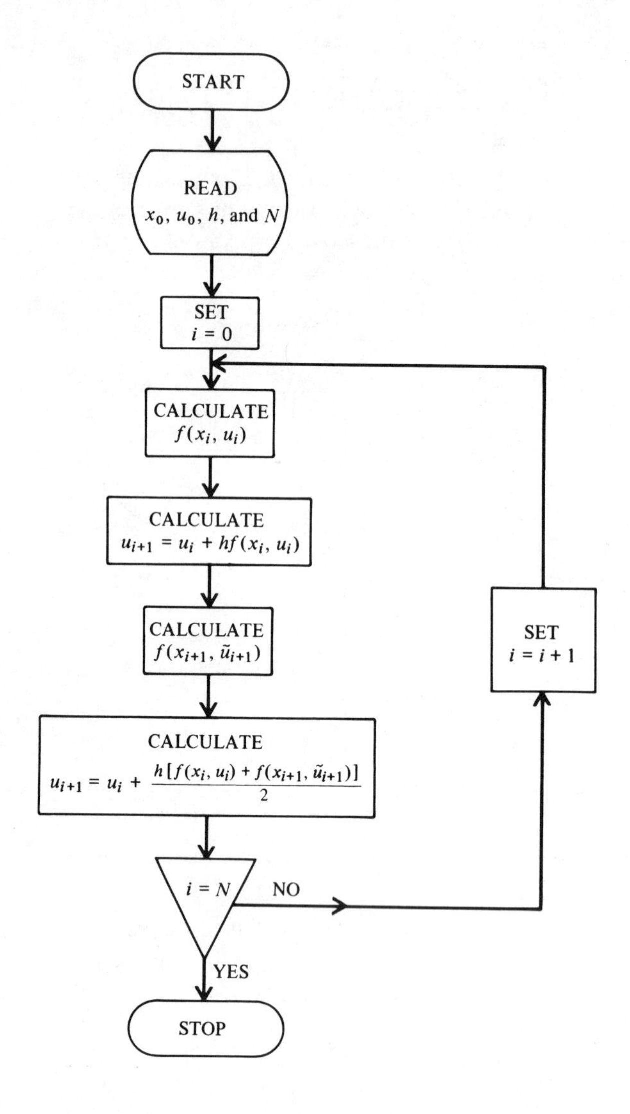

Figure 15.7

EXAMPLE

Solve the IVP $y' - y = 1$, $y(0) = 0$ on the interval $0 \leq x \leq 1$ by the modified Euler method, using the mesh widths $h_1 = .5$ and $h_2 = .25$. Evaluate the accuracy of these approximations and compare them with the Euler method. Retain two-decimal-place accuracy.

solution. For the DE of this example, $f(x, y) = y + 1$. Therefore the modified Euler difference equation, given by Eq. (4), is

$$u_{i+1} = u_i + \left[\frac{u_i + 1 + \tilde{u}_{i+1} + 1}{2}\right]h, \tag{5}$$

where $\tilde{u}_{i+1}$ is given by the Euler difference formula, Eq. (3), as

$$\tilde{u}_{i+1} = u_i + (u_i + 1)h. \tag{6}$$

We insert Eq. (6) into Eq. (5). After rearranging the terms we obtain

$$u_{i+1} = \frac{h}{2}(2 + h) + \left[1 + \frac{h}{2}(2 + h)\right]u_i. \tag{7}$$

For the mesh with $h = h_1 = .5$, the mesh points are $x_0 = 0$, $x_1 = .5$, and $x_2 = 1$. In addition, $(h/2)(2 + h) = .625$. By rounding this number to two decimal places and inserting the result in Eq. (7) we get the difference equation

$$u_{i+1} = .63 + 1.63u_i. \tag{8}$$

Since $u_0 = 0$, we can evaluate u_1 and u_2 by successively setting $i = 0$ and $i = 1$ in Eq. (8). The results are summarized in Table 15.6.

TABLE 15.6 Numerical Solution Using Modified Euler Method with $h = .5$ and Two-Decimal-Place Accuracy

x_i	u_i	$y(x_i)$	*Percent Error*	*Percent Error by Euler's Method*
0.0	0.0	0.0	0.0	0.0
0.5	0.63	0.649	2.93	22.96
1.0	1.66	1.718	3.38	27.24

The last two columns in Table 15.6 clearly illustrate the superior accuracy of the modified Euler method.

For the mesh with $h = h_2 = .25$, the difference equation, given by Eq. (7), is

$$u_{i+1} = .28 + 1.28u_i. \tag{9}$$

The values of u_i as successively evaluated from Eq. (9) are summarized in Table 15.7.

TABLE 15.7 Numerical Solution Using Modified Euler Method with $h = .25$

x_i	u_i	$y(x_i)$	*Percent Error*	*Percent Error by Euler's Method*
0.0	0.0	0.0	0.0	0.0
0.25	0.28	0.284	1.41	11.97
0.50	0.64	0.649	1.39	13.71
0.75	1.10	1.117	1.52	14.95
1.00	1.69	1.718	1.63	16.18

Thus by reducing the mesh width to half its value, the error is reduced by more than half of its value, as we can see from the percent error columns in Tables 15.6 and 15.7. A similar reduction is not achieved for the Euler method, as we see from the last column in these tables. The superiority of the modified Euler method is apparent again.

Exercises

1. Obtain numerical solutions of the IVPs
   (a) $y' + 3y = x^2, \quad y(0) = 1$ (b) $y' + y = \sin x, \quad y(0) = 0$
   (c) $y' + 2y(y - 1) = 0, \quad y(0) = \frac{1}{2}$
   on the interval $0 \leq x \leq 1$ by the modified Euler method. Use mesh widths of $h_1 = .5$ and $h_2 = .25$. and two-decimal-place accuracy. Use the exact solution to determine the error in the numerical solution. Compare these results with the numerical results obtained in the Exercises in Section 15.2 by the Euler method.

2. Use the results of Exercise 1 to determine the two-mesh Richardson approximation. Use the exact solution to calculate the errors in the Richardson approximation. Compare with the Richardson approximations obtained in Exercise 1 of Section 15.4.

3. Repeat Exercises 1 and 2 by calculating with three-decimal-place accuracy. Compare the results with the results of Exercises 1 and 2.

4. We showed in Section 15.1 that in Euler's method we approximate the derivative y' in the DE $y' = f(x, y)$ at the mesh point x_i by the forward difference approximation $y'(x_i) \approx (u_{i+1} - u_i)/h$. Another difference approximation of $y'(x_i)$ is given by

$$y'(x_i) \approx \frac{u_{i+1} - u_{i-1}}{2h}.$$

   It is called the centered difference approximation of $y'(x_i)$ because it involves the approximation u_{i+1}, which corresponds to the mesh point $x_{i+1} = x_i + h$ immediately to the right of x_i, and u_{i-1}, which corresponds to the mesh point $x_{i-1} = x_i - h$ immediately to the left. By using the centered difference approximation of y' in the DE show that

$$u_{i+1} = u_{i-1} + f(x_i, u_i)2h, \qquad i = 1, 2, \ldots, N - 1. \tag{i}$$

This is called the *centered difference method.* Since the difference equation (i) gives the value of u_{i+1} in terms of the two preceding values u_i and u_{i-1}, it is called a two-step method. Therefore if we set $i = 0$ in Eq. (i), the right side contains u_{-1}, which is not defined. Thus Eq. (i) can apply only for $i = 1, 2, \ldots, N - 1$. By setting $i = 1$ in Eq. (i) we get u_2 in terms of u_1 and u_0. The initial data give $u_0 = a$. Therefore we require a procedure for determining u_1. The Euler method or the modified Euler method applied to the point $i = 1$ is frequently used to give an approximation for u_1. This is called a starting method. Give a geometric interpretation for the centered difference method.

5. Repeat Exercises 1 and 2 using the centered difference method instead of the modified Euler method.

15.6 POPULATION DYNAMICS WITH IMMIGRATION

The population dynamics model that was described in Section 15.1 requires the solution of the IVP

$$y' = \alpha(A - y)y + \beta e^{-ky}, \qquad y(0) = a_0. \tag{1}$$

For $\beta = 0$, Eq. (1) is reduced to the IVP for the logistic equation. The explicit solution of the logistic IVP was obtained in Section 14.5-1. The solution of the IVP given by Eq. (1) is not known explicitly.

For simplicity of presentation we shall consider the special case of Eq. (1) with $\alpha = A = k = 1$. Thus we shall consider the IVP

$$y' = (1 - y)y + \beta e^{-y}, \qquad y(0) = a_0. \tag{2}$$

We use the modified Euler method to obtain numerical solutions of the IVP in Eq. (2) for fixed values of the parameters β and a_0. Then we shall compare these results with the solution of the logistic equation ($\beta = 0$) to determine the effects of immigration on the growth of the population. Specifically, we shall use the typical parameter values, $\beta = a_0 = .1$.

The modified Euler difference equations of the IVP are

$$\tilde{u}_{i+1} = u_i + f(x_i, u_i)h, \tag{3}$$

$$u_{i+1} = u_i + \frac{h}{2}[f(x_i, u_i) + f(x_{i+1}, \tilde{u}_{i+1})], \tag{4}$$

$$u_0 = a_0 = .1, \tag{5}$$

and

$$f(x, u_i) = (1 - u_i)u_i + .1e^{-u_i}. \tag{6}$$

In the examples studied in Section 15.5 it was convenient to eliminate $\tilde{u}_{i+1}$ from Eqs. (8) and (9) because of the simple form of $f(x, y)$. Although it is possible to do this in the present problem, the resulting expressions are quite complicated. It is more convenient to work with the difference equations directly in the form given by Eqs. (3) through (6). Thus if u_i is known, we determine u_{i+1} first by evaluating $\tilde{u}_{i+1}$ from Eq. (3). Then by calculating $f(x_{i+1}, \tilde{u}_{i+1})$ from Eq. (6) and inserting the result into Eq. (4), we obtain u_{i+1}.

```
      PROGRAM PEOPLE (INPUT, OUTPUT)
      DIMENSION U(101), X(101)
      EFF IS FUNCTION OF X AND Z, IN THIS CASE NOT A FUNCTION OF X
      EFF(Z) = (1 - Z) * Z + BETA * EXP(-Z)
      READ 1000, H, BETA, A, M
1000  FORMAT (F6.3, I6)
      N = 1
      U(N) = A
      X(N) = 0.
  10  N = N + 1
      X(N) = X(N - 1) + H
      UTIL = U TILDE
      UTIL = U(N - 1) + EFF(U(N - 1)) * H
      U(N) = U(N - 1) + .5 * H * (EFF(U(N - 1)) + EFF(UTIL))
      IF(N.LT.M) GO TO 10
      PRINT 1001, (J, X(J), U(J), J = 1, M)
1001  FORMAT (1X, I6, 4X, F6.3, F11.7)
      STOP
      END
```

Figure 15.8

A sample digital computer code for solving the difference equations given by Eqs. (3) through (6) is shown in Figure 15.8. It is written in the Fortran IV computer language for the CDC 6600 computer. Codes for other computers may differ slightly from the code in the figure. Numerical solutions were obtained by using this program on the interval $0 \leq x \leq 10$ with the mesh width $h = .1$. The numerical results at $x = 1, 2, \ldots, 10$ are shown in Table 15.8. A graph of these results is given in Figure 15.9. The exact solution of the logistic IVP (the logistic curve) is also shown in the figure. The numerical solutions required .7 second of computer time.

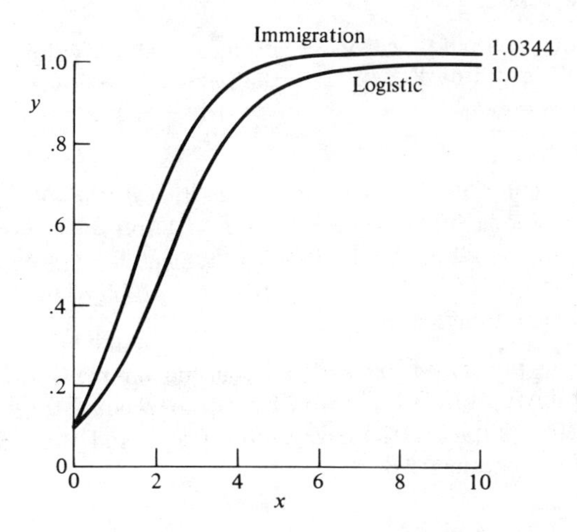

Figure 15.9

The change from the convex or accelerating portion of the solution curve to the decelerating portion in Figure 15.9 occurs for $x \approx 1$. For the logistic curve, it occurs for $x \approx 2.25$. The logistic curve approaches its asymptote $y = 1$ as $x \to \infty$. The numerical results approach the asymptote $y = 1.0344$ as $x \to \infty$. Thus by permitting immigration into the population, the equilibrium state is increased from $y = 1$ to $y = 1.0344$. This is an increase of 3.44 percent. The approach to equilibrium is similar for the two curves, as the figures show. The largest difference in the two populations occurs for x approximately equal to 2.8.

TABLE 15.8 Numerical Solution for Population Dynamics

x	u_i	x	u_i
0.0	0.1000	6.0	1.0272
1.0	0.3427	7.0	1.0320
2.0	0.6459	8.0	1.0336
3.0	0.8669	9.0	1.0341
4.0	0.9726	10.0	1.0343
5.0	1.0131		

Exercises

1. (a) Use the digital computer code in Figure 15.8 to solve the IVP given by Eq. (2) with $\beta = a_0 = .1$, using $h = .1$. (b) Compare your result with Table 15.8.

2. Use the digital computer code given in Figure 15.8 (or write your own code) to solve the IVP given by Eq. (2) with the following choices of β and a_0.
(a) $\beta = .1$, $a_0 = .5$
(b) $\beta = 1$, $a_0 = .1$
(c) $\beta = .1$, $a_0 = 2$

3. (a) To determine the dependence of the solution of the IVP given by Eq. (2) on the parameter a_0, fix the value of β at $\beta = .1$ and solve the IVP for $a_0 = .1$, .3, .5, .7, and .9. (b) Use the results of (a) to sketch a graph of the asymptote of the solution as a function of a_0. (c) Sketch a graph of the inflection point x_1 $(y''(x_1) = 0)$ as a function of a_0.

4. (a) Determine the dependence of the solution on the constant β by fixing $a_0 = .1$ and solving the IVP given in Eq. (2) for β equal to .1, .5, 1.0, 1.5, and 2.0. (b) Sketch graphs of the asymptote of $y(x)$ and the inflection point as functions of the parameter β.

chapter 16

THE LAPLACE TRANSFORM

16.1 INTRODUCTION

In many applications it is necessary to solve IVPs with discontinuous inhomogeneous terms. A suddenly applied or removed voltage in an electric network (as in a power failure), a drug that is injected into the circulatory system, and a suddenly applied or removed force in a mechanical system are examples of discontinuous forcing functions. In this brief chapter we introduce the Laplace transform and show how it is used to solve IVPs for linear DEs with constant coefficients. No attempt will be made to motivate or rigorously justify it.[1]

The Laplace transform of a function $f(x)$ is denoted by $\mathcal{L}\{f(x)\}$. The following is a precise definition of $\mathcal{L}\{f(x)\}$.

[1] Two useful references on the Laplace transform are: R. V. Churchill, *Operational Mathematics*, McGraw-Hill, New York, 1972; and I. H. Sneddon, *The Use of Integral Transforms*, McGraw-Hill, New York, 1972. These books also discuss other transform methods and their applications to differential equations.

definition.

Let $f(x)$ be defined for all $x \geq 0$. The Laplace transform of f is defined by

$$\mathcal{L}\{f(x)\} = \int_0^\infty e^{-sx} f(x)\, dx \tag{1}$$

for every value of the real number s for which the improper integral exists.[2]

As the parameter s varies, the value of the integral varies. Thus the Laplace transform is a function of s, which we denote by its corresponding capital letter; for example, $\mathcal{L}\{f(x)\} = F(s)$. The parameter s is usually called the transform variable.

In Sections 16.2 and 16.3 we derive and discuss some elementary properties of the Laplace transform. Applications to the solution of IVPs are given in Section 16.4. In sections 16.5 we consider discontinuous forcing functions.

16.2 EXAMPLES AND ELEMENTARY PROPERTIES OF LAPLACE TRANSFORMS

We first obtain the Laplace transforms of some elementary functions $f(x)$ by using the definition

$$\mathcal{L}\{f(x)\} = \int_0^\infty e^{-sx} f(x)\, dx = \lim_{b\to\infty} \int_0^b e^{-sx} f(x)\, dx.$$

EXAMPLE 1

Find the Laplace transform of the function, $f(x) = 1$ for $x \geq 0$.

solution. From the definition we have

$$\mathcal{L}\{1\} = \int_0^\infty e^{-sx}\, dx$$

for all values of s for which the integral exists. If $s = 0$, then

$$\int_0^\infty e^{-sx}\, dx = \int_0^\infty dx = \infty,$$

[2] In more advanced treatments of Laplace transforms, it is important to consider s as a complex number. Improper integrals are reviewed in Appendix III. We wish to emphasize that not every function has a Laplace transform. Thus mathematicians have derived conditions on $f(x)$ to ensure that its Laplace transform exists; c.f. Section 16.6.

and the integral diverges. Thus the Laplace transform of $f(x) = 1$ does not exist for $s = 0$. If $s \neq 0$, then we have

$$\mathcal{L}\{1\} = \lim_{b\to\infty} \int_0^b e^{-sx}\,dx = \lim_{b\to\infty} \frac{e^{-sx}}{-s}\bigg|_0^b = \lim_{b\to\infty}\left[\frac{e^{-sb}}{-s} + \frac{1}{s}\right]$$

$$= \frac{1}{s} - \frac{1}{s}\lim_{b\to\infty} e^{-sb}. \tag{1}$$

If $s < 0$, then $e^{-sb} \to \infty$ as $b \to \infty$, and the integral is divergent. If $s > 0$, then $e^{-sb} \to 0$ as $b \to \infty$. Thus the improper integral in Eq. (1) converges and yields

$$\mathcal{L}\{1\} = \frac{1}{s}, \qquad \text{for } s > 0.$$

EXAMPLE 2

Find $\mathcal{L}\{e^{ax}\}$, where a is a specified constant.

solution.

$$\mathcal{L}\{e^{ax}\} = \int_0^\infty e^{-sx}e^{ax}\,dx = \int_0^\infty e^{-(s-a)x}\,dx.$$

If $s - a \leq 0$, then the reader can show, just as in Example 1, that the integral does not exist. However, for $s > a$, we have

$$\mathcal{L}\{e^{ax}\} = \lim_{b\to\infty}\int_0^b e^{-(s-a)x}\,dx = \lim_{b\to\infty}\left[\frac{e^{-(s-a)x}}{-(s-a)}\right]_0^b = \frac{1}{s-a}.$$

EXAMPLE 3

Find $\mathcal{L}\{\cos ax\}$, where a is a given constant.

solution.

$$\mathcal{L}\{\cos ax\} = \lim_{b\to\infty}\int_0^b e^{-sx}\cos ax\,dx.$$

For $s = 0$, $\mathcal{L}\{\cos ax\} = \int_0^\infty \cos ax\,dx$ is divergent. For $s \neq 0$, we have, by integrating by parts twice,

$$\int_0^b e^{-sx}\cos ax\,dx = \frac{e^{-sb}}{a^2}(a\sin ab - s\cos ab) + \frac{s}{a^2} - \frac{s^2}{a^2}\int_0^b e^{-sx}\cos ax\,dx.$$

We transfer the integral from the right side of this equation to the left and then take the limit of the resulting equation as $b \to \infty$. For $s > 0$ this gives

$$\lim_{b\to\infty}\left(1 + \frac{s^2}{a^2}\right)\int_0^b e^{-sx}\cos ax\,dx = \frac{s}{a^2}$$

because $\lim e^{-sb} = 0$, as $b \to \infty$, for $s > 0$. For $s < 0$ the limit does not exist because $e^{-sb}(a\sin ab - s\cos ab)$ does not have a limit as $b \to \infty$. Thus, using the definition of the Laplace transform we, have

$$\mathcal{L}\{\cos ax\} = \frac{s/a^2}{1 + (s^2/a^2)} = \frac{s}{a^2 + s^2} \qquad \text{for } s > 0.$$

The Laplace transforms of the functions considered in Examples 1 through 3 and the Laplace transforms of other frequently occurring functions are summarized in Table 16.1. More extensive tabulations may be found in the texts mentioned in Section 16.1.

TABLE 16.1 Laplace Transforms

$f(x)$	$F(s) = \mathcal{L}\{f(x)\}$	
1	$\frac{1}{s}$,	$s > 0$
x^n, n is a positive integer	$\frac{n!}{s^{n+1}}$,	$s > 0$
e^{ax}	$\frac{1}{s-a}$,	$s > a$
$\sin ax$	$\frac{a}{s^2+a^2}$,	$s > 0$
$\cos ax$	$\frac{s}{s^2+a^2}$,	$s > 0$
$e^{ax} \sin bx$	$\frac{b}{(s-a)^2+b^2}$,	$s > a$
$e^{ax} \cos bx$	$\frac{s-a}{(s-a)^2+b^2}$,	$s > a$
$e^{ax}x^n$, n is a positive integer	$\frac{n!}{(s-a)^n}$,	$s > a$

We now discuss two properties of the Laplace transform, which will be used in later sections to solve DEs.

theorem 1 (linearity of the Laplace transform)

If the Laplace transforms of the functions $f(x)$ and $g(x)$ exist for $s > s_0$, then

$$\mathcal{L}\{c_1 f(x) + c_2 g(x)\} = c_1\mathcal{L}\{f(x)\} + c_2\mathcal{L}\{g(x)\}, \qquad \text{for } s > s_0, \tag{2}$$

where c_1 and c_2 are arbitrary constants.

proof. For any finite $b > 0$ we have

$$\int_0^b e^{-sx}[c_1 f(x) + c_2 g(x)]\, dx = c_1 \int_0^b e^{-sx} f(x)\, dx + c_2 \int_0^b e^{-sx} g(x)\, dx.$$

Since $\mathcal{L}\{f(x)\}$ and $\mathcal{L}\{g(x)\}$, exist we know that both terms on the right side of this equation have limits as $b \to \infty$. Thus the left side has a finite limit as $b \to \infty$, and

$$\begin{aligned}\lim_{b\to\infty} \int_0^b e^{-sx}[c_1 f + c_2 g]\, dx &= c_1 \lim_{b\to\infty} \int_0^b e^{-sx} f(x)\, dx + c_2 \lim_{b\to\infty} \int_0^b e^{-sx} g(x)\, dx \\ &= c_1 \mathcal{L}\{f\} + c_2 \mathcal{L}\{g\},\end{aligned}$$

which proves the theorem.

By repeated application of this property, we can obtain

$$\mathcal{L}\{c_1 f_1(x) + c_2 f_2(x) + \cdots + c_n f_n(x)\} = c_1 \mathcal{L}\{f_1(x)\} + \cdots + c_n \mathcal{L}\{f_n(x)\}. \tag{3}$$

We now show that the Laplace transform of a derivative of a function is easily related to the Laplace transform of the function itself. This property is central for the use of Laplace transforms in solving DEs. The result for the first derivative is contained in the next theorem.

theorem 2

If $f'(x)$ is a continuous function for $x \geq 0$ and there is a real number s_0 such that $\mathcal{L}\{f\}$ exists for $s > s_0$, and

$$\lim_{b\to\infty} e^{-sb} f(b) = 0, \qquad \text{for all } s > s_0,$$

then

$$\mathcal{L}\{f'(x)\} = s\mathcal{L}\{f(x)\} - f(0), \qquad \text{for } s > s_0. \tag{4}$$

proof. By the definition of the Laplace transform of $f'(x)$, we must show that

$$\lim_{b\to\infty} \int_0^b e^{-sx} f'(x)\, dx$$

exists. Integration by parts yields

$$\int_0^b e^{-sx} f'(x)\, dx = e^{-sb} f(b) - f(0) + s \int_0^b e^{-sx} f(x)\, dx.$$

Taking the limits on both sides as $b \to \infty$ and using the hypothesis that $e^{-sb} f(b) \to 0$ as $b \to \infty$ for $s > s_0$, we get

$$\lim_{b\to\infty} \int_0^b e^{-sx} f'(x)\, dx = 0 - f(0) + s \lim_{b\to\infty} \int_0^b e^{-sx} f(x)\, dx.$$

Since $\mathcal{L}\{f(x)\}$ exists, we conclude that the limit on the left side exists for $s > s_0$. Thus $\mathcal{L}\{f'(x)\}$ exists and Eq. (4) is verified.

By successive application of this property, we can show that

$$\mathcal{L}\{f''(x)\} = s^2\mathcal{L}\{f(x)\} - sf(0) - f'(0), \qquad \text{for } s > s_0, \tag{5}$$

if $f''(x)$ is continous, $e^{-sb}f(b) \to 0$ as $b \to \infty$, and $e^{-sb}f'(b) \to 0$ as $b \to \infty$.

Exercises

1. Find the Laplace transforms of the following functions. Indicate the values of s for which they exist.
   (a) $\sin ax$, a real
   (b) $e^{ax}\cos bx$, a and b real
   (c) $e^{ax}\sin bx$, a and b real
   (d) xe^{ax}, a real

2. Use the definitions

$$\cosh at = \frac{e^{at} + e^{-at}}{2} \qquad \text{and} \qquad \sinh at = \frac{e^{at} - e^{-at}}{2},$$

where a is any real number, to show that

$$\mathcal{L}\{\cosh at\} = \frac{s}{s^2 - a^2} \qquad s > s_0 = |a|$$

and

$$\mathcal{L}\{\sinh at\} = \frac{a}{s^2 - a^2} \qquad s > s_0 = |a|.$$

3. The Laplace transform of $f(x) = x^n$ is given by

$$\mathcal{L}\{x^n\} = \int_0^\infty e^{-sx}x^n\,dx.$$

By changing the dummy variable x by $sx = u$, show that

$$\mathcal{L}\{x^n\} = \frac{1}{s^{n+1}}\int_0^\infty e^{-u}u^n\,du, \qquad s > 0.$$

Hence by using Exercise 3 of Appendix III show that

$$\mathcal{L}\{x^n\} = \frac{n!}{s^{n+1}}, \qquad s > 0,\ n \text{ a positive integer.}$$

4. (a) Use integration by parts to derive Eq. (5). (b) Apply Theorem 2 with f replaced by f' to establish this result.

5. If $f^{(n)}(x)$ is a continuous function and if there exists an s_0 such that $e^{-sb}f(b)$, $e^{-sb}f'(b)$, $\ldots$, $e^{-sb}f^{(n-1)}(b)$ all tend to zero as $b \to \infty$ for $s > s_0$, then show by

repeated integration by parts that $\mathcal{L}\{f^{(n)}(x)\}$ exists for $s > s_0$, and that

$$\mathcal{L}\{f^{(n)}(x)\} = s^n\mathcal{L}\{f(x)\} - s^{n-1}f(0) - s^{n-2}f'(0) - \cdots - f^{(n-1)}(0),$$

for $s > s_0$.

Exercises 6 and 7 discuss the Laplace transform for periodic functions.

6. Let $f(x)$ be a periodic function with period T. (a) Show that if $A > 0$, then

$$\mathcal{L}\{f(x)\} = \frac{\int_0^T e^{-sx}f(x)\,dx}{1 - e^{-sT}}.$$

(b) Use this method to show that $\mathcal{L}\{\sin ax\} = a/(s^2 + a^2)$.

7. Use Exercise 6(a) to find the Laplace transform of each of the following periodic functions. In all cases A is a positive constant. Sketch the function in each case.

$$f(x) = \begin{cases} A & 0 \le x < 1 \\ -A & 1 \le x < 2 \end{cases} \quad \text{and } f(x+2) = f(x), \qquad \text{for all } x \ge 0.$$

This function, which occurs frequently in electronics, is called a square wave. It is not a continuous function, but its Laplace transform is defined.
(b)

$$f(x) = \begin{cases} A & 0 \le x < 1 \\ 0 & 1 \le x < 2 \end{cases} \quad \text{and } f(x+2) = f(x), \qquad \text{for all } x \ge 0.$$

(c) $f(x) = Ax$, for $0 \le x < T$ and $f(x + T) = f(x)$, for all $x \ge 0$. This function is called a sawtooth wave.
(d)

$$f(x) = \begin{cases} A \sin x & 0 \le x < \pi \\ 0 & \pi \le x < 2\pi \end{cases} \quad \text{and } f(x+2\pi) = f(x), \qquad \text{for all } x \ge 0.$$

This function arises in the study of electric circuits. A half-wave rectifier converts an alternating sinusoidal signal into the given function. See Figure 16.1.
(e) A full-wave rectifier converts the signal $A \sin x$ into

$$f(x) = \begin{cases} A \sin x & 0 \le x < \pi \\ -A \sin x & \pi \le x < 2\pi \end{cases} \quad \text{and } f(x+2\pi) = f(x), \qquad \text{for all } x \ge 0.$$

8. (a) Show that if $a > 0$ then

$$\mathcal{L}\{f(ax)\} = \frac{1}{a}F\left(\frac{s}{a}\right), \qquad \text{for } s > as_0,$$

where $F(s) = \mathcal{L}\{f(x)\}$, $s > s_0$. (b) Find the Laplace transforms of $\sin \omega t$, $\cos \omega t$, and $e^{\omega t}$, where ω is a constant, using the result of (a).

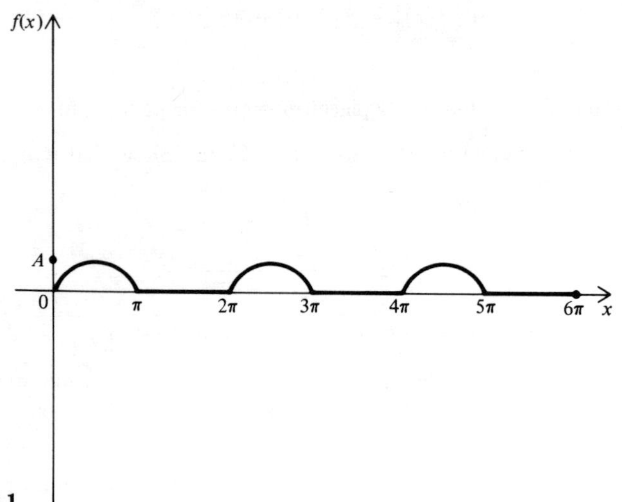

Figure 16.1

9. Show that if $f(x)$ is a continuous function with a Laplace transform $F(s)$ for $s > s_0 > 0$, then

$$\mathcal{L}\left\{\int_0^x f(z)\,dz\right\} = \frac{1}{s}\,F(s), \qquad \text{for } s > s_0.$$

16.3 THE INVERSE LAPLACE TRANSFORM

In Section 16.2 we showed how to determine the Laplace transform $F(s)$ of a given function $f(x)$. In the applications of Laplace transforms to solve DEs, we must solve the inverse problem: Given a function $F(s)$, determine the function $f(x)$ whose Laplace transform is $F(s)$. The process of solving the inverse problem is called the inversion of the Laplace transform and $f(x)$ is then called the *inverse Laplace transform*. It is denoted by $f(x) = \mathcal{L}^{-1}\{F(s)\}$.

The theory of inverse transforms requires knowledge of functions of a complex variable. Consequently, we shall not discuss it here. Instead, we shall proceed in an operational way. We assume without proof that no two continuous functions have the same Laplace transform. Thus if $F(s)$ is the Laplace transform of a continuous function, then there is only one continuous function $f(x) = \mathcal{L}^{-1}\{F(s)\}$ that is the inverse transform of $F(s)$. However, there may be other functions, which are not continuous, that have $F(s)$ as their Laplace transform (see Exercise 3). This result is proved in more advanced treatments of the Laplace transform. With this property we can use Table 16.1 to determine inverse Laplace transforms. That is, if a given $F(s)$ appears in the right-hand column of Table 16.1, we can determine uniquely the continuous inverse transform $f(x)$ from the corresponding entry in the left-hand column.

Finally, we state a property of the inverse transform that is useful in the applications of Table 16.1. A proof is sketched in Exercise 2.

theorem (linearity of the inverse transform)

If $f(x)$ and $g(x)$ are the continuous inverse transforms of $F(s)$ and $G(s)$, respectively, that is, if $f(x) = \mathcal{L}^{-1}\{F(s)\}$ and $g(x) = \mathcal{L}^{-1}\{G(s)\}$, then the continuous inverse transform of $c_1F(s) + c_2G(s)$, where c_1 and c_2 are arbitrary constants, is given by

$$\mathcal{L}^{-1}\{c_1F(s) + c_2G(s)\} = c_1\mathcal{L}^{-1}\{F(s)\} + c_2\mathcal{L}^{-1}\{G(s)\}. \tag{1}$$

EXAMPLE 1

Determine $f(x)$ if its Laplace transform is

$$F(s) = \frac{s}{s^2 + 9}, \qquad \text{for } s > 0.$$

solution. We observe from Table 16.1 that $s/(s^2 + a^2)$ is the Laplace transform of $\cos ax$. Thus $\cos ax$ is the inverse transform of $s/(s^2 + a^2)$; that is, $\cos ax = \mathcal{L}^{-1}\{s/(s^2 + a^2)\}$. Thus, by setting $a = 3$, we conclude that

$$f(x) = \cos 3x = \mathcal{L}^{-1}\left\{\frac{s}{s^2 + 9}\right\}.$$

EXAMPLE 2

Find the inverse transform $f(x)$ of

$$F(s) = \frac{1}{s^2 - 3s + 2}, \qquad \text{for } s > 2.$$

solution. This transform does not appear in the right-hand column of Table 16.1. We shall now rewrite $F(s)$ so that Table 16.1 can be used. The technique that we shall employ is the *method of partial fractions*, which was studied in calculus. As we shall see, partial-fraction decompositions of Laplace transforms are frequently used in solving DEs. Since $s^2 - 3s + 2 = (s - 1)(s - 2)$, we have

$$F(s) = \frac{1}{(s - 1)(s - 2)}.$$

By partial fractions we obtain

$$F(s) = \frac{1}{s - 2} - \frac{1}{s - 1}.$$

Then using the linearity property of $\mathcal{L}^{-1}$ (see the theorem), we get

$$f(x) = \mathcal{L}^{-1}\{F(s)\} = \mathcal{L}^{-1}\left\{\frac{1}{s-2} - \frac{1}{s-1}\right\} = \mathcal{L}^{-1}\left\{\frac{1}{s-2}\right\} - \mathcal{L}^{-1}\left\{\frac{1}{s-1}\right\}.$$

Since $e^{ax} = \mathcal{L}^{-1}\{1/(s-a)\}$, we get $f(x) = e^{2x} - e^{x}$.

Exercises

1. For each of the following expressions find the inverse Laplace transform by using Table 16.1.

   (a) $\dfrac{12}{2s-8}, \quad s > 4$ (b) $\dfrac{16}{s^2+8}, \quad s > 0$

   (c) $\dfrac{2}{s^2-5s+6}, \quad s > 3$ (d) $\dfrac{2}{s^3-s^2}, \quad s > 0$

   (e) $\dfrac{2}{s^3}, \quad s > 0$ (f) $\dfrac{s-6}{s^2-s-2}, \quad s > 2$

   (g) $\dfrac{s^2+2}{(s^2+1)(s^2+4)}, \quad s > 0$ (h) $\dfrac{2s^2+5s-1}{s^2(s^2+1)}, \quad s > 0$

2. Let $\mathcal{L}\{f(x)\} = F(s)$ and $\mathcal{L}\{g(x)\} = G(s)$. Then using the linearity property of the Laplace transform (Theorem 1 of Section 16.2), we have $\mathcal{L}\{c_1 f(x) + c_2 g(x)\} = c_1 F(s) + c_2 G(s)$. By taking the inverse Laplace transform on both sides, show that

$$\mathcal{L}^{-1}\{c_1 F(s) + c_2 G(s)\} = c_1 \mathcal{L}^{-1}\{F(s)\} + c_2 \mathcal{L}^{-1}\{G(s)\}.$$

3. (a) Show that the functions $f_1(x) = 1$ and

$$f_2(x) = \begin{cases} 1 & x \neq 2 \text{ and } x \neq 4 \\ 0 & x = 2 \text{ and } x = 4 \end{cases}$$

   have the same Laplace transform, namely $1/s$. Thus $\mathcal{L}^{-1}\{1/s\}$ is not unique. Observe that f_2 is not continuous for $x = 2$, and $x = 4$.

   (b) Consider $N(x) = f_2(x) - f_1(x)$. This function is zero except at $x = 2$ and $x = 4$. Show that

$$\int_0^x N(z)\, dz = 0 \tag{i}$$

   for *all* $x \geq 0$. Any function $N(x)$ satisfying (i) for all $x \geq 0$ is called a null function. A function that is zero except at a finite number of points is an example of a null function. It can be shown in general that if two functions have the same Laplace transform, then they must differ by a null function.

16.4 SOLUTION OF INITIAL-VALUE PROBLEMS

In this section we shall use the Laplace transform to solve IVPs. As we shall see, it is a compact procedure that is applicable to both homegeneous and inhomogeneous DEs. It gives the solution to the IVP directly without determining a general solution. The following examples illustrate the procedure.

EXAMPLE 1

Find the solution of the IVP

$$y' + y = x, \tag{1}$$

$$y(0) = 1. \tag{2}$$

solution. We multiply DE (1) by e^{-sx} and integrate with respect to x. This yields

$$\int_0^\infty y'e^{-sx}\,dx + \int_0^\infty ye^{-sx}\,dx = \int_0^\infty xe^{-sx}\,dx$$

or, equivalently,

$$\mathcal{L}\{y'\} + \mathcal{L}\{y\} = \mathcal{L}\{x\}. \tag{3}$$

Since $\mathcal{L}\{y'\} = s\mathcal{L}\{y\} - y(0)$ (see Eq. (4) of Section 16.2) and since $\mathcal{L}\{x\} = 1/s^2$ (see Table 16.1), Eq. (3) is reduced to

$$s\mathcal{L}\{y\} - y(0) + \mathcal{L}\{y\} = \frac{1}{s^2}.$$

We use the notation $Y(s) = \mathcal{L}\{y\}$ for the Laplace transform of y and solve this algebraic equation for $Y(s)$. Since $y(0) = 1$, this gives

$$Y(s) = \frac{1 + (1/s^2)}{s+1} = \frac{s^2+1}{s^2(s+1)}.$$

Thus we have obtained the Laplace transform of the solution of the IVP given by Eqs. (1) and (2). To determine the solution $y(x)$, we must find the inverse transform of $y(x) = \mathcal{L}^{-1}\{Y(s)\}$. First we use partial fractions to obtain

$$Y(s) = \frac{s^2+1}{s^2(s+1)} = -\frac{1}{s} + \frac{1}{s^2} + \frac{2}{s+1}.$$

Using the linearity property of $\mathcal{L}^{-1}$, we get

$$\begin{aligned}\mathcal{L}^{-1}\{Y(s)\} &= \mathcal{L}^{-1}\left\{-\frac{1}{s} + \frac{1}{s^2} + \frac{2}{s+1}\right\}\\ &= -\mathcal{L}^{-1}\left\{\frac{1}{s}\right\} + \mathcal{L}^{-1}\left\{\frac{1}{s^2}\right\} + 2\mathcal{L}^{-1}\left\{\frac{1}{s+1}\right\}.\end{aligned}$$

Since $1/s$, $1/s^2$, and $1/(s+1)$ are the Laplace transforms of 1, x, and e^{-x}, respectively, we have

$$y(x) = \mathcal{L}^{-1}\{Y(s)\} = -1 + x + 2e^{-x}$$

as the solution of the IVP.

EXAMPLE 2

Find the solution of the IVP

$$y'' + y = \cos 3x, \tag{4}$$

$$y(0) = 0,\ y'(0) = 0. \tag{5}$$

solution. We take the Laplace transform of both sides of DE (4) and get $\mathcal{L}\{y'' + y\} = \mathcal{L}\{\cos 3x\}$. We now use the linearity property of $\mathcal{L}$ and Equation (5) of Section 16.2 for $\mathcal{L}\{y''\}$ to get

$$\mathcal{L}\{y'' + y\} = s^2Y(s) - sy(0) - y'(0) + Y(s),$$

where $Y(s) = \mathcal{L}\{y\}$. Since $\mathcal{L}\{\cos 3x\} = s/(s^2 + 9)$, we finally obtain

$$s^2Y(s) - sy(0) - y'(0) + Y(s) = \frac{s}{s^2 + 9}.$$

By using the initial conditions given by Eq. (5) and solving this algebraic equation for $Y(s)$, we find that

$$Y(s) = \frac{s}{(s^2 + 9)(s^2 + 1)}.$$

Taking inverse Laplace transforms on both sides and using a partial fraction decomposition of Y, we obtain

$$y(x) = \mathcal{L}^{-1}\{Y(s)\} = \mathcal{L}^{-1}\left\{\frac{1}{8}\frac{s}{s^2 + 1} - \frac{1}{8}\frac{s}{s^2 + 9}\right\}$$

$$= \frac{1}{8}\mathcal{L}^{-1}\left\{\frac{s}{s^2 + 1}\right\} - \frac{1}{8}\mathcal{L}^{-1}\left\{\frac{s}{s^2 + 9}\right\}$$

We now use Table 16.1 and find

$$y(x) = \frac{1}{8}(\cos x - \cos 3x). \tag{6}$$

The two preceding examples exhibit the essential features of the application of Laplace transforms to the solution of IVPs. The DE for the unknown solution $y(x)$ is transformed into an algebraic equation for the Laplace transform. This equation is easily solved for $Y(s)$. The solution of the IVP is then obtained by finding the inverse transform of $Y(s)$. That is, we find the function $y(x)$ whose Laplace transform is $Y(s)$. This last step presents the main difficulty in the application of the method, and complicated techniques have been developed to find inverse transforms. However, in many cases of interest, Table 16.1 is sufficient. We note that in many instances $Y(s)$ must be decomposed by partial fractions as a preliminary to the use of Table 16.1.

We now use the transform method to solve an IVP for a system of two first-order ODEs.

EXAMPLE 3

Find the solution $y_1(x)$ and $y_2(x)$ of the following system of first-order DEs,

$$2y_1' - 3y_2' = 2e^{2x},$$

$$y_1' - 2y_2' = 0,$$

with the initial conditions $y_1(0) = 2$ and $y_2(0) = 1$.

solution. We define $Y_1(s) = \mathcal{L}\{y_1(x)\}$ and $Y_2(s) = \mathcal{L}\{y_2(x)\}$. We take the Laplace transforms of both sides of the DEs and use the linearity property of $\mathcal{L}$ to get

$$2\mathcal{L}\{y_1'\} - 3\mathcal{L}\{y_2'\} = 2\mathcal{L}\{e^{2x}\}$$

and

$$\mathcal{L}\{y_1'\} - 2\mathcal{L}\{y_2'\} = 0.$$

We now express $\mathcal{L}\{y_1'\}$, and $\mathcal{L}\{y_2'\}$ in terms of $\mathcal{L}\{y_1\} = Y_1(s)$ and $\mathcal{L}\{y_2\} = Y_2(s)$, respectively, and use $\mathcal{L}\{e^{2x}\} = 1/(s-2)$ in these equations to obtain

$$2[sY_1(s) - y_1(0)] - 3[sY_2(s) - y_2(0)] = \frac{2}{s-2}$$

and

$$[sY_1(s) - y_1(0)] - 2[sY_2(s) - y_2(0)] = 0.$$

By substituting the initial conditions into the equations and simplifying, we obtain

$$2sY_1(s) - 3sY_2(s) = \frac{s}{s-2},$$

and

$$sY_1(s) - 2sY_2(s) = 0.$$

The solution of these two algebraic equations in $Y_1(s)$ and $Y_2(s)$ is

$$Y_1(s) = \frac{2}{s-2} \quad \text{and} \quad Y_2(s) = \frac{1}{s-2}.$$

Since $\mathcal{L}\{e^{2x}\} = 1/(s-2)$, the solution of our problem is

$$y_1(x) = \mathcal{L}^{-1}\{Y_1(s)\} = 2e^{2x},$$

and

$$y_2(x) = \mathcal{L}^{-1}\{Y_2(s)\} = e^{2x}.$$

EXAMPLE 4

Find the solution of the IVP

$$y'' - 6y' + 10y = 0, \qquad y(0) = 0,\ y'(0) = 1.$$

solution. By taking the Laplace transform of both sides of the DE, using the linearity property of $\mathcal{L}$, and the expressions for the Laplace transforms of y' and y'', we get

$$Y(s) = \frac{1}{s^2 - 6s + 10}.$$

Since the factors of $s^2 - 6s + 10$ are complex numbers, the partial fraction decomposition of $Y(s)$ involves a substantial amount of manipulations with complex numbers. Therefore we shall use an alternative technique to find $y(x) = \mathcal{L}^{-1}\{Y(s)\}$. First we complete the square in the denominator of Y; that is, $s^2 - 6s + 10 = (s-3)^2 + 1$. Hence

$$Y(s) = \frac{1}{(s-3)^2 + 1}.$$

From Table 16.1, we have

$$\mathcal{L}\{e^{ax} \sin bx\} = \frac{b}{(s-a)^2 + b^2}.$$

Thus

$$y = \mathcal{L}^{-1}\{Y(s)\} = \mathcal{L}^{-1}\left\{\frac{1}{(s-3)^2 + 1}\right\} = e^{3x} \sin x$$

is the solution of the IVP.

The inverse transform $Y(s)$ in the above example can also be found by using the following shift theorem.

theorem

If $\mathcal{L}\{f(x)\} = F(s)$ for $s > s_0$, then

$$\mathcal{L}\{e^{ax} f(x)\} = F(s-a), \qquad \text{for } s > s_0 + a. \tag{7}$$

proof. By definition,

$$\mathcal{L}\{e^{ax} f(x)\} = \int_0^\infty e^{-sx} e^{ax} f(x)\, dx$$

$$= \int_0^\infty e^{-(s-a)x} f(x)\, dx = F(s-a).$$

This integral exists for $s - a > s_0$ because

$$\mathcal{L}\{f(x)\} = \int_0^\infty e^{-sx} f(x)\,dx$$

exists for $s > s_0$. This establishes the theorem.

corollary

$$\mathcal{L}^{-1}\{F(s-a)\} = e^{ax}f(x) = e^{ax}\mathcal{L}^{-1}\{F(s)\}. \tag{8}$$

The proof of the corollary follows directly from Eq. (7).

In Example 4 we used Table 16.1 to find $\mathcal{L}^{-1}\{1/[s-3)^2+1]\}$. We now use the corollary to evaluate it. Let $F(s) = 1/(s^2+1)$. Since $\mathcal{L}^{-1}\{1/(s^2+1)\} = \sin x$, an application of the corollary with $a = 3$ yields

$$\mathcal{L}^{-1}\left\{\frac{1}{(s-3)^2+1}\right\} = \mathcal{L}^{-1}\{F(s-3)\} = e^{3x}\mathcal{L}^{-1}\{F(s)\} = e^{3x}\sin x.$$

Exercises

1. Solve the following IVPs by using Laplace transforms.
   (a) $y'' - 3y' + 2y = 0, \quad y(0) = 1,\ y'(0) = -1$
   (b) $y'' + 3y' - 4y = 0, \quad y(0) = 1,\ y'(0) = 0$
   (c) $y^{(4)} - y = 0, \quad y(0) = 1,\ y'(0) = 0,\ y''(0) = 0,\ y'''(0) = 0$
   (d) $y'' + 9y = 4\sin x, \quad y(0) = 0,\ y'(0) = 0$
   (e) $y'' + 5y' + 6y = 3\sin 4x, \quad y(0) = 0,\ y'(0) = 1$
   (f) $y''' - 6y'' + 11y' - 6y = e^{4x}, \quad y(0) = 1,\ y'(0) = 1,\ y''(0) = 0$
   (g) $y_1' + y_1 + 2y_2 = 0, \quad y_1(0) = 1,\ y_2(0) = 2$
   $y_2' + 3y_1 + 2y_2 = 0,$
   (h) $y_1' + y_2 = e^{3x}, \quad y_1(0) = 0,\ y_2(0) = 0$
   $y_2' - y_1 = e^{2x},$
   (i) $y'' - 8y' + 25y = 0, \quad y(0) = 1,\ y'(0) = 0.$
   (j) $y''' - 5y'' + 12y' - 8y = e^{2x}, \quad y(0) = 0,\ y'(0) = 0,\ y''(0) = 1$
   (k) $y'' + 2y' + 2y = \cos 2x, \quad y(0) = 0,\ y'(0) = 0$
2. Find the solution of the following IVPs.
   (a) $y' + 2y = x, \quad y(1) = 0$. (*Hint*: Change the independent variable by the transformation $x = 1 + t$ so that the initial point $x = 1$ transforms to $t = 0$.)
   (b) $y'' - 3y' + 2y = e^x, \quad y(1) = 0,\ y'(1) = 0$
   (c) $y'' + 4y' + 3y = x, \quad y(-1) = 0,\ y'(-1) = 2$

16.5 DISCONTINUOUS FUNCTIONS

Discontinuous forcing functions and discontinuous coefficients occur frequently in the applications of DEs, as we have mentioned in the introduction to this chapter. Several such functions were given in Exercise 7 of Section 16. 2.

The Heaviside function $H(x)$, which is defined by

$$H(x) = \begin{cases} 0, & x < 0 \\ 1, & x \geq 0, \end{cases} \tag{1}$$

is a discontinuous function that is important in certain applications. It is discontinuous at $x = 0$; see Figure 16.2. This function is also called the unit step function.

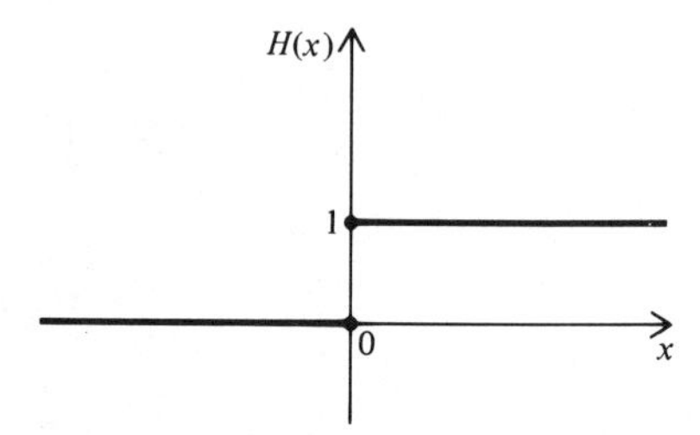

Figure 16.2

In order to study discontinuous forcing functions, we first recall some concepts from calculus. A function $f(x)$ is continuous at $x = x_0$ if $f(x_0)$ exists and $\lim_{x \to x_0} f(x) = f(x_0)$. If this limit does not exist, then $f(x)$ is discontinuous at $x = x_0$. We now use this definition to establish the geometrically obvious fact that $H(x)$ is discontinuous at $x = 0$. We observe that $H(x) = 1$ for all $x > 0$. Therefore $H(x) \to 1$ as $x \to 0$ only through positive values of x. However $H(x) = 0$ for all values of $x < 0$. Therefore $H(x) \to 0$ as $x \to 0$ through negative values of x. Since $H(x)$ approaches different values depending on whether $x \to 0$ from the left or the right, the function $H(x)$ has no limit at $x = 0$. Finally, we observe that $H(x)$ is finite for all values of x and it jumps in its value from 0 to 1 as x passes through $x = 0$ from left to right.

It is clear from the foregoing discussion that in studying discontinuous functions it is convenient to define a one-sided limit of a function. A right limit of $f(x)$ as $x \to x_0$, which we indicate by

$$\lim_{x \to x_0^+} f(x),$$

means take the limit as $x \to x_0$ through values of $x > x_0$. Similarly the left limit,

$$\lim_{x \to x_0^-} f(x),$$

means take the limit as $x \to x_0$ through values of $x < x_0$. Thus in terms of one-sided limits we have, for $f(x) = H(x)$,

$$\lim_{x \to 0^-} H(x) = 0 \quad \text{and} \quad \lim_{x \to 0^+} H(x) = 1.$$

If the one-sided limits are equal, then $f(x)$ is continuous at $x = x_0$. If

$$\lim_{x \to x_0^-} f(x) \quad \text{and} \quad \lim_{x \to x_0^+} f(x)$$

are finite numbers, which are not equal, then $f(x)$ has a *finite discontinuity* or *jump discontinuity* at x_0. A function $f(x)$ is *piecewise continuous* in an interval if $f(x)$ is continuous in that interval except at a finite number of points where it has jump discontinuities. Thus $H(x)$ defined by Eq. (1) is a piecewise continuous function. As another example we consider the function $f(x)$ that is defined by

$$f(x) = \begin{cases} x & 0 \le x < 1, \\ 3/2 & 1 \le x < 2, \\ x & x \ge 2. \end{cases}$$

See Figure 16.3 for the graph of this function. Here the left limit of $f(x)$ at $x = 1$ is given by

$$\lim_{x \to 1^-} f(x) = \lim_{x \to 1^-} x = 1$$

and the right limit is given by

$$\lim_{x \to 1^+} f(x) = \lim_{x \to 1^+} 3/2 = 3/2.$$

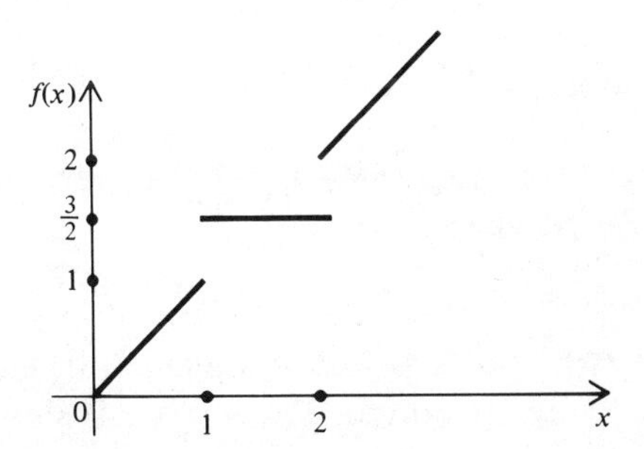

Figure 16.3

Since the left and the right limits of $f(x)$ at $x = 1$ exist and are not equal, the function $f(x)$ has a jump discontinuity at $x = 1$. Similarly, there is a jump discontinuity at $x = 2$. Moreover, $f(x)$ is continuous at all other points. Hence the function $f(x)$ is piecewise continuous for all $x \geq 0$.

We now consider the function $H(x - a)$. Since $H(x) = 0$ for $x < 0$ and $H(x) = 1$ for $x \geq 0$, the function $H(x - a) = 0$ for $x - a < 0$ and $H(x - a) = 1$ for $x - a \geq 0$. Thus

$$H(x - a) = \begin{cases} 0, & x < a, \\ 1, & x \geq a. \end{cases} \tag{2}$$

The graph of the function $y = H(x - a)$ with $a > 0$ is shown in Figure 16.4. It represents a shift of the function $H(x)$ by a units to the right.

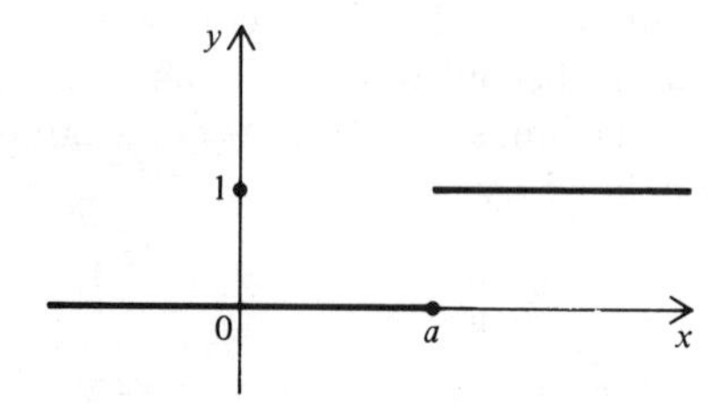

Figure 16.4

EXAMPLE 1

Find the Laplace transform of the unit step function $H(x - a)$.

solution. By the definition of the Laplace transform,

$$\mathcal{L}\{H(x - a)\} = \int_0^\infty e^{-sx} H(x - a)\, dx.$$

Since $H(x - a)$ is defined by Eq. (2), $H(x - a) = 0$ for $x < a$ and $H(x - a) = 1$ for $x > a$. Thus

$$\mathcal{L}\{H(x - a)\} = \int_a^\infty e^{-sx}\, dx = \frac{e^{-sx}}{-s}\bigg|_a^\infty = \frac{e^{-sa}}{s}, \qquad \text{for } s > s_0 = 0. \tag{3}$$

If $s \leq 0$, then the integral in Eq. (3) is unbounded.

If we set $a = 0$ in Eq. (3) we find that $\mathcal{L}\{H(x)\} = 1/s$, and thus Eq. (3) can be rewritten as

$$\mathcal{L}\{H(x - a)\} = e^{-sa}\mathcal{L}\{H(x)\}. \tag{4}$$

Since $H(x - a)$, with $a > 0$, is just a shift or translation of the function $H(x)$ to the right by an amount a, Eq. (4) gives a simple relation between the Laplace transforms of $H(x)$ and of its translation $H(x-a)$. We now show that the Laplace transforms of any function $f(x)$ defined for $x \geq 0$ and its translation are related as in Eq. (4). As we shall see, this is a useful result for finding inverse transforms. The analytic expression for the translation of a function $f(x)$ is

$$H(x-a)f(x-a) = \begin{cases} 0 & x < a \\ f(x-a) & x \geq a \end{cases}$$

theorem

If $\mathfrak{L}\{f(x)\} = F(s)$ for $s > s_0$, then for any $a \geq 0$,

$$\mathfrak{L}\{H(x-a)f(x-a)\} = e^{-sa}\mathfrak{L}\{f(x)\} = e^{-sa}F(s), \tag{5}$$

for $s > s_0$.

proof. From the definition of a Laplace transform, we have

$$\mathfrak{L}\{H(x-a)f(x-a)\} = \int_0^\infty e^{-sx}H(x-a)f(x-a)\,dx.$$

Since $H(x-a) = 0$ for $x < a$ and $H(x-a) = 1$ for $x \geq a$, the last equation simplifies to

$$\mathfrak{L}\{H(x-a)f(x-a)\} = \int_a^\infty e^{-sx}f(x-a)\,dx.$$

In this integral we change the variable of integration by $x - a = t$. Then

$$\mathfrak{L}\{H(x-a)f(x-a)\} = \int_0^\infty e^{-s(t+a)}f(t)\,dt$$

$$= e^{-sa}\int_0^\infty e^{-st}f(t)\,dt = e^{-sa}F(s)$$

for $s > s_0$. This proves the theorem.

corollary

$$\mathfrak{L}^{-1}\{e^{-sa}F(s)\} = H(x-a)f(x-a). \tag{6}$$

The proof of the corollary follows directly by taking the inverse transform of both sides of Eq. (5).

This theorem gives the relation between the Laplace transform of a function and the Laplace transform of a shifted function. We illustrate its application in the following example.

EXAMPLE 2

Find the solution of the IVP

$$y'' - 3y' + 2y = H(x-6), \qquad y(0) = y'(0) = 0.$$

solution. In the usual way, we take the Laplace transform of both sides of the DE. Then using the linearity of the Laplace transform and the expressions for the Laplace transforms of the derivatives of y, we have

$$(s^2 - 3s + 2)Y(s) = \frac{e^{-6s}}{s}.$$

Here we have used $\mathcal{L}\{H(x-6)\} = (1/s)e^{-6s}$ and $\mathcal{L}\{y(x)\} = Y(s)$. Solving for $Y(s)$, we have

$$Y(s) = \frac{1}{(s^2 - 3s + 2)s} e^{-6s}. \tag{7}$$

We observe that $s^2 - 3s + 2 = (s-2)(s-1)$. Therefore, by using partial fractions, we can rewrite Eq. (7) as

$$Y(s) = \frac{1}{2}\frac{e^{-6s}}{s-2} - \frac{e^{-6s}}{s-1} + \frac{1}{2}\frac{e^{-6s}}{s}.$$

Taking the inverse transform on both sides of this equation and using the linearity property of $\mathcal{L}^{-1}$, we get

$$y(x) = \mathcal{L}^{-1}\{Y(s)\} = \frac{1}{2}\mathcal{L}^{-1}\left\{\frac{e^{-6s}}{s-2}\right\} - \mathcal{L}^{-1}\left\{\frac{e^{-6s}}{s-1}\right\} + \frac{1}{2}\mathcal{L}^{-1}\left\{\frac{e^{-6s}}{s}\right\}. \tag{8}$$

From Table 16.1 we get $\mathcal{L}\{e^{bx}\} = 1/(s-b)$. Thus, from the corollary, we have

$$\mathcal{L}^{-1}\left\{e^{-6s}\frac{1}{s-b}\right\} = H(x-6)e^{b(x-6)}.$$

Using this fact in Eq. (8) for $b = 2$, 1, and 0, we obtain

$$y = H(x-6)[\tfrac{1}{2}e^{2(x-6)} - e^{x-6} + \tfrac{1}{2}]. \tag{9}$$

This is the required solution of the IVP. We observe that the only possible solution of the IVP for $0 < x < 6$ is $y = 0$ since the initial conditions are zero and the inhomogeneous term $H(x-6) = 0$ for $x < 6$. The reader should verify that $y''(x)$ is discontinuous at $x = 6$.

In order to study DEs with discontinuous inhomogeneous terms, we must extend our definition of a solution to include functions that do not have the same number of derivatives as the order of the equation at all points x. Thus the solution of Example 2, given by Eq. (9), does not have a second derivative at $x = 6$.

Exercises

1. Find the Laplace transforms of the following functions.

(a) $f(x) = \begin{cases} 1 & 0 \leq x < 1 \\ 2 & x \geq 1 \end{cases}$ (b) $f(x) = \begin{cases} 1 & 0 \leq x < 1 \\ 0 & 1 \leq x < 2 \\ 1 & x \geq 2 \end{cases}$

2. Solve the following by using the Laplace transform and the corollary in the text.

(a) $y'' + y' - 6y = H(x - 4), \quad x > 0$
$y(0) = 0, \; y'(0) = 1$

(b) $y'' + y = H(x - \pi), \quad x > 0$
$y(0) = 1, \; y'(0) = 0$

(c) $y'' + 6y' + 8y = H(x - \pi), \quad x > 0$
$y(0) = 0, \; y'(0) = 0$

(d) $y' + y = r(x), \; y(0) = 1,$
where

$$r(x) = \begin{cases} 1 & 0 \leq x < 1 \\ 2 & x \geq 1 \end{cases}$$

(e) $y'' + y = r(x), \quad y(0) = 0, \; y'(0) = 1,$
where

$$r(x) = \begin{cases} 1 & 0 \leq x < 1 \\ 0 & 1 \leq x < 2 \\ 1 & x \geq 2 \end{cases}$$

(f) $y'' + 4y = r(x), \quad y(0) = 0, \; y'(0) = 0,$
where

$$r(x) = \begin{cases} x & 0 \leq x < 2 \\ 1 & x \geq 2 \end{cases}$$

3. The Heaviside function can be used to write discontinuous functions such as those of Exercises 1 in more compact form. For example,

$$f(x) = \begin{cases} 1 & 0 \leq x < 1 \\ 2 & x \geq 1 \end{cases}$$

can be written as $H(x) + H(x - 1)$. Express $r(x)$ in Exercises 2(d)–2(f) in terms of Heaviside functions.

*4. (a) Suppose that f has a jump at $x = x_0$. Let

$$[f]_{x_0} = \lim_{x \to x_0^+} f(x) - \lim_{x \to x_0^-} f(x)$$

be the jump in f at $x = x_0$. Show that

$$\mathcal{L}\{f'\} = s\mathcal{L}\{f\} - f(0) - [f]_{x_0}e^{-sx_0}.$$

(b) Compute the Laplace transform of f' if f is equal to $H(x)$ and $H(x-1) + H(x)$.

5. Consider the IVP $\theta'' + 4\theta' + 3\theta = F(t)$, $\theta(0) = \theta'(0) = 0$, where $F(t) = F_0[H(t) - H(t-1)]$. This IVP describes the motion of a pendulum in a resisting medium that is acted upon by a force F_0 for $0 \leq t < 1$. At $t = 1$ the force is removed. (a) Determine the motion $\theta(t)$ of the pendulum. (b) Sketch $\theta(t)$. (*Hint*: Use the corollary of the theorem in Section 16.5.)

6. An undamped pendulum is acted upon by a force

$$F(t) = \begin{cases} F_0 t & 0 \leq t < \pi \\ 0 & \pi \leq t \end{cases}.$$

If the pendulum starts from rest, then the IVP to describe its motion is $\theta'' + \sqrt{g/l}\theta = F(t)$, $\theta(0) = \theta'(0) = 0$. (a) Determine the motion of the pendulum. (b) Does the steady-state solution depend on F_0?

We recall that the DE for the charge $q(t)$ in a RCL circuit is

$$Lq'' + Rq' + \frac{1}{C}q = E(t).$$

Here R, L, and C are the resistance, inductance, and capacitance of the circuit, respectively, and $E(t)$ is the electromotive force of applied voltage. The current I and the charge q are related by $I = q'$.

7. (a) Consider a series RC circuit with $E(t) = E_0[H(t-1) - H(t-2)\}$, assuming that the capacitor is initially uncharged and that the initial current in the circuit is zero. Solve this problem. (b) Sketch both the current and the charge.

8. (a) Show that if $y' + ky = r(x)$, $y(0) = 0$ then the Laplace transform $Y(s)$ of y is given by $Y(s) = T(s)R(s)$ where $R(s)$ is the Laplace transform of the characteristic polynomial of the associated homogeneous DE with m replaced by s. Here k is a constant. (b) Generalize this result to $y'' + 2by' + cy = r(x)$ $y(0) = y'(0) = 0$, with b and c constants. The functions $T(s)$ and $R(s)$ are called the transfer and excitation functions respectively.

*9. We have seen in Exercise 10 that in solving IVPs the Laplace transform of the solution $Y(s)$ appears in a product form.

We shall now describe a technique for inverting Laplace transforms when they are in the product form. If the transform is given as a product of two transforms $F(s)$ and $G(s)$, and the inverse transforms $f(x)$ and $g(x)$ respectively are known, then the inverse transform of the product is given by the *convolution formula*:

$$\mathcal{L}^{-1}\{F(s)G(s)\} = \int_0^x f(u)g(x-u)\,du. \tag{i}$$

By taking the Laplace transform of both sides of Eq. (i), we obtain

$$\mathcal{L}\left\{\int_0^x f(u)g(x-u)\,du\right\} = F(s)G(s). \tag{ii}$$

The integral

$$\int_0^x f(u)g(x-u)\,du$$

is called the convolution integral of f and g, and is denoted by $f * g$.

A rigorous proof of the validity of the convolution formula requires advanced mathematical techniques. However, an intuitive understanding can be obtained as follows:
(a) Show that

$$\mathcal{L}\{f * g\} = \int_0^\infty e^{-sx}\,dx \int_0^\infty f(u)g(x-u)H(x-u)\,du. \tag{iii}$$

where $H(x)$ is the step function. (b) Change the order of integration in (iii) and use the properties of the step function to get

$$\mathcal{L}\{f * g) = \int_0^\infty f(u)\,du \int_u^\infty e^{-sx}g(x-u)\,dx.$$

(c) In the inner integral let $x - u = t$ to derive (i).

*10. Solve the following IVPs using the result of Exercise 11.
(a) $y'' + y = \cos 3x, \quad y(0) = y'(0) = 0$
(b) $y'' - 3y' + 2y = H(x-6), \quad y(0) = y'(0) = 0$
(c) $y'' + y = f(x), \quad y(0) = 1,\ y'(0) = 0$

16.6 FUNCTIONS OF EXPONENTIAL ORDER AND EXISTENCE OF THE LAPLACE TRANSFORM

The Laplace transform is an improper integral. In Appendix III we show by example that improper integrals and hence Laplace transforms may be divergent. For example, it can be shown that the function $f(x) = e^{x^2}$ does not have a Laplace transform; see Exercise 2. We shall now give sufficient conditions on $f(x)$ to ensure that its Laplace transform exists. These conditions are concerned with the smoothness of $f(x)$ and its behavior for large values of x.

definition

A function $f(x)$ is of exponential order if there exist constants $M > 0$, $A > 0$ and a, such that

$$|f(x)| < Me^{ax}, \qquad \text{for all } x > A. \tag{1}$$

The definition states that $f(x)$ is of exponential order if, for all sufficiently large x ($x > A$), $f(x)$ is always smaller than the exponential e^{ax}. An equivalent definition of exponential order a is that $|f(x)|/e^{ax}$ is bounded as $x \to \infty$.

EXAMPLE 1

Show that $f(x) = x$ is of exponential order.

solution. If $a > 0$, then by using L'Hospital's rule we have

$$\lim_{x\to\infty} \frac{x}{e^{ax}} = \lim_{x\to\infty} \frac{1}{ae^{ax}} = 0.$$

Thus, for large x, $x < e^{ax}$. In particular, if a is sufficiently large (for example, $a = 2$), then $x < e^{ax}$ for all $x > 0$. Thus in the definition we have $M = 1$, A is any nonnegative number and $a = 2$.

EXAMPLE 2

Show that $f(x) = e^{x^2}$ is not of exponential order.

solution. For any a we have $e^{x^2}/e^{ax} = e^{x(x-a)}$. Since

$$\lim_{x\to\infty} e^{x(x-a)} = \infty$$

for any a, then e^{x^2}/e^{ax} is not bounded as $x \to \infty$ for some a. This violates the definition of exponential order.

If the Laplace transform

$$\mathcal{L}\{f(x)\} = \int_0^\infty e^{-sx} f(x)\, dx$$

is to exist, then, as we have already mentioned in Section 16.1, $f(x)$ cannot grow faster as $x \to \infty$ than e^{ax} for some $a < \infty$. Thus, as is indicated in the following theorem, the concept of a function of exponential order is natural for the existence of the Laplace transform.

theorem

If $f(x)$ is a piecewise continuous function on the interval $0 \le x \le x_0$ for all $x_0 > 0$ and $f(x)$ is a function of exponential order with constants $M > 0$, $A > 0$ and a, then the Laplace transform $\mathcal{L}\{f(x)\} = F(s)$ exists for all $s > a$.

proof: The proof of this theorem is sketched in Exercise 3.

Exercises

1. Determine which of the following functions are of exponential order. Give values of M, A, and a for each function that is of exponential order.

(a) $6x^2$ (b) $\sin x$ (c) e^{6x}
(d) x^n, for n any integer (e) e^{x^3} (f) $e^{\sin x}$

2. Show that the Laplace transform of the function e^{x^2} does not exist. (*Hint:* $\mathcal{L}\{e^{x^2}\} = \int_0^\infty e^{x(x-s)}\, dx$. Observe that the integrand is positive for $x > 0$ and is an increasing function for $x > s/2$.)

3. Prove the theorem using the following approach: Use the definition of exponential order to show that

$$\int_A^b e^{-sx} \, | f(x) | \, dx \leq \frac{M}{s-a} e^{-(s-a)A}$$

and thus $\int_A^\infty e^{-sx} \, | f(x) | \, dx$ converges. Use the inequality $0 \leq | f(x) | - f(x) \leq 2 \, | f(x) |$ to show that $\int_A^\infty e^{-sx} f(x) \, dx$ exists. Finally, show that $\int_0^\infty e^{-sx} f(x) \, dx$ exists.

appendix I

MODEL FORMULATION

In this Appendix we continue the discussion of Sections 1.4, 1.4-1, and 1.4-2 and derive mathematical models of some elementary mechanics and electric circuit problems.

I.1 THE FALLING PARTICLE

We now consider the vertical motion of a particle of constant mass m that is released at $t = 0$ from an altitude h with a given initial velocity v_0. We let $y(t)$ equal the distance the particle has descended at the time t from the initial altitude h; see Figure I.1. The velocity of the particle $v(t)$ is the rate of change of the displacement,

$$v = y'. \tag{1}$$

The acceleration $a(t)$ of the particle is the rate of change of velocity,

$$a = v'. \tag{2}$$

The momentum of the particle is defined as, momentum $= mv$.

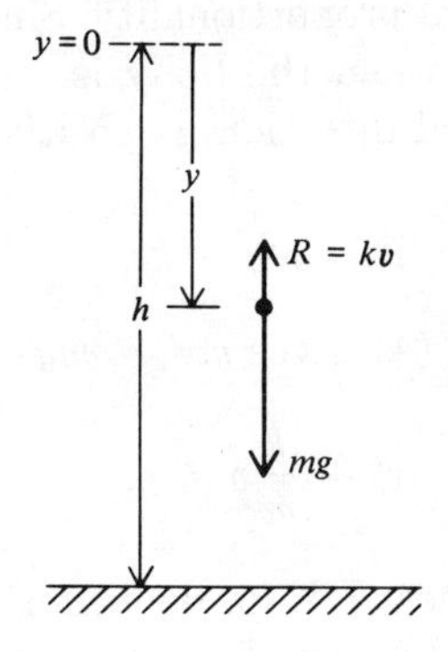

Figure I.1

Newton's laws of motion are fundamental in studying the motion of particles. We use in particular Newton's second law: *The total external force applied to a particle is equal to the rate of change of momentum of the particle.* That is,

$$F = (mv)' \tag{3}$$

where F is the total external force. If m is a constant, then we get from Eq. (3)

$$F = mv' \tag{4}$$

or, equivalently, by using Eq. (2) for the acceleration of a particle,

$$F = ma. \tag{5}$$

This is the familiar form of Newton's law: Force equals mass times acceleration.

The forces acting on the particle are the gravitational attraction of the earth and the frictional resistance of the air. The gravitational force acts downward, as we have indicated in Figure I.1. Its magnitude is equal to the weight of the particle, mg, where g is the gravitational acceleration. The weight of the particle and thus the gravitational force vary with the particle's distance from the center of the earth. We shall neglect this variation and assume that g is a constant. This constant is approximately 980 cm/sec^2 or 32 ft/sec^2. This assumption is accurate if h is small compared to the radius of the earth; that is, if we do not go far from the earth's surface. However, in problems of space travel we must account for the variation of the earth's gravitational attraction.

It has been observed experimentally that for sufficiently small velocities the resistance force R is proportional to the velocity. Thus we have

$$R = kv$$

where $k > 0$ is the constant of proportionality. Since the resistance opposes the motion, it acts upward when the body is falling as shown in Figure I.1. If we take the downward direction as positive, the total force on the particle is

$$F = mg - kv.$$

The equation of motion Eq. (4) gives $mv' = mg - kv$ or

$$v' + \frac{k}{m}v = g. \tag{6}$$

This is a first-order linear DE for determining the particle's velocity. Since the initial velocity of the particle is given as

$$v(0) = v_0, \tag{7}$$

the IVP for the velocity of the falling particle is to determine solutions of DE (6) that satisfy the initial condition given by Eq. (7). When this IVP is solved the displacement for any value of t is determined from the IVP

$$y' = v, \qquad y(0) = 0. \tag{8}$$

The DE in Eq. (8) is the definition of the velocity. The initial condition in Eq. (8) is the mathematical statement that the particle starts the motion at the height h because y is the distance the particle has descended from the height h.

Exercises

1. A body of mass m is dropped from the top of a building of height h.
   (a) Write an IVP for the velocity v of the body, assuming that the air resistance is proportional to the velocity of the body.
   (b) Write an IVP for the distance the mass has fallen.
   (c) Write an IVP for the distance the mass is above the earth.

2. For bodies moving at, or slightly above, the speed of sound the force of air resistance is proportional to the square of the velocity. Assume that a body is projected downward from a height of 10,000 feet with an initial velocity v_0 that is close to the speed of sound. Find an IVP for the velocity v.

3. (a) Neglecting air resistance, show that an IVP for the distance fallen by a body that is dropped from a height h is

   $$y'' = g, \qquad y(0) = y'(0) = 0.$$

   (b) Solve this IVP by integrating both sides of the DE twice.

4. Suppose that a body is projected upward from the earth's surface with velocity v_0. Neglecting air resistance, derive the IVP.

5. Use Eqs. (1) and (6) to derive a single DE for the displacement $y(t)$. What are the appropriate initial conditions?

I.2 THE VIBRATING PENDULUM

The problem of mathematically predicting the motion of the vibrating pendulum shown in Figure I.2 was studied as early as the eighteenth century by L. Euler and other scientists. It was motivated by the use of pendulums in clocks and other devices and by the simplicity of conducting experiments with pendulums. It is still of interest because it has many features common to other phenomena where oscillations take place.

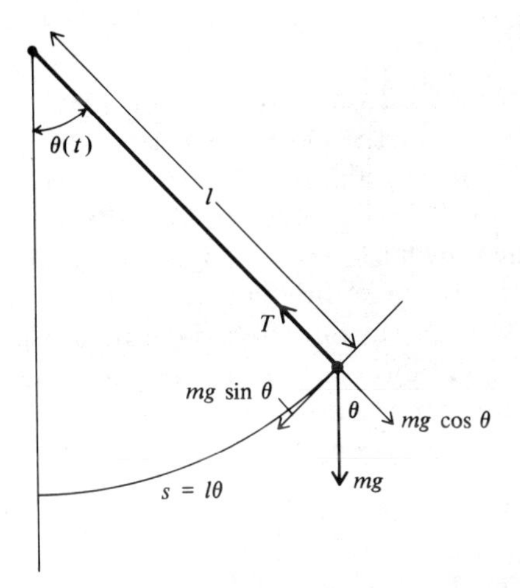

Figure I.2

The pendulum consists of a weight of constant mass m, attached to a slender rod of length l. The opposite end of the rod is connected to a pivot. The pendulum swings in the plane of the paper, as shown in Figure I.2. We wish to derive a DE to describe the motion of the pendulum due to a given initial angular displacement and a given initial angular velocity. To obtain a simple theory, we neglect the flexibility of the rod; that is, we assume that it is a rigid body. We also assume that the distributed mass is equivalent to a point mass of the same weight as we have shown in the figure. For many pendulums, the weight of the rod is small compared to the weight of the mass. To simplify our derivation we shall completely neglect the rod's weight. Furthermore, we shall neglect any resistance to the pendulum's motion due to air resistance, friction in the pivot, or any other source.

Since the rod is rigid, it does not change length, and thus the point mass must move along an arc of the circle with center at the pivot and radius l. If we denote the angular displacement of the pendulum in radians by $\theta(t)$, as shown in Figure I.2, then the distance $S(t)$ that the mass moves along the arc from its vertical position at time t is $S = l\theta$. The angle θ is positive if the pendulum is to the right of the vertical, and negative if it is to the left of the vertical. We assume that the motion of the point mass or particle is adequately described by Newton's second law. This states that the total force acting on a particle is equal to the change in momentum of the particle in the direction that the force is acting. Since the particle has constant mass, this gives the familiar equation

$$F = ma. \tag{1}$$

Here F is the force and a is the acceleration of the mass in the direction of F.

For any value of t we apply Newton's law, Eq. (1), to the mass in the tangent direction of the path along which it moves. The forces acting on the mass are shown in Figure I.2. Here T is the reaction force of the rod on the mass. Since T is along the radius of the arc of motion, it is perpendicular to the tangent to the arc. Thus, only the gravity force mg has a component in the tangent direction. It is $F = -mg \sin \theta$. The negative sign is required because the force acts opposite to the direction of increasing θ when θ is positive. Since $S(t)$ is the length along the arc of motion, the acceleration in this direction is given by $a = S'' = l\theta''$. Thus from Newton's law we have $-mg \sin \theta = ml\theta''$ or, equivalently,

$$\theta'' + \frac{g}{l} \sin \theta = 0. \tag{2}$$

At the initial time t_0 the pendulum is given an initial angular displacement θ_0 and an initial angular velocity θ_1. Then the IVP for the pendulum is to determine solutions of Eq. (2) that satisfy the initial conditions

$$\theta(t_0) = \theta_0, \qquad \theta'(t_0) = \theta_1. \tag{3}$$

This IVP is a mathematical model of the motion of the pendulum. Several important assumptions have been made in its derivation. Although these assumptions seem to be reasonable, it is not clear in advance whether the IVP has a solution and whether it has only one solution. Physical experiments show that the vibrating pendulum will execute a well-defined motion for any choice of θ_0 and θ_1. If the IVP is to be a good mathematical model of the pendulum, it certainly should have one, and only one, solution.

Equations (2) and (3) can be solved explicitly. Furthermore, it has been shown that there is only one solution. The solution predicts that the pendulum oscillates indefinitely. Experiments with pendulums show that their motions are accurately approximated by this solution for small values of t. However, for large values of t, experiments show that the pendulum executes successively smaller swings and eventually comes to rest. This is true no matter how carefully the pendulum is made or how thoroughly the pivot is lubricated. The resistance to the motion, because of effects such as friction in the pivot and air resistance, ultimately stops the pendulum. Thus, the mathematical model given in Eqs. (2) and (3) is appropriate only for relatively small values of t.

If we wish to discuss the motion for large t, we must modify the DE (2) to account for frictional resistance. Experiments suggest that the frictional resistance R is proportional to the velocity $v = l\theta'$ and that R is always opposite to the direction of motion. That is,

$$R = -\mu l\theta' \tag{4}$$

where the proportionality constant $\mu > 0$ is called the friction coefficient. When θ is increasing, $\theta' > 0$ and hence $R < 0$. When θ is decreasing, $\theta' < 0$ and $R > 0$. Thus the frictional resistance force R given by Eq. (4) always opposes the motion. To modify DE (2), we again apply Newton's law, $F = ma$, to the mass but we now include R. This gives $-mg \sin\theta - \mu l\theta' = ml\theta''$ or, equivalently,

$$\theta'' + \left(\frac{\mu}{m}\right)\theta' + \left(\frac{g}{l}\right)\sin\theta = 0. \tag{5}$$

The initial conditions for Eq. (5) are given in Eq. (3).

It is shown in more advanced texts that for $\mu > 0$ the solution of Eq. (5) approaches zero as $t \to \infty$, as is observed in experiments. Thus, the mathematical model given by Eqs. (5) and (3) more accurately describes the motion of he pendu um than the model given by Eqs. (2) and (3). Al hough many properties of the solution of the IVP given by Eqs. (5) and (3) have been derived, an explicit solution has not been obtained.

If the initial values θ_0 and θ_1 are small, then it is reasonable to expect that the solution $\theta(t)$ will be small for all t. By using the Taylor series expansion for $\sin\theta$, we have

$$\sin\theta = \theta - \frac{\theta^3}{3!} + \frac{\theta^5}{5!} - \cdots.$$

For θ small we can neglect the θ^3 term and all higher powers of θ in this equation, and we get $\sin\theta \approx \theta$. By substituting this approximation for

$\sin\theta$ into the DE (5) we have

$$\theta'' + \left(\frac{\mu}{m}\right)\theta' + \left(\frac{g}{g}\right)\theta = 0. \tag{6}$$

This is a linear, second-order equation, as the reader can easily verify. It is used to describe motions of the pendulum where the value of $\theta(t)$ is always small. Such motions are called *small-amplitude* motions. As we shall show in Chapters 4 and 5, we can explicitly solve this DE.

Finally, if a force $r(t)$ is applied to the mass and acts tangent to the circular arc, then the modified equation of motion corresponding to Eq. (6) is

$$\theta'' + \left(\frac{\mu}{m}\right)\theta' + \left(\frac{g}{l}\right)\theta = \frac{r(t)}{ml}. \tag{7}$$

Equation (7) describes the small-amplitude forced vibrations of the pendulum. The inhomogeneous term $r(t)$ is called the forcing function. This DE will be studied in Chapters 7 and 8.

Exercises

1. A pendulum that is in the vertical position $\theta = 0$ at $t = 0$ is started in motion with an initial angular velocity θ_1. Write the IVP that describes the small-amplitude motion, assuming no air resistance.

2. List all the assumptions that were made in deriving the mathematical model given by Eq. (2).

3. Compare the values of $\sin\theta$ and θ for small angles and determine approximately how large θ can be and still satisfy $|\sin\theta - \theta| < 0.01$.

*4. Suppose that a pendulum moves through a resisting medium, which exerts a force on the pendulum that is proportional to the square of the velocity. Derive the DE for the vibrating pendulum? (*Hint:* $(\theta')^2 = \pm\theta'|\theta'|$).

1.3 THE SIMPLE ELECTRIC CIRCUIT

The flow of electricity and other electromagnetic phenomena are mathematically described by a complicated set of partial differential equations known as Maxwell's equations. However, a simpler mathematical model can be obtained for many problems concerning the flow of electricity in a conductor by assuming that the electrical properties of the conductor, which may vary along the length, can be discretized and assigned specific values for the entire conductor. For example, when a current passes through a wire, the flow of electricity is resisted by the wire along its entire length. Indeed, the resistance may vary from point to point in the wire. However, a single quantity is assigned to the wire as its resistance. It is determined by measuring the current i that passes from one end of the wire to the other end due to a potential difference of e_R volts between the ends of the wire. Then the resistance R is given by Ohm's law as

$$R = \frac{e_R}{i}. \tag{1}$$

Here i is measured in amperes and R is given in ohms.

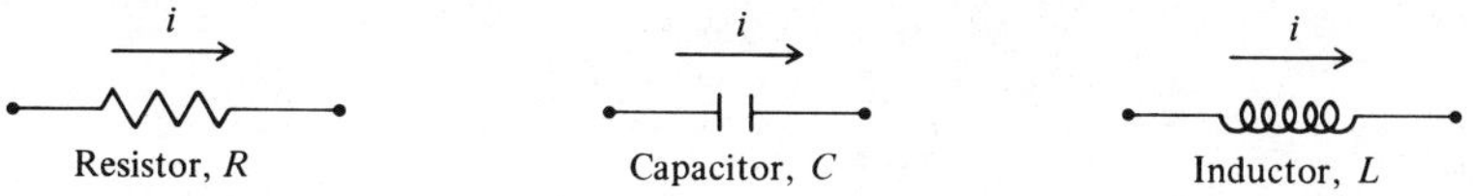

Resistor, R Capacitor, C Inductor, L

Figure I.3

The discretization of the properties of an electric circuit is frequently called the *lumped-parameter* description of the circuit. In the lumped-parameter description the basic parameters of a simple electric circuit are resistance, capacitance, and inductance. In diagrams of electric cicuits they are usually indicated by the symbols shown in Figure I.3. The three lumped parameters or circuit elements are called resistors, capacitors, and inductors. The three basic quantities that describe the flow of electricity in a circuit are the current i, the voltage e, and the charge y. They are all functions of time t. The current in a wire is the time rate of change of the charge flowing in the wire, or

$$i = y'. \tag{2}$$

From Ohm's law, Eq. (1), we conclude that the voltage drop e_R across a resistor due to the flow of current i is given by

$$e_R = Ri. \tag{3}$$

By using the other elementary laws of electricity, we can show that the voltage drops, e_C across a capacitor and e_L across an inductor, due to the flow of current i are

$$e_C = \frac{1}{C} y \tag{4}$$

$$e_L = Li'. \tag{5}$$

The quantity C in Eq. (4) is called the capacitance of the circuit and the quantity L in Eq. (5) is called the inductance. The units of C and L that we shall use are farads and henries, respectively. In general, R, C, and L

depend on t and may also depend on the current i. They are given functions of these variables. For example, the resistance of wire may change due to heating caused by the flow of current. However, for many applications R, C, and L can be approximated by constants.

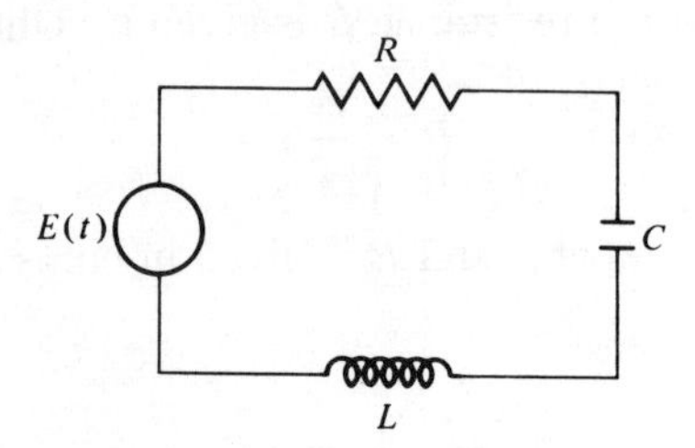

Figure I.4

A closed electric circuit is represented in Figure I.4. An inductor L, a resistor R, and a capacitor C are connected in series with a voltage source $E(t)$. The voltage source can be, for example, a battery or an electric generator. The basic law that results from the conservation of energy and describes the flow of electricity through the circuit was discovered by the German physicist and mathematician G. Kirchhoff. It is: *The algebraic sum of all voltage drops around a closed circuit is zero for all t.* Since the voltage drops across the individual circuit elements are given by Eqs. (3), (4), and (5), by applying Kirchhoff's law to the circuit in Figure I.4, we obtain

$$Li' + Ri + \frac{1}{C}y - E = 0. \tag{6}$$

In Eq. (6) the minus sign reflects the fact that the voltage supplied by the source is a voltage increase, and not a voltage drop.

Equations (2) and (6) are a system of two first-order DEs for the current i and the charge y. We can eliminate either i or y and obtain a single DE. For example, we eliminate i by inserting Eq. (2) into (6) and get

$$Ly'' + Ry' + \frac{1}{C}y = E \tag{7}$$

This is a second-order DE for the charge y. If L, R, and C are functions of t only, then Eq. (7) is a linear DE. If $E(t) = 0$ for all t, then Eq. (7) is a homogeneous DE.

To determine the charge y in the circuit for all time, we must specify in addition to DE (7), the charge and current in the circuit at the initial time t_0. Since $y' = i$, the initial conditions are

$$y(t_0) = a_0, \qquad y'(t_0) = a_1, \tag{8}$$

where a_0 and a_1 are given numbers. The IVP for the RCL electric circuit is to determine a solution of DE (7) that satisfies the initial conditions given in Eq. (8).

If the circuit is such that $L = 0$, then DE (7) is reduced to

$$Ry' + \frac{1}{C}y = E, \tag{9}$$

which is a first-order DE. A circuit with $L = 0$ is called an RC circuit. The IVP for the RC circuit consists of DE (9) and the initial condition

$$y(t_0) = a_0. \tag{10}$$

If the capacitor is removed from the circuit in Figure I.4, then DE (7) is reduced to $Ly'' + Ry' = E$. However, by using the definition given in Eq. (2) for i in this DE, we obtain

$$Li' + Ri = E. \tag{11}$$

This is a first-order DE for the current i. Such a circuit is called an RL circuit. Then the IVP consists of DE (11) and the initial condition

$$i(t_0) = a_1. \tag{12}$$

Exercises

1. (a) A capacitor with initial charge a_0 is discharged through a resistor of resistance R. What is the IVP for the charge on the capacitor? Find a DE for the current $i(t)$ in the circuit.
2. Write the IVP that describes the current in an LC circuit.
3. Kirchhoff's law could also be stated as: The sum of all the voltage drops around a closed circuit is equal to the applied electromotive force or the source voltage $E(t)$. Show that this statement leads to DE (7) for the RCL circuit.
4. Derive an IVP for the current in an RCL circuit.

   In deriving Eq. (6) we have neglected the effects described in Exercise 5.

5. The resistor in an RC circuit heats up due to the current flowing through it. The change in temperature of the resistor alters its resistance and thus the resistance is a function of the current. What is the DE for the current i if $R = R_0 + \alpha i$, where R_0 and α are constants? Is it a linear DE?

appendix II

THE COMPLEX-VALUED EXPONENTIAL FUNCTION

We assume that the reader has prior experience with the manipulation of complex numbers. In this Appendix we establish the properties of the complex-valued exponential function that were stated in Section 4.3. We first consider the complex number e^C that is defined for any complex number $C = A + iB$ by

$$e^C = e^A (\cos B + i \sin B); \tag{1}$$

see Eq. (3) in Section 4.3.

theorem 1

Let $C_1 = A + iB$ and $C_2 = a + ib$ be any two complex numbers. Then

$$e^{C_1+C_2} = e^{C_1}e^{C_2}. \tag{2}$$

proof. If $B = b = 0$, that is, C_1 and C_2 are real numbers, then Eq. (2) is a well-known property of the exponential function. If C_1 and C_2 are complex numbers, then the proof follows directly from the definition given in Eq. (1). Since $C_1 + C_2 = (A + a) + i(B + b)$, we have from Eq. (1) that

$$e^{C_1+C_2} = e^{(A+a)+i(B+b)} = e^{(A+a)}[\cos (B + b) + i \sin (B + b)]. \tag{3}$$

By substituting the trigonometric identities,

$$\cos(B+b) = \cos B \cos b - \sin B \sin b,$$

$$\sin(B+b) = \sin B \cos b + \cos B \sin b$$

in Eq. (3), and rearranging the terms in the resulting equation, we get

$$e^{C_1+C_2} = e^{(A+a)}(\cos B + i \sin B)(\cos b + i \sin b). \tag{4}$$

Since A and a are real, $e^{(A+a)} = e^A e^a$. We use this result in Eq. (4) to get

$$e^{C_1+C_2} = [e^A(\cos B + i \sin B)][e^a(\cos b + i \sin b)]. \tag{5}$$

Since the first factor in Eq. (5) is e^{C_1} and the second factor is e^{C_2}, Eq. (5) gives the required result (2).

theorem 2

If $C = A + iB$ is any complex number, then $e^C \neq 0$.

proof. From Eq. (1) $e^C = e^A(\cos B + i \sin B)$. Since $\cos^2 B + \sin^2 B = 1$, $\cos B$ and $\sin B$ cannot be zero simultaneously. Furthermore, $e^A \neq 0$ for any real A. Thus we conclude $e^C \neq 0$.

We shall now study complex-valued functions of the real variable x that are defined on the interval (x_1, x_2). Let $f(x)$ denote a function defined on (x_1, x_2). If, for at least one value of x in this interval the corresponding value of $f(x)$ is a complex number, then $f(x)$ is a complex-valued function of the real variable x. Thus for each x in the interval we may express the complex-valued function $f(x)$ as,

$$f(x) = u(x) + iv(x), \tag{7}$$

where for each fixed x, $u(x)$, and $v(x)$ are real numbers. Hence $u(x)$ and $v(x)$ are real-valued functions of x. They are called the real and imaginary parts of $f(x)$ respectively. If $v(x) = 0$ for all values of x, then $f(x) = u(x)$ is a real-valued function. An example of a complex-valued function is $f(x) = 2x + \sqrt{1 - x^2}$ for any interval containing points $|x| > 1$. Another example is e^{mx}, where m is a complex number; see Eq. (5) of Section 4.3. Other examples are given in Exercise 1.

The product of a complex number $C = A + iB$ and a complex-valued function $f(x)$ is given by

$$\begin{aligned} Cf(x) &= (A + iB)[u(x) + iv(x)] \\ &= [Au(x) - Bv(x)] + i[Bu(x) + Av(x)]. \end{aligned}$$

definition 1

Let $f(x)$ be a complex-valued function of the real variable x. If

$$f'(x) = \lim_{h \to 0} \frac{f(x+h) - f(x)}{h} \tag{8}$$

exists, then $f(x)$ is differentiable at the point x and its derivative is denoted by $f'(x)$.

This definition of the derivative, in terms of the limit of a difference quotient, is analogous to the definition for real-valued functions.

We substitute Eq. (7) into Eq. (8) and rearrange terms to get,

$$f'(x) = \lim_{h \to 0} \left\{ \frac{u(x+h) - u(x)}{h} + i \frac{v(x+h) - v(x)}{h} \right\}. \tag{9}$$

We recall from calculus that the limit of the sum of two functions is the sum of the limits of the individual functions provided that these limits exist. The same result can be established for complex-valued functions of a real variable. Thus we can write Eq. (9) as

$$f'(x) = \lim_{h \to 0} \frac{u(x+h) - u(x)}{h} + i \lim_{h \to 0} \frac{v(x+h) - v(x)}{h}. \tag{10}$$

The two expressions in brackets on the right side of Eq. (10) are merely the difference quotients of the real-valued functions $u(x)$ and $v(x)$. The following theorem is then a direct consequence of Eq. (10):

theorem 3

If $u(x)$ and $v(x)$ are differentiable functions, then $f(x)$ is a differentiable function and

$$f'(x) = u'(x) + iv'(x). \tag{11}$$

The reader should should verify by using Theorem 3 that

$$[Cf(x)]' = Cf'(x), \tag{12}$$

where C is a constant. Many of the other differentiation formulas for real-valued functions are also valid for complex-valued functions; see Exercise 10.

The derivative of the complex-valued exponential function is given in

theorem 4

If $m = \alpha + i\beta$ is any complex number, then

$$(e^{mx})' = me^{mx} \tag{13}$$

proof. We replace C in the definition given in Eq. (1) of e^C by mx and obtain,

$$e^{mx} = e^{\alpha x}(\cos \beta x + i \sin \beta x). \tag{14}$$

By applying Eq. (11) for the derivative of a complex-valued function to Eq. (14) we get

$$(e^{mx})' = [e^{\alpha x} \cos \beta x]' + i[e^{\alpha x} \sin \beta x]'. \tag{15}$$

We perform the indicated differentiations on the right side of Eq. (15) and obtain,

$$(e^{mx})' = (\alpha + i\beta)e^{\alpha x} \cos \beta x + (-\beta + i\alpha)e^{\alpha x} \sin \beta x. \tag{16}$$

Since $\alpha + i\beta = m$ and $-\beta + i\alpha = i(\alpha + i\beta) = im$, Eqs. (16) and (14) yield

$$(e^{mx})' = m(e^{\alpha x} \cos \beta x + ie^{\alpha x} \sin \beta x) = me^{mx},$$

which proves the theorem.

Exercises

1. For what values of x, if any, are each of the following functions not real valued:
   (a) $\sqrt{1 - x}$ (b) $3x + i\sqrt{1 - x}$
   (c) $\sin x + i \cos x$ (d) $e^{\sqrt{1-x^2}}$

2. Show that $e^{i\pi/2} = i$, $e^{i\pi} = -1$ and $e^{2n\pi i} = 1$ for $n = 0, \pm 1, \pm 2, \ldots$.

3. Prove, using Definition 1, that $e^{C_1 - C_2} = e^{C_1}/e^{C_2}$.

4. For complex numbers $C = A + iB$, we consider the A, B plane and introduce the polar coordinates r, θ by the transformation $A = r \cos \theta$ and $B = r \sin \theta$. Show that $C = r(\cos \theta + i \sin \theta) = re^{i\theta}$ where $r = | C | = \sqrt{A^2 + B^2}$ and $\theta = \tan^{-1}(B/A)$. This is called the polar representation of a complex number. Represent the following complex numbers in polar form: $C = i$; $C = -i$; $C = 1 + i$; $C = 1 - i$ and $C = i^2$. Show that $\bar{C} = re^{-i\theta}$, where $\bar{C}$ is the complex conjugate of C. That is, if $C = A + iB$, then $\bar{C} = A - iB$.

5. Let $C_1 = r_1 e^{i\theta_1}$ and $C_2 = r_2 e^{i\theta_2}$ be the polar representations of the complex numbers C_1 and C_2. Show that

$$C_1 C_2 = r_1 r_2[\cos(\theta_1 + \theta_2) + i \sin(\theta_1 + \theta_2)],$$

$$C_1/C_2 = (r_1/r_2)[\cos(\theta_1 - \theta_2) + i \sin(\theta_1 - \theta_2)].$$

6. (a) Use the results of Exercise 5 to prove De Moivre's formula,

$$(\cos \theta + i \sin \theta)^n = \cos n\theta + i \sin n\theta,$$

   where n is any integer.
   (b) Use De Moivre's formula to show that

$$C^n = r^n(\cos n\theta + i \sin n\theta). \tag{i}$$

7. De Moivre's formula and the result of Exercise 6 can be summarized by writing

$$C^n = r^n e^{in\theta}.$$

Evaluate i^2, $(1+i)^3$, $(3+2i)^2$, $(i)^{-2}$, $(1+i)^{-3}$ using the results of Exercise 6.

8. Let n be a positive integer and C be a complex number. The nth root of C, $C^{1/n}$, is any complex number K that satisfies the equation $K^n = C$. If $K = ke^{i\phi}$ and $C = re^{i\theta}$, use Exercise 6b to show that $k = r^{1/n}$ and $n\phi = \theta + 2m\pi$ ($m = 0, \pm 1, \pm 2, \ldots$). Thus show that

$$C^{1/n} = r^{1/n}\left[\cos\left(\frac{\theta}{n} + \frac{2m\pi}{n}\right) + i\sin\left(\frac{\theta}{n} + \frac{2m\pi}{n}\right)\right]$$

is a formula for the n, nth roots of the number C. Use this formula to find the roots:

$$\sqrt{i},\ (1+i)^{1/3},\ \sqrt[3]{1},\ \sqrt[4]{4},\ (3+7i)^{1/4}.$$

In each case show that $\sqrt[p]{C}$ has p distinct values.

9. Calculate $f'(x)$ where
(a) $f(x) = e^{\alpha x}\cos\beta x + ie^{\alpha x}\sin x$, where α and β are constants,
(b) $f(x) = \ln(x) + i\sin\beta x$.

10. If $f_1(x)$ and $f_2(x)$ are two differentiable complex-valued functions of x, use Eq. (11) to show that the following familiar differentiation formulas are valid.
(a) $(f_1f_2)' = f_1f_2' + f_2f_1'$
(b) $(f_1/f_2)' = (f_2f_1' - f_1f_2')/f_2^2$, if $f_2 \neq 0$.

appendix III

IMPROPER INTEGRALS AND THE LAPLACE TRANSFORM

The Laplace transform involves an integral over the range from zero to infinity. An integral over an infinite interval, such as

$$\int_a^\infty f(x)\,dx,$$

is an improper integral. We review some of the basic properties of improper integrals. We give a definition of this improper integral in terms of a limit of ordinary definite integrals.

definition

The improper integral

$$\int_a^\infty f(x)\,dx$$

exists (or converges) if the definite integral

$$\int_a^b f(x)\,dx$$

exists for all $b > a$ and if

$$\lim_{b\to\infty} \int_a^b f(x)\,dx$$

exists. When the improper integral exists, its value is given by

$$\int_a^\infty f(x)\,dx = \lim_{b\to\infty} \int_a^b f(x)\,dx.$$

If the limit does not exist, then the improper integral does not exist (diverges).

EXAMPLE 1

Show that the improper integral

$$\int_1^\infty x^{-2}\,dx$$

exists, and determine its value.

solution. We use the Definition Thus we first consider the ordinary integral

$$\int_1^b x^{-2}\,dx.$$

Then

$$\int_1^b x^{-2}\,dx = \frac{x^{-1}}{-1}\Bigg|_1^b = -\frac{1}{b} + 1.$$

The improper integral is then given by

$$\int_1^\infty x^{-2}\,dx = \lim_{b\to\infty} \int_1^b x^{-2}\,dx = \lim_{b\to\infty}\left(-\frac{1}{b} + 1\right) = 1.$$

Since the limit exists, the improper integral exists and its value is 1.

EXAMPLE 2

Show that the improper integral

$$\int_1^\infty x^{-1}\,dx$$

diverges.

solution.

$$\int_1^\infty x^{-1}\,dx = \lim_{b\to\infty} \int_1^b x^{-1}\,dx = \lim_{b\to\infty} [\ln x]_1^b = 0$$

$$= \lim_{b\to\infty} [\ln b - \ln 1] = \infty.$$

Therefore the integral diverges.

EXAMPLE 3

Show that

$$\int_0^\infty \sin x \, dx$$

diverges.

solution.

$$\int_0^\infty \sin x \, dx = \lim_{b\to\infty} \int_0^b \sin x \, dx = -\lim_{b\to\infty} \cos b + 1.$$

Although $\cos b$ is bounded for all b, the limit of $\cos b$ does not exist because $\cos b$ oscillates between $+1$ and -1 infinitely often as $b \to \infty$. Therefore the integral is divergent.

Exercises

1. Determine whether the following integrals converge or diverge. Evaluate the ones that converge.

   (a) $\int_1^\infty x^{-5/2} \, dx$ (b) $\int_0^\infty e^{-x} \, dx$

   (c) $\int_0^\infty \frac{dx}{x^2+1}$ (d) $\int_0^\infty x^2 e^{-x} \, dx$

   (e) $\int_0^\infty e^{-x} \sin x \, dx$

2. For what values of n does

$$\int_1^\infty x^n \, dx$$

   converge?

3. By using L'Hospital's rule, show that $\lim b^n e^{-b} = 0$, as $b \to \infty$, for any positive integer n. Then prove that

$$\int_0^\infty e^{-x} x^n \, dx = n!.$$

ANSWERS TO EXERCISES

section 1.1

1. (a), (c), (e), (g) ordinary, second order

section 1.2

1. (a) all x **(c)** $x \neq -4$ **2. (a)** all c, all x **(c)** all c_1, c_2; all x **(e)** all c, $x \neq 0$ **4. (a)** $m = \pm 1$ **(c)** $m = \pm 2, 1$ **5. (a)** $n = -3, n = 1$ **6.** $n = -2, A = \pm\sqrt{6}, 0$ **10. (a)** $1 + e^x = 1\cdot(e^x) + \frac{1}{2}\cdot(2)$ **(c)** $x + 1 = -1\cdot(x - 1) + 2\cdot(x)$

section 1.3

1. (a) linear, inhomogeneous **(c)** linear, homogeneous **(e)** linear, inhomogeneous **(g)** nonlinear

section 1.4-1

1. (a) IVP **(c)** BVP **(e)** BVP **(g)** IVP **2. (a)** $y = 2e^{-x}$ **3. (a)** $y = \sin x$ **(c)** an infinite number

section 1.4-2

1. $y' = y/10$, $y(0) = 10{,}000$ **3. (a)** $y' = ky$ **(c)** no **4. (a)** $y' = ky^2$, $y(0) = a_0$ **(b)** $k < 0$ **5.** $y' = (m(t) - n)y$ **6.** $y' = -\alpha y$, $\alpha > 0$, $y(0) = 20$ **8.** $P'(t) = \alpha(S - D)$, $\alpha < 0$ **9.** $N'(t) = -\alpha N(t)$, $\alpha > 0$, $N(0) = 1{,}000{,}000$

section 2.2

1. (a) $y = ce^{-x^2/2}$ **(c)** $y = ce^{x^2/2}$ **2. (a)** $y = 2e^{-\sin x}$, all x **(c)** $y = 3e^{1-e^x}$, all x **4. (a)** $y = \frac{1}{2}e^{3(\cos x - 1)}$

6. **(a)** discontinuous at $n\pi$, $n = 0, \pm 1, \ldots$; $y = \sin x$, $x \neq n\pi$, $n = 0, \pm 1, \ldots$ **(c)** $x = 0$, $y = 2e^{\{(1/x)+1\}}$, $(-\infty, 0)$
(e) discontinuous at $x = -1$, $y = (1 + x)^{-1}$, $(-1, \infty)$.

section 2.3-1

1. **(a)** $y = ce^{-x^2/2}$ **(c)** $y = ce^{-\sin x + x \cos x}$ **2.** **(a)** $y = -\csc x$, $0 < x < \pi$
(c) $y = x^{-1}$, $0 < x < \infty$ **(e)** $y = (1/x)e^{2-2x}$, $0 < x < \infty$.
4. **(a)** $y = 0$, one solution **(c)** $y = c_1 \cos x$, infinite number of solutions if $a_0 = 0$ and none if $a_0 \neq 0$

section 2.4-1

1. $\sim$544. **3.** $R = .406$, $t_2 = 1.7$ hr, 3 hr **5.** 1.95 billion **6.** 2.7 billion

section 2.4-2

1. **(a)** $y = 10e^{-.030543t}$ **(c)** $\sim$75 min **3.** **(a)** $y = 15e^{-.916291t}$ **(c)** 2.51 hr
7. $s(t) = 800e^{-t/25}$, $s(25) = 294.3$ lb **9.** $t = (1/k) \ln s$

section 2.5

1. $y(t) = \exp\left(-t + \frac{1}{2}\cos t - \frac{1}{2}\right)$ **3.** $i(t) = a[L_0/(L_0 + \beta^2 t)]^{R/\beta^2}$
5. **(a)** $y = \exp\left[-\frac{1}{2}x + \frac{1}{4}\sin 2x\right]$ **(b)** no, π **(c)** $\frac{1}{2}$ **(d)** yes

section 3.1

1. **(a)** $y = xe^{-x} + e^{-x}$ **(c)** $y = 2e^{-(x^2-1)/2} + 1 + e^{-(x^2-1)/2} \int_1^x e^{(z^2-1)/2}\, dz$
3. **(a)** $y = ce^{bt} + e^{st}/(s - b)$, $b \neq s$, $y = te^{bt} + ce^{bt}$, $b = s$
4. **(c)** $\exp\left[\int^x g(z)\, dz\right]$ **6.** **(a)** discontinuous at $x = 0$, $y = \frac{1}{4}x$, all x
(c) discontinuous at $x = n\pi$, $n = 0, \pm 1, \ldots$, $y = x \sin x - (\pi/2) \sin x$, all x
(e) discontinuous at $x = 0$, $y = x^{-2} + x^{-2} \ln |x|$, $x \neq 0$.

section 3.2

1. **(a)** $y_c = ce^{-3x}$, $y_p = 2$ **(c)** $y_c = ce^{-x}$, $y_p = e^{-x} \tan^{-1}(e^x)$
2. **(a)** $H(x) = 2e^{1-x}$, $F(x) = (x - 1)e^{-x}$, $y_t = 2e^{1-x} - e^{-x} + xe^{-x}$, $y_s(x) = 0$
(c) $H(x) = e^{(x^2-1)/2}$, $F(x) = 2e^{(x^2-1)/2} - 2$, $y_t(x) = 0$, $y_s(x) = 3e^{(x^2-1)/2} - 2$
(e) $H(x) = 0$, $F(x) = -(x/k) - (1/k^2) + (1/k^2)e^{kx}$, $y_t(x) = 0$, $y_s(x) = F(x)$
3. $y = a_0 e^{kx} - (A/k)(1 - e^{kx})$
4. **(a)** $y = e^{-x} + [A/(1 + \omega^2)]\{\sin \omega x - \omega \cos \omega x + \omega e^{-x}\}$ **(c)** no
5. **(a)** $y = a_0 e^{kx} + [A/(k^2 + \omega^2)]\{-k \sin \omega x - \omega \cos \omega x + \omega e^{kx}\}$
7. **(a)** $y = 1$, yes **9.** **(a)** $y_p = \frac{2}{3}$ **(c)** $y_p = 3$

section 3.3-1

1. $A = (\ln 2)/10 \approx .0693$
3. $A_1 = (\ln 4)/10 \approx .1386$, $A_2 = (\ln 2)/10 \approx .0693$

5. (a) $y = M\left(1 - \left(\frac{1}{1+t}\right)^{A_0}\right)$ **7. (a)** $y = \frac{AM}{A+B}[1 - e^{-(A+B)t}]$

section 3.3-2

1. (a) $y = a_0e^{-3t} + (1000/3)(1 - e^{-3t})$
2. (a) $y = 500[e^{-t/10} + 1]$ **(c)** 500 **3. (a)** $y = a_0e^{kt} + (A/k)(e^{kt} - 1)$
5. (a) 23,284, 2005 **6. (a)** $t = 10 \ln [(2a_0 + 10A)/(a_0 + 10A)]$
7. if $A = a_0k$.

section 3.3-3

1. $v = 94e^{-16t/3} + 6$

2. $v = \left[100 - \frac{150+P}{25}\right]e^{-(25g/(150+P))t} + \frac{150+P}{25}$

3. (a) $v = -60 + 32t$, $t = 1.875$ sec, 56.25 ft
4. (a) $t = (m/k) \ln [(kv_0 + mg)/mg]$

6. (a) $v = v_0(1+t)^{-\mu_0/m} + \frac{g}{\mu_0/m + 1}[1 + t - (1+t)^{-\mu_0/m}]$

7. (a) $v' + (k_1/m)v = g$ in air, $v(0) = 0$; $v' + (k_2/m)v = g$, in water
8. (a) $v' = g(\sin \alpha - \mu \cos \alpha)$, $v(0) = v_0$

(c) $$t = -\frac{v_0}{g(\sin \alpha - \mu \cos \alpha)} + \sqrt{\frac{v_0^2}{g^2(\sin \alpha - \mu \cos \alpha)^2} + \frac{200}{g(\sin \alpha - \mu \cos \alpha)}}$$

9. (a) $y = \frac{v_0^2}{2g(\sin \alpha - \mu \cos \alpha)}$

section 3.3-4

1. (a) $s(t) = E_0C(1 - e^{-t/RC})$ **(b)** $E_0C, 0$ **(c)** yes **(d) (e)** $RC \ln 2$

2. (a) $s(t) = \frac{A}{[10^6 + (120\pi)^2]}\left\{10 \cos 120\pi t - 10e^{-10^3t} + \frac{6\pi}{5} \sin 120\pi t\right\}$

(c) $\lim_{t\to\infty} i(t) = \frac{48\pi A}{[10^6 + (120\pi)^2]}\{-25 \sin 120\pi t + 3 \cos 120\pi t\}$

4. (a) $$s(t) = \begin{cases} a_0e^{-t/RC}, & 0 \le t < T, \\ a_0e^{-t/RC} + CE_0(1 - e^{(T-t)/RC}), & t > T \end{cases}$$

$$i(t) = \begin{cases} -a_0/(RC)\, e^{-t/RC} & 0 \le t < T \\ -a_0/(RC)e^{-t/RC} + (E_0/R)e^{(T-t)/RC}, & t > T \end{cases}$$

5. **(a)** $i(t) = (E_0/\pi)(1 - e^{-Rt/L})$.

section 4.2

1. **(a)** $m^2 - 4m + 3 = 0$, e^{3x}, e^x **(c)** $m^2 - 4m + 4 = 0$, e^{2x}
(e) $m^2 - m + 3 = 0$, $e^{\{(1+i\sqrt{11})/2\}x}$, $e^{\{(1-i\sqrt{11})/2\}x}$
2. **(a)** $y = (5/4)e^x - (1/4)e^{5x}$ **(c)** $y = 2e^{5(x-1)}/3 + e^{-7(x-1)}/3$
5. $c < 0$. **7.** e^{-2bx}, e^0 **9.** $y = a_1(e^{4x} - e^{2x})/2$

section 4.3

1. **(a)** $y = c_1e^{(-1+3i/2)x} + c_2e^{(-1-3i/2)x}$ **(c)** $y = c_1e^{2x} + c_2e^{-4x}$
(e) $y = c_1e^{(-1+2i)x} + c_2e^{(-1-2i)x}$
3. $y = c_1e^{(-2+\sqrt{4-a})x} + c_2e^{(-2-\sqrt{4-a})x}$, $a < 4$; $y = e^{-2x}(c_1 \cos \sqrt{a-4}x + c_2 \sin \sqrt{a-4}x)$, $a > 4$ **4.** **(a)** $y = e^{-x}[\cos(3x/2) + 2\sin(3x/2)]$, 0
(c) $y = \cos x + 2\sin x$, does not exist

5. $y = \begin{cases} \cos\sqrt{c}x, & c > 0 \\ \frac{1}{2}e^{\sqrt{-c}x} + \frac{1}{2}e^{-\sqrt{-c}x}, & c < 0; \\ 1, & c = 0 \end{cases}$ $\quad y \approx \begin{cases} \cos\sqrt{c}x, & c > 0 \\ \frac{1}{2}e^{\sqrt{-c}x}, & c < 0 \\ 1, & c = 0 \end{cases}$

9. $y = c_1e^{\sqrt{2}(1-i)x/2} + c_2e^{-\sqrt{2}(1-i)x/2}$; $y = c_1e^{\sqrt{2}(1+i)x/2} + c_2e^{-\sqrt{2}(1+i)x/2}$.

section 4.4

1. **(a)** e^x, xe^x **(c)** $\cos\sqrt{7}x$, $\sin\sqrt{7}x$ **(e)** e^{2x}, xe^{2x} **2.** **(a)** $y = xe^x$
(c) $y = (3 - 2x)e^{3(x-1)}$ **3.** $y = (c_1 + c_2x)e^{-bx}$; $\lim_{x\to\infty} y = 0$, for $b > 0$; $\pm\infty$, for $b < 0$; $\pm\infty$, for $b = 0$ and $c_2 \neq 0$; c_1, for $b = 0$ and $c_2 = 0$

section 4.5

1. **(a)** $y = c_1e^{\sqrt{5}x} + c_2e^{-\sqrt{5}x}$ **(c)** $y = c_1e^{-x/2} + c_2e^{5x/2}$ **(e)** $y = c_1 + c_2e^{-6x}$
3. **(c)** $y = c_1e^x + c_2e^{-x}$, $y = c_1e^x + c_2 \sinh x$, $y = c_1e^x + c_2\cosh x$,
$y = c_1e^{-x} + c_2\sinh x$, $y = c_1e^{-x} + c_2\cosh x$ **5.** $a_0 = -a_1$

section 4.5-1

1. **(a)** $W(x) = (a_2 - a_1)e^{(a_1+a_2)x}$ **(c)** $W(\sin 2x, \cos 2x) = -2$
2. **(a)** $W(x) = 0$, dependent **(c)** $W(x) = 0$, dependent **(e)** independent
5. **(a)** $W(x) = -2/x$ **(c)** $W(x) = (-x^{-3} + x^{-2})e^{1/x}$

section 5.1

1. **(a)** $y = 5(e^{3x} - e^x)$, $\lim_{x\to\infty} y(x) = \infty$
(c) $y = [(-2 + 3a_0)e^{2x} + (2 + 2a_0)e^{-3x}]/5$; as $x \to \infty$,

$y(x) \to [(-2 + 3a_0)/5]e^{2x}$ **(e)** $y = \frac{1}{2}\sin 2x$
2. (a) $y = \sqrt{10}e^{-x}\cos(x - \phi)$, $\phi = \tan^{-1}(3)$; amplitude $= \sqrt{10}e^{-x}$; phase $= \tan^{-1}(3)$; period $= 2\pi$; frequency $= 1$
(c) $y = \sqrt{2}e^{-3x}\cos(\frac{1}{2}x - \frac{1}{4}\pi)$; amplitude $= \sqrt{2}e^{-3x}$; phase $= -\pi/4$, period $= 4\pi$, frequency $= 1/2$
3. (a) $y = 2\sin x$, **(c)** $y = 10e^{4x}[\cos(3x - \phi)]$, $\phi = \tan^{-1}(4/3)$
9. $a_1 = m_2a_0$; $a_1 = m_1a_6$ **11.** $a_0 = b_0$, $a_1 = b_1$

section 5.2
1. $y = \sqrt{4901} \times 10^2[\cos(2t - \phi_1)]$, $\phi_1 = \tan^{-1}(49/50)$
$t = \frac{1}{2}\tan^{-1}(49/50) + \pi/4 \approx 3\pi/8$; $z = \sqrt{9802} \times 10^2[\cos(2t - \phi_2)]$, $\phi_2 = \tan^{-1}(99)$, $t = \frac{1}{2}\tan^{-1}(99) + (\pi/4) \approx \pi/2$
3. $\bar{y} = c_1\cos\sqrt{\alpha_1\beta_1}t + c_2\sin\sqrt{\alpha_1\beta_1}t$,

$$\bar{z} = c_1\frac{\beta_2}{\alpha_2}\sqrt{\frac{\alpha_1}{\beta_1}}\sin\sqrt{\alpha_1\beta_1}t - \frac{\beta_2c_2}{\alpha_2}\sqrt{\frac{\alpha_1}{\beta_1}}\cos\sqrt{\alpha_1\beta_1}t$$

The predators and prey vary periodically with period $2\pi/\sqrt{\alpha_1\beta_1}$ about the equilibrium solution.
5. Equilibrium solutions are $(\bar{y}, \bar{z}) = (0, 0)$, $(0, \beta_1/\beta_2)$, $(\alpha_1/\alpha_2, 0)$.
$\bar{y} = c_1e^{2t}$, $\bar{z} = c_2e^{t}$, unstable about $(0, 0)$
$\bar{y} = c_1e^{t}$, $\bar{z} = c_2e^{-t} - (1 + c_1)e^{t}$, unstable about $(0, 1)$
$\bar{z} = c_1e^{-t}$, $\bar{y} = c_2e^{-2t} + 2(1 + c_1)e^{t}$, stable about $(2, 0)$

section 5.3
1. $\theta = (1/10)\cos 4t$, amplitude $= 1/10$, period $= \pi/2$, frequency $= 4$, phase $= 0$ **3. (a)** $\theta = (\sqrt{17}/8)\cos(8t + \phi)$, $\phi = \tan^{-1}(1/4)$
(b) $t = [2n\pi - (\pi/3) - \phi]/8$, $n = 1, 2, 3, \ldots$
(c) $t = [2n\pi + (\pi/3) - \phi]/8$, $n = 0, 1, 2, 3, \ldots$ **5.** $E = \frac{1}{2}A^2\omega^2$
9. $a_1 < 2\sqrt{10}/\sqrt{3}$

section 5.4
1. $\theta(t) = e^{-2t} - \frac{1}{2}e^{-4t}$ **3.** $\theta_1 < \sqrt{5}\exp[(\pi/2 + \phi)/2]$, $\phi = \tan^{-1}(1/2)$
5. $\delta = 2\pi$ **7.** $s \approx .1846$

section 5.5
1. (a) $s = e^{-4t}[\cos(8\sqrt{6}t) + 1/(2\sqrt{6})\sin 8\sqrt{6}t]$,
$i = e^{-4t}[-(25\sqrt{6}/3)\sin 8\sqrt{6}t]$, both underdamped
(c) $t = (\pi/2 + \phi)/(8\sqrt{6})$, $\phi = \tan^{-1}[1/(2\sqrt{6})]$
3. underdamped, $(R^2/L^2) - (4/LC) < 0$; overdamped, $(R^2/L^2) - (4/LC) > 0$; critically damped, $(R^2/L^2) - (4/LC) = 0$
5. $s = s_0\cos\sqrt{1/LC}t$; $i = s' = -(s_0/\sqrt{LC})\sin(t/\sqrt{LC})$

section 6.1

1. (a) $y = [1/(e - 1)](e^{x+1} - e^{2x})$ **(c)** $y = \frac{1}{4}e^{3x} \sin 4x$
(e) $y = \sinh (kx)/[k^2 \cosh (k)]$
3. $y = 4[\cos 2x + (1 - \cos 2l) \sin 2x/\sin 2l]$, exactly one solution, $l \neq n\pi/2$; $y = 4(\cos 2x + c_2 \sin 2x)$, infinitely many solutions, $l = n\pi$; no solution when $l = (n + \frac{1}{2})\pi$.

5. $$y = \frac{1}{7 \sin (2) - 5 \cos (2)} \{[10 \cos (2) + 60 \sin (2) - 1] \cos x + [1 + 10 \sin (2) - 60 \cos (2)] \sin x\}$$

7. If $\alpha \neq n\pi/q$, then $$y = e^{-bx}\left[A \cos qx + \frac{Be^{b\alpha} - A \cos q\alpha}{\sin q\alpha} \sin qx\right]$$
is the only solution for all A and B.
9. $y = \sin x/\cos 1$

section 6.1-1

1. $\sim$44 ft

3. $$y = e^{ux/2}\left[y_0 \cosh qx + \frac{y_1 e^{-L/2} - y_0 \cosh qL}{\sinh qL} \sinh qx\right], \quad q = \frac{1}{2}\sqrt{u^2 + 4\alpha^2}$$

5. $$y = a_0 \frac{\sinh \sqrt{12/T}x}{\sinh \sqrt{k/T}l}, \text{ maximum deflection } = y(L) = a_0$$

7. $$y = \frac{a_1\sqrt{T/k}}{\cosh \sqrt{k/T}L} \sinh (\sqrt{k/T}x)$$

section 6.2

1. (a) $y_n = a_n \sin \sqrt{k_n}x$, $k_n = (2n - 1)^2\pi^2/4$, $n = 1, 2, \ldots$
(c) $y = a_n \sin \sqrt{k}x$, where k is any solution of $\cot \sqrt{k} = -1/\sqrt{k}$; $y = b \sinh \sqrt{k}x$, where k is the solution of $\tanh \sqrt{-k} = \sqrt{-k}$
(e) $y = a$, $k = 0$; $y = a_n \cos (\sqrt{k_n}x) + b_n \sin (\sqrt{k_n}x)$, $k_n = 4n^2\pi^2$, $n = 0, 1, \ldots$ **5.** no eigenvalues **7. (a)** no eigenvalues **(c)** $y = a$, $k = 0$

section 6.2-1

1. $k_1 = \pi^2/L^2$ **5.** $y = a_n \sin \sqrt{k_n}x$, $k_n = n^2\pi^2$

section 7.1

1. (a) $y = c_1 \cos x + c_2 \sin x + x$ **(c)** $y = c_1e^{4x} + c_2e^{-x} - x^2 + 3x/2 - 13/8$
3. $y = c_1e^{-x} + c_2 - xe^{-x}$

section 7.2

1. (a) $y_p(x) = \frac{1}{2}e^{2x}$ **(c)** $y_p(x) = \frac{1}{2}xe^x$ **(e)** $y_p(x) = 3e^x$
(g) $y_p(x) = -\frac{1}{8}\sin 3x$ **(i)** $y_p(x) = -\frac{1}{2}x\cos x$
(k) $y_p(x) = [1/(\rho^2 - \alpha^2)]\sin \alpha t$
(m) $y_p(x) = -x^4/4 - x^3/3 - x^2/3 - 2x/9$
(o) $y_p(x) = (x^2/3 - x/3 + 1/9)e^x$
2. (a) $y(x) = e^{-3x/2}[c_1\cos(\sqrt{5}/2)x + c_2\sin(\sqrt{5}/2)x] + Ax^2e^x + Bxe^x + Ce^x + De^x\sin x + Ee^x\cos x + Fx^3e^{2x}\sin 3x + Gx^2e^{2x}\sin 3x + Hxe^{2x}\sin 3x + Ie^{2x}\sin 3x + Jx^3e^{2x}\cos 3x + Kx^2e^{2x}\cos 3x + Lxe^{2x}\cos 3x + Me^{2x}\cos 3x$ **3. (a)** $y(x) = e^{2x} - e^{3x} + xe^{-x}$
(c) $y(x) = 3\pi\cos 2x + \frac{1}{2}\sin 2x - x\cos 2x + x\sin 2x$

section 7.3

1. (a) $y = c_1e^x + c_2e^{-x} + xe^x/2$
(c) $y = c_1\cos x + c_2\sin x + \cos x\ln\cos x + x\sin x$
(e) $y = c_1e^{-2x} + c_2e^x - \frac{1}{3}e^{-2x}\int^x e^{2z}\ln z\,dz + \frac{1}{3}e^x\int^x e^{-z}\ln z\,dz$
(g) $y = c_1e^{-x} + c_2e^{-2x} + e^{-x}\ln(1 + e^x) + e^{-2x}[\ln(1 + e^x)]$
(i) $y = c_1e^x + c_2e^{-x} + \frac{1}{4}[e^x\ln(1 - e^{-2x}) - e^{-x}\ln(e^{2x} - 1)]$
3. $y = c_1\sin x + c_2\cos x + \sin x\int_0^x r(\xi)\cos\xi\,d\xi + \cos x\int_0^x r(\xi)\sin\xi\,d\xi$

section 8.1

1. (a) $y = \frac{2}{3}\sin x - \frac{1}{3}\sin 2x$, $y_s = \frac{2}{3}\sin x - \frac{1}{3}\sin 2x$, $y_t = 0$, periodic, 2π, oscillatory

$$\text{(c) } y = \frac{1}{40b^2 - 48b + 85}\left\{e^{-bx}\left[\frac{6b^2 - 6b + 27}{\sqrt{1 - b^2}}\sin\sqrt{1 - b^2}x - 6(1 - b)\cos\sqrt{1 - b^2}x\right] + e^{-x}[6(1 - b)\cos 3x - (7 + 2b)\sin 3x]\right\}$$

For $0 < b < 1$, $y_s = 0$, $y_t = y$, not periodic, $\lim_{x\to\infty} y(x) = 0$; for $b = 0$, $y_s = (27/85)\sin x - (6/85)\cos x$, $y_t = e^{-x}[(6/85)\cos 3x - (7/85)\sin 3x]$, not periodic, oscillatory **2. (a)** $y = \cosh x + (1 - \cosh 1)\sinh x/\sinh 1 - 1$
(c) $y = -8e^{-x} + (8 - 6e)xe^{-x} + x^2 - 4x + 9$
4. (a) $H(x) = 2\sin x + \cos x$, $F(x) = k(1 - \cos x)$
(c) $H(x) = (m\alpha_1/\mu)(1 - e^{-(\mu/m)x})$, $F(x) = (m^2g^2/\mu^2)(e^{-(\mu/m)x} - 1) + (mg/\mu)x$

$$\text{5. (a) } y = \begin{cases} \sin x + 1 - x, & 0 < x \le 1 \\ [1 - \cos 1]\sin x + \sin 1\cos x, & x > 1 \end{cases}$$

$$\text{6. (a) } y = \begin{cases} \frac{2}{3}\sin x - \frac{1}{3}\sin 2x, & 0 \le x \le \pi/2 \\ \frac{2}{3}\sin x - \frac{2}{3}\cos x, & x > \pi/2 \end{cases}$$

section 8.2

1. (a) $y(t) = \frac{1}{3}(29{,}000 + 59{,}600t)e^{-t} + 400 - (200/3)(\cos t + 3 \sin t)$
(b) $y_t = \frac{1}{3}(29{,}000 + 59{,}600t)e^{-t}$, $y_s = 400 - (200/3)(\cos t + 3 \sin t)$
(c) 2π, $A_{\max} = 400 + 200\sqrt{10}/3$ **(d)** yes **(e)** $S_0/2$
(2) $y_s = S_0[(9/7) - (7/2) \cos t - (9/2) \sin t]/2$,
$z_s = S_0[1 - (7/2) \sin t]/2$, $S_0 \leq 400/9$
3. (a) $y_{s_{\max}} = (51/14)S_0$, $z_{s_{\max}} = (9/4)S_0$, 2π, $\phi_y = \tan^{-1}(9/7)$, $\phi_z = 0$
(b) yes; for example, when $\sin t > 2/7$ **(5)** no steady-state solution
(7) $z_{\max} = (3a_0 - 200)e^{-\{(3a_0+200)/(3a_0-200)\}}/3$

section 8.3

1. (a) $\theta = (1/25) \cos 4t + (4/25) \cos t$
(b) $t = 2n\pi/5$, $n = 0, 1, 2, \ldots$, $\theta_{\max} = 24/5$ (*Hint*: Use $\sin \theta = \sin \alpha$ implies that $\theta = (-1)^n\alpha + n\pi$) **(3)** $\theta = (1/16) \sin 2t - (t/8) \cos 2t$
(5) $\theta = (\theta_1/\omega) \sin \omega t - k(\cos \omega t - \cos at)/(\omega^2 - a^2)$
(7) $\theta = (1/10) \cos t + (1/24) \cos 2t - (17/120) \cos 4t$

section 8.3-1

1. (a) $\theta = (1/10) \cos t$ **(b)** $\theta_p = (1/10) \cos t$, 2π
3. (a) $\theta = (1/64) \sin 4t + (1/10) \cos 4t - (t/16) \cos 4t$
(b) $\theta_p = -(t/16) \cos 4t$, no
(c) as $t \to \infty$, the solution oscillates with increasing amplitude

section 9.2

1. (a) $y = c_1e^x + c_2e^{\sqrt{2}x} + c_3e^{-\sqrt{2}x}$ **(c)** $y = c_1 + c_2 \sin x + c_3 \cos x$
(e) $y = c_1e^{-2x} + e^{2x}(c_2 \sin x + c_3 \cos x)$ **2. (a)** $y = (1 - \cos 3x)/9$
(c) $y = [3e^{2x} + (\sqrt{3}/3) \sin \sqrt{3}x + 4 \cos \sqrt{3}x)]/7$
3 (a) $y = e^{(\sqrt{2}/2)x}[c_1 \sin(\sqrt{2}x/2) + c_2 \cos(\sqrt{2}x/2)] +$
$e^{-(\sqrt{2}/2)x}[c_3 \sin(\sqrt{2}x/2) + c_4 \cos(\sqrt{2}x/2)]$
(c) $y = c_1 \cos 2x + c_2 \sin 2x + \cos x\,(c_3e^{\sqrt{3}x} + c_5e^{-\sqrt{3}x}) +$
$\sin x\,(c_4e^{\sqrt{3}x} + c_6e^{-\sqrt{3}x})$

section 9.3

1. (a) $y = c_1e^x + c_2xe^{-2x} + c_3e^{-2x}$ **(c)** $y = c_1 + c_2x + c_3e^{-x} + c_4xe^{-x}$
(e) $y = (c_1 + c_2x + c_3x^2) \sin x + (c_4 + c_5x + c_6x^2) \cos x$
2. (a) $y \doteq [1 + x + (x^2/2)]e^{-x}$ **(c)** $y = 1$
5. (a) $y = (c_1 + c_2x + c_3x^2 + c_4x^3)e^x$
(b) $y = (c_1 + c_2x + c_3x^2 + \cdots + c_px^{p-1})e^{ax}$

section 9.4

1. (a) $y = c_1 + c_2 \sin 2x + c_3 \cos 2x$, $W(x) = -8$
(c) $y = c_1 + c_2x + (c_3 + c_4x)e^x$, $W(x) = e^{2x}$
2. (a) $W(x) = -12e^{4x}$ **(c)** $W(x) = 32\sqrt{3}$
3. (a) $c_1 = c_2 = -c_4$, $c_3 = 0$, c_1 arbitrary **5. (c)** $a_1 = a_2$, a_0 arbitrary

section 9.5

1. (a) $y = c_1e^{-x} + c_2e^{2x} + c_3e^{3x} + \frac{3}{4}e^x$
(c) $y = c_1 + c_2 \sin 2x + c_3 \cos 2x - \frac{1}{3} \cos x$
(e) $y = c_1 + c_2x + c_3 \sin x + c_4 \cos x + (x^4/12) + (x^3/3) - x^2$
(g) $y = c_1e^{-5x} + c_2e^{\sqrt{3}x} + c_3e^{-\sqrt{3}x} - (5/72)e^x - (xe^x/12)$
2. (a) $y = (1/6)(e^{-x} - e^x) + (1/12)(e^{2x} - e^{-2x}) = -(1/6)(2 \sinh x - \sinh 2x)$ **(c)** $y = (e^{-x}/4) - (7e^x/12) - (e^{-2x}/60) + (7e^{2x}/20) + (\sin x)/10$
3. (a) $y_p = Ae^x + x^2(B \sin x + C \cos x)$
(c) $y_p = Ax^3e^x + Bx^3 + Cx^2 + Dx + E$
(e) $y_p = (Ax^4 + Bx^3 + Cx^2 + Dx)e^x$
7. (b) $y_p = (B_0x + B_1x^2 + \cdots + B_mx^{m+1})e^{ax}$, e^{ax} a solution; $y_p = (B_0 + B_1x + \cdots + B_mx^m)e^{ax}$, e^{ax} not a solution
(c) $y_p = (B_0x^j + B_1x^{j+1} + \cdots + B_mx^{j+m})e^{ax}$, $(1 + x + \cdots + x^{j-1})e^{ax}$ a solution; $y_p = (B_0 + B_1x + \cdots + B_mx^m)e^{ax}$, $(1 + x + \cdots + x^{j-1})e^{ax}$ not a solution
9. (a) $y_p = \ln|\sin x| + \cos x \ln|\csc x + \cot x| - 1$ **(c)** $y_p = xe^x/10$

section 9.6-1

1. (a) stable; $y = (c_1 + c_2x + c_3x^2)e^{-x}$
(c) not stable; $y = c_1e^{-x} + c_2e^{-2x} + c_3e^x$

section 9.6-2

1. $y = (1/EI)(12/\pi)^4 \sin(\pi x/12)$; $x = 6$; $y_{\max} = (1/EI)(12/\pi)^4$
3. $y = ((25)^6/180EI) - ((25)^4/120EI)(x - 25)^2 + [(x - 25)^6/360EI]$; $x = 25$; $y_{\max} = (25)^6/180EI$ **5.** $y = (q_0/EI)[(L^2x^2/4) - (Lx^3/3) + (x^2/24)]$

7. $y = \dfrac{47}{300}x - \dfrac{x^3}{600} + \dfrac{1}{10}\left(\dfrac{e^x - e^{-x}}{e^{10} - e^{-10}}\right)$

9. $y = \dfrac{q_0}{T}\left[-1 + \dfrac{L}{2}x - \dfrac{x^2}{2} + \dfrac{1}{e^L + 1}(e^x + e^{L-x})\right]$

section 9.6-3

1. $\lambda = \lambda_n$, where $\cos \sqrt[4]{\lambda_n}L \cosh \sqrt[4]{\lambda_n}L = 1$; $\lambda \leq 0$ is not an eigenvalue;

$$y_n = a_n\left[\frac{\cos\sqrt[4]{\lambda_n}x - \cosh\sqrt[4]{\lambda_n}x}{\cos\sqrt[4]{\lambda_n}L - \cosh\sqrt[4]{\lambda_n}L} - \frac{\sin\sqrt[4]{\lambda_n}x - \sinh\sqrt[4]{\lambda_n}x}{\sin\sqrt[4]{\lambda_n}L - \sinh\sqrt[4]{\lambda_n}L}\right]$$

2. (a) $\lambda \leq 0$ is not an eigenvalue, $\lambda = \lambda_n > 0$, $\cos \sqrt[4]{\lambda_n}L \cosh \sqrt[4]{\lambda_n}L = -1$;

$$y = a_n\left[\frac{\cosh\sqrt[4]{\lambda_n}x - \cos\sqrt[4]{\lambda_n}x}{\cosh\sqrt[4]{\lambda_n}L + \cos\sqrt[4]{\lambda_n}L} - \frac{\sinh\sqrt[4]{\lambda_n}x - \sin\sqrt[4]{\lambda_n}x}{\sinh\sqrt[4]{\lambda_n}L + \sin\sqrt[4]{\lambda_n}L}\right]$$

(c) $\lambda_n = n^2\pi^2$, $n = 1, 2, 3, \ldots$, $y_n = a_n \sin n\pi x$

section 10.1

1. (a) $y' = z$, $z' = -y + \sin x$; $y = c_1 \sin x + c_2 \cos x - \frac{1}{2}x \cos x$, $z = c_1 \cos x - c_2 \sin x - \frac{1}{2} \cos x + \frac{1}{2}x \sin x$
(c) $y' = w$, $w' = z$, $z' = -3z - w - y + x$; $y = (c_1 + c_2x + c_3x^2)e^{-x} + x - 3$, $w = [(c_2 - c_1) + (2c_3 - c_2)x - c_3x^2]e^{-x} + 1$, $z = [(2c_3 - 2x_2 + c_1) + (c_2 - 4c_3)x + c_3x^2]e^{-x}$

section 10.2

1. (a) $y = c_1e^x + c_2e^{3x}$, $z = \frac{1}{3}(c_2e^{3x} - c_1e^x)$
(c) $y = e^{2x}(c_1 \sin x + c_2 \cos x)$, $z = e^{2x}(-c_1 \cos x + c_2 \sin x)$
(e) $y = (c_1 + c_2x)e^{2x}$, $z = (c_1 - c_2 + c_2x)e^{2x}$
(g) $z = e^{(3/14)x}[c_1 \sin(\sqrt{215}x/14) + c_2 \cos(\sqrt{215}x/14)]$, $y = -(5/12)e^{(3/14)x}[(c_1 + \sqrt{215}c_2/5) \sin(\sqrt{215}x/14) + (c_2 - \sqrt{215}c_1/5) \cos(\sqrt{215}x/14)]$
(i) $z = c_1e^{2x}$, $w = 3c_1xe^{2x} + c_2e^{2x}$, $y = [(3c_1x/5) - (3c_1/25) + c_2/5)]e^{2x} + c_3e^{-3x}$
2. (a) $y_c = e^{x/2}(c_1e^{(\sqrt{57}/2)x} + c_2e^{-(\sqrt{57}/2)x})$, $y_p = -[x + (69/14)]/14$, $z_c = (e^{x/2}/8)[(\sqrt{57} - 3)c_1e^{(\sqrt{57}/2)x} - (\sqrt{57} + 3)c_2e^{-(\sqrt{57}/2)x}]$, $z_p = -[3x - (31/14)]/14$
(c) $y_c = 3c_1e^{-x} - 2c_2e^{4x}$, $y_p = \frac{1}{4}[2x^2 + x + (25/4)]$, $z_c = c_1e^{-x} + c_2e^{4x}$, $z_p = \frac{1}{4}[x^2 - (5x/2) + (19/8)]$
3. (a) $y = (1/2\sqrt{57})(\sqrt{57} + 7)e^{(5+\sqrt{57})x/2} + (\sqrt{57} - 7)e^{(5-\sqrt{57})x/2}$, $z = (2/\sqrt{57})[e^{(5+\sqrt{57})x/2} - e^{(5-\sqrt{57})x/2}]$

section 10.3

1. (a) $y = e^{2x}(c_1 \sin 3x + c_2 \cos 3x)$, $z = (e^{2x}/2)[(3c_1 - c_2) \cos 3x - (3c_2 - c_1) \sin 3x]$
(c) $y = c_1 \sin 2\sqrt{bx} + c_2 \cos 2\sqrt{bx}$, $z = -(2/\sqrt{b})[c_1 \cos 2\sqrt{bx} - c_2 \sin 2\sqrt{bx}]$, $b > 0$; $y = c_1$, $z = 4c_1x + c_2$, $b = 0$; $y = c_1e^{2\sqrt{-b}x} + c_2e^{-2\sqrt{-b}x}$, $z = -(2\sqrt{-b}/b)[c_1e^{2\sqrt{-b}x} - c_2e^{-2\sqrt{-b}x}]$, $b < 0$
3. $z = c_1e^{-x} + c_2e^{2x}$, $y = -c_1e^{-x} + 2c_2e^{2x}$ **7. (b)** $y_p = 2$, $z_p = -x$

9.
$$y_p = y_1(x) \int^x \frac{r_1(\xi)z_2(\xi) - r_2(\xi)y_2(\xi)\, d\xi}{W(\xi)} + y_2(x) \int^x \frac{r_2(\xi)y_1(\xi) - r_1(\xi)z_1(\xi)}{W(\xi)}\, d\xi,$$

$$z_p = z_1(x) \int^x \frac{r_1(\xi)z_1(\xi) - r_2(\xi)y_2(\xi)}{W(\xi)}\, d\xi + z_2(x) \int^x \frac{r_2(\xi)y_1(\xi) - r_1(\xi)z_1(\xi)}{W(\xi)}\, d\xi$$

11. (a) $y_p = (1/(1 - a^2))[a \cos ax - \sin ax]$,

$z_p = (1/1 - a^2)[a \cos ax + \sin ax]$
(b) $y_p = (Ax + B) \cos x + (Cx + D) \sin x$,
$z_p = (Ex + F) \cos x + (Gx + H) \sin x$

section 10.4
1. (a) $\theta = \frac{1}{2}[(\theta_0 + \psi_0) \cos 4t + (\theta_0 - \psi_0) \cos \sqrt{21}t]$, $\psi = \frac{1}{2}[(\theta_0 + \psi_0) \cos 4t - (\theta_0 - \psi_0) \cos \sqrt{21}t]$ **(b)** no **(c)** $\theta = \psi = \theta_0 \cos 4t$, $\theta = -\psi = \theta_0 \cos \sqrt{21}t$
3. (a) $\theta = (1/2\omega)(\theta_1 + \psi_1) \sin \omega t + (1/2p)(\theta_1 - \psi_1) \sin pt$,
$\psi = (1/2\omega)(\theta_1 + \psi_1) \sin \omega t - (1/2p)(\theta_1 - \psi_1) \sin pt$
(b) $n\omega = mp$ for some integers n and m
7. (a) $\theta = \psi = [R_0/(\omega^2 - a^2)](\cos at - \cos \omega t)$, $\omega \neq \pm a$;
$\theta = \psi = (R_0/2\omega)t \sin \omega t$, $\omega = \pm a$ **(b)** $\omega = \pm a$ **(c)** $\omega = \pm a$ and $n\omega = ma$ for some integers m and n

section 10.5
1. (a) $i_1 = (c_1 e^{-t}/4) + 4c_2 e^{-16t} + E_0$, $i_2 = c_1 e^{-t} + c_2 e^{-16t} + E_0$,
$i_3 = 3[c_2 e^{-16t} - (c_1 e^{-t}/4)]$ **(b)** $i_1 = i_2 = E_0$, $i_3 = 0$
3. (a) $(R_1 + R_2 + R_3)i' + [(1/c_1) + (1/c_2)]i = E(t)$,
(c) $R = R_1 + R_2 + R_3$, $1/c = (1/c_1) + (1/c_2)$
5. $i_1 = c_1 e^{-3t} + c_2 e^{-t/2} + (31 \sin t - 17 \cos t)/50$,
$i_2 = (\sqrt{2}/3)c_1 e^{-3t} - (\sqrt{2}/2)c_2 e^{-t/2} + \sqrt{2}(\cos t + 7 \sin t)/50$

section 11.1
2. (a) $y = e^{2x}$
3. either $-(P/2) + \sqrt{(P^2/4) - Q}$ or $-(P/2) - \sqrt{(P^2/4) - Q}$ is a constant **5. (a)** no discontinuities **(c)** no discontinuities
6. (a) $y = c_1 \ln x + c_2$
(c) $y = [(a_0 - a_1) \ln x + a_1 \ln x_0 - a_0 \ln x_1]/(\ln x_0 - \ln x_1)$

section 11.3
1. $y = c_1 e^{2x} + c_2(x + 1)$ **3.** $y = c_1 x e^x + c_2 x$

section 11.4
1. (a) $y = c_1 x + c_2 x^2$ **(c)** $y = c_1 x + c_2 x^5$
2. (a) $y = c_1 \sqrt{x} + c_2 x$ **(c)** $y = (1/\sqrt{x})(c_1 + c_2 \ln x)$
6. (a) $y = (1/x^4)(c_1 x^{\sqrt{15}} + c_2 x^{-\sqrt{15}})$
(c) $y = x^3[c_1 \sin (4 \ln x) + c_2 \cos (4 \ln x)]$

7. $$z = c_1 \exp\left[\left(\frac{1 - b_1}{2}\right) + \sqrt{\left(\frac{1 - b_1}{2}\right)^2 - b_2}\right]t$$

$$+ c_2 \exp\left[\left(\frac{1 - b_1}{2}\right) - \sqrt{\left(\frac{1 - b_1}{2}\right)^2 - b_2}\right]t,$$

$[(1 - b_1)/2]^2 - b_2 > 0$;

$$z = (c_1 + c_2 t)e^{b_2 t},\ [(1 - b_2)/2]^2 - b_2 = 0;$$

$$z = e^{[(1-b_1)/2]t}\left(c_1 \sin\sqrt{b_2 - \left(\frac{1-b_1}{2}\right)^2}\,t + c_2 \cos\sqrt{b_2 - \left(\frac{1-b_1}{2}\right)^2}\,t\right),$$

$$\left(\frac{1-b_1}{2}\right)^2 - b_2 < 0;$$

$$y = x^{(1-b_1)/2}\left[c_1 x^{\sqrt{[(1-b_1)/2]^2 - b_2}} + c_2 x^{-\sqrt{[(1-b_1)/2]^2 - b_2}}\right],\ \left(\frac{1-b_1}{2}\right)^2 - b_2 > 0;$$

$$y = (c_1 + c_2 \ln x)x^{b_2},\ [(1 - b_1)/2]^2 - b_2 = 0;$$

$$y = x^{(1-b_1)/2}\left[c_1 \sin\left(\sqrt{b_2 - \left(\frac{1-b_1}{2}\right)^2}\ln x\right)\right.$$

$$\left.+ c_2 \cos\left(\sqrt{b^2 - \left(\frac{1-b_1}{2}\right)^2}\ln x\right)\right],\ \left(\frac{1-b_1}{2}\right)^2 - b_2 < 0$$

section 12.1

1. **(a)** $x - x^3/6 + \cdots$
(c) $e + 2e(x-1) + 3e(x-1)^2 + 10e(x-1)^3/3 + \cdots$
(e) $1 + x + x^2 + x^3 + \cdots$
2. **(a)** $y = ce^{-x^3} = c(1 - x^3 + x^6/2 - x^9/6 + \cdots)$
(c) $y = ce^{-x-x^2/2-x^3/3} = c(1 - x + x^4/4 - x^5/20 + \cdots)$
5. **(a)** $y = 2(1 - 2x^3/3 + 2x^6/9 - 4x^9/81 + \cdots)$
(c) $y = 2(1 + x + x^2/2! + x^3/3! + x^4/4! + \cdots)$
7. $y = c[1 + (x - x_0) + (x - x_0)^2/2! + (x - x_0)^3/3! + \cdots]$

section 12.2

1. **(a)** $R = 1$ **(c)** $R = 4$ **(e)** $R = \infty$

section 12.2-1

1. **(a)** $f'(x) = f''(x) = \sum_{n=0}^{\infty} (x^n/n!)$ for all x
(c) $f'(x) = \sum_{n=0}^{\infty} (-1)^n x^n$; $f''(x) = \sum_{n=0}^{\infty} (-1)^{n+1}(n+1)x^n$, $-1 < x < 1$
4. **(a)** $\sum_{m=0}^{\infty} (m^2 + 3m + 5)x^m$ **(c)** $\sum_{m=0}^{\infty} (m^2 + 6m + 6)x^m$
7. $\sum_{n=0}^{\infty} x^{2n}$

section 12.3

1. **(a)** $b_{m+1} = -2b_m/(m+1)$, $R = \infty$, $y = c\sum_{m=0}^{\infty} [(-1)^m 2^m x^m/m!]$
(c) $b_{m+1} = -(b_m + b_{m-2})/(m+1)$, $R = \infty$; $y = c(1 - x + x^2/2 - x^3/2 + 3x^4/8 - \cdots)$ **3.** **(a)** $y = 3\sum_{m=0}^{\infty} [(-1)^m x^m/m!]$

section 12.3-1

1. (a) $y = c_1 \sum_{n=0}^{\infty} [x^{2n}/(2n)!] + c_2 \sum_{n=0}^{\infty} [x^{2n+1}/(2n+1)!]$, $R = \infty$

(c) $y = c_1 x + c_2 \left[1 - \frac{x^2}{2} - \sum_{n=2}^{\infty} \frac{1\cdot 3\cdot 5\cdot \cdots \cdot (2n-3)}{2\cdot 4\cdot 6\cdot \cdots \cdot 2n} \right]$, $R = 1$

(e) $y = c_1 \sum_{n=0}^{\infty} \frac{(-1)^n x^{2n}}{2^n n!} + c_2 \sum_{n=0}^{\infty} \frac{(-1)^n 2^n n! x^{2n+1}}{(2n+1)!}$, $R = \infty$

2. (c) $y = b_1 x + b_0 \left\{ 1 - x^2 - 4 \sum_{n=2}^{\infty} \frac{(2n-3)!}{(2n)!(n-2)!} x^{2n} \right\}$

4. (a) $y = b_1 x + b_0 \left\{ 1 - \sum_{n=1}^{\infty} \frac{x^{2n}}{2n-1} \right\}$

7. $y = b_0[1 - (x+1)^2/2 - (x+1)^3/2 - 5(x+1)^4/12 - \cdots]$
$+ b_1[(x+1) + (x+1)^2/2 + (x+1)^3/6 + \cdots]$

section 12.4

1. (a) $x = 0$ **(c)** $x = 0$

3. (a) $y = b_0 \left[1 + \sum_{n=1}^{\infty} (-1)^n \frac{[2\cdot 5\cdot 8\cdot \cdots \cdot (3n-1)]^2}{(3n)!} \left(\frac{x}{2}\right)^{3n} \right]$

$+ b_1 \sum_{n=0}^{\infty} (-1)^n \frac{8[1\cdot 4\cdot 7\cdot \cdots \cdot (3n+1)]^2}{(3n+2)!} \left(\frac{x}{2}\right)^{3n+2}$

(c) $y = b_1(x + x^2)$, no general solution

5. (a) $x = 0$ **(c)** no singular points

6. (a) $y = a_0(1 + x^2/2 - x^3/6 + x^4/12 - \cdots)$
$+ a_1(x + x^3/6 - x^4/12 - \cdots)$

section 13.1

1. (a) $y = cx^{-1}$, $k = -1$

(c) $y = c \exp [-\int^x (\cos t)/t\, dt] = cx^{-1}(1 + x^2/2 + x^4/48 + \cdots)$

section 13.2

1. (a) $x = 0$, regular **(c)** $x = 1$, regular

2. (a) $y = c_1 \sum_{n=0}^{\infty} \frac{(-1)^n x^n}{(2n)!} + c_2 |x|^{1/2} \sum_{n=0}^{\infty} \frac{(-1)^n x^n}{(2n+1)!}$

(c) $y = c_1 \sum_{n=0}^{\infty} \frac{n! x^n}{(2n)!} + c_2 |x|^{-1/2} \sum_{n=0}^{\infty} \frac{x^n}{2^{2n} n!}$

4. (a) $y = c_1 x^{-1}(1 + x/5 + x^2/30 + x^3/90 - \cdots)$
$+ c_2 |x|^{5/2}(1 - x/9 + x^2/198 - x^3/7722 + \cdots)$

7. (a) $x = 0$: irregular singular point
8. (a) $x = 0$: regular singular point for $N = 1, 2$; irregular singular point for $N > 2$

(c) $y = cx \sum_{n=0}^{\infty} \dfrac{(-1)^n x^n}{n!(n+1)!}$

section 13.3

1. $y = c_1 |x|^{1/2} \sum_{n=0}^{\infty} \dfrac{(-1)^n x^{2n}}{(2n+1)!} + c_2 |x|^{-1/2} \sum_{n=0}^{\infty} \dfrac{(-1)^n x^{2n}}{(2n)!}$

2. (a) $y = cx^{-1} \sum_{n=0}^{\infty} \dfrac{(-1)^n x^{2n}}{2^n n!}$ **3. (a)** $y = (c_1 + c_2 x^2) \sum_{n=0}^{\infty} \dfrac{x^n}{n!}$

(c) $y = c_1 x^2(1 + x + x^2/5 - 2x^3/15 + \cdots)$
$\quad + c_2 x^{-1}(1 + x + 2x^2 - 2x^4 \cdots)$

4. (a) $y_1 = c_1 \sum_{n=1}^{\infty} nx^n = x(1-x)^{-2}$, $y_2 = c_2 y_1 \ln |x| + \sum_{n=0}^{\infty} x^n$

$= c_2 y_1 \ln |x| + (1-x)^{-1}$ **(c)** $y_1 = c_1 x^{-1}$,
$y_2 = c_2 y_1 \ln |x| + x^{-1} + (2x)^{-2} \{1 - \sum_{n=2}^{\infty} [(2x)^n / n!(n-1)!]\}$

section 14.1

1. (a) separable **(c)** separable **(e)** separable **(g)** not separable
3. (a) $r(x) = cq(x)$, c = constant
4. (a) r, a, and b differ by a multiplicative constant **(5)** no

section 14.2

1. (a) $y = \pm \sqrt{(x^4/2) + c}$ **(c)** $y = \cos^{-1}[-(e^x + c)]$
(e) $y = \pm \sqrt{4 - (x-c)^2}$ **(g)** $y = \tan[-(x^3/3) + c]$
3. (a) $y = -1 + 2\sqrt{x^3 + x + 1}$ for all x, where $x^3 + x + 1 > 0$
(c) $y = -1 + \sqrt{x^2 + 2x + 16}$ for all x **6. (a)** $(\sqrt{\pi}/2)\,\mathrm{erf}\,(y) = x$
8. (a) $y = \pm \sqrt{5x^2 + c} - 2x$ **(c)** $y = xe^{cx}$

section 14.3

1. (a) $\phi(x, y) = x^2 + y^2 - xy - c = 0$ **(c)** not exact
(e) $\phi(x, y) = 2xy + (y^3/3) - c = 0$ **(g)** $\phi(x, y) = \frac{1}{2}(xy)^2 - xy - c = 0$
2. (a) $\phi(x, y) = y^2 - xy - c = 0$
3. (a) $\phi(x, y) = 4xy + x^2 + 3x + (y^2/2) = 17/2$
(c) $\phi(x, y) = (x^3/3) - x^2 y - \cos y + \cos 1 = 0$ **5. (a)** never exact
(7) $H'(y) = P(y)$; $G'(x) = Q(x)$
10. (a) $\phi(x, y) = (1/x) + (1/y) + c = 0$
(c) $\phi(x, y) = xy^2 - (x^4/4) - c = 0$

section 14.4-1
3. (b) $y/x = c$ **4. (a)** $(0, 0)$ **(c)** $(n\pi, 0)$, $n = 0, \pm 1, \pm 2, \ldots$

section 14.4-2

1. (a) $y = xc + c^2$ **2. (c)** $\left(\frac{c}{2}, \frac{c^2}{4}\right)$ **3. (b)** $y = [\frac{3}{4}(x + c)]^{4/3}$

(c) $$y = \begin{cases} 0, & \text{for } 0 \le x \le x_1 \\ [\frac{3}{4}(x - x_1)]^{4/3}, & \text{for } x > x_1 \end{cases}$$

5. (c) IVP has no solution if P is a negative number because $y'(0)$ is infinite **(d)** $P > 1$
7. (a) $-\infty < x < 1/2a_0^2$ **(c)** $-\infty < x < \infty$

section 14.5-1
1. (a) $y = 6/(1 + e^{-12t})$ **7.** $y(3) = 19.06$; $y(4) = 19.67$
8. (a) $y = (a_0^{-n} - nkt)^{-1/n}$ **9. (b)** $y = L \exp(ce^{-kt})$

section 14.5-2

1. (a) $t = \dfrac{\ln (999)}{1000\alpha} = \dfrac{6{\cdot}9 \times 10^{-3}}{\alpha}$

3. (a) $t = -\ln [a_0(A + a_0)/(A - a_0)^2]/A\alpha$ **4. (b)** no
5. (a) 999 **(b)** $t = (1/999) \ln [2(999 - a_0)/(999 - 2a_0)]$
(c) $t = \ln [(999/a_0) - 1]/999$ **7.** $a_0 = 30$, $k = 1/300$
11. (b) $y = 0$, $z = \alpha + \beta$ **12. (a)** $k = -.9$, $v = -.5$
13. $t = (a_0 + k \ln 4)/2v$

section 14.5-3
1. (a) $y + 2x = c$ **(c)** $y^2 - x^2 = c$ **(e)** $3y^2 + x^2 = c$
3. (a) $y^2 + 2x^2 = c$ **(c)** $y = cx^4$

section 14.5-4
1. (a) $8\sqrt{2}$ ft/sec **3. (b)** $v = [v_0^{1-n} + (1 - n/m)t]^{1/(1-n)}$ **4. (a)** $n < 1$
(b) $mv_0^{1-n}/[k(1 - n)]$ **(c)** $n < 1$ **7. (a)** 11.3 ft/sec, 110 ft approximately
8. (a) 35.8 ft/sec, 923 ft approximately

section 15.1
1. (a) $f'(0) = f'(1/4) = f'(1/2) = f'(3/4) = 1$ for approximate and actual
(c) approximate: $f'(0) = .9896$, $f'(1/4) = .9281$, $f'(1/2) = .8088$,

$f'(3/4) = .6393$; actual: $f'(0) = 1, f'(1/4) = .9689, f'(1/2) = .8776$, $f'(3/4) = .7317$

2. (a)

x	*Diff. Quot.*	$f'(x)$
0	.1	0
.1	.3	.2
.3	.7	.6
.5	1.1	1.0
.7	1.5	1.4
.9	1.9	1.8

5. $e(x_i, h) = (y''(\xi)/2)h$, where ξ is a point in the interval $x_{i-1} < \xi < x_i$

section 15.2

1. $y = 25e^{-3x}/27 + x^2/3 - 2x/9 + 2/27$

(a) $h = .5$:

x	$u(x)$	$y(x)$	$E(x)$
0.0	1.0	1.0	0%
.5	− .50	.25290	297
1.0	.38	.23128	− 64

(c) $h = .1$:

x	$u(x)$	$y(x)$	$E(x)$
0.0	1.0	1.0	0.0%
.1	.70	.74113	5.5
.5	.19	.25290	24.9
1.0	.20	.23123	13.5

(3) $y = \frac{1}{2}(e^{-x} + \sin x - \cos x)$

(a) $h = .5$ (two decimal places and three decimal places)

x	$u(x)$	$y(x)$	$E(x)$
.5	0.0	.10419	100.0%
1.0	.240	.33452	28.3

(c) $h = .1$ (two decimal places)

x	$u(x)$	$y(x)$	$E(x)$
.1	0.0	.00483	100.0%
.5	.09	.10419	13.6
1.0	.32	.33452	4.3

$h = .1$ (three decimal places)

x	$u(x)$	$y(x)$	$E(x)$
.1	0.0	.00483	100.0%
.5	.089	.10419	14.6
1.0	.320	.33452	4.3

7. $u_{i+1} = u_i + f(x_{i+1}, u_{i+1})h$

9. $u_{i+1} = u_i + f(x_i, u_i)h + \frac{1}{2}\left[\frac{\partial f}{\partial x}(x_i, u_i) + \frac{\partial f}{\partial y}(x_i, u_i)f(x_i, u_i)\right]h^2$

section 15.3

1. (a) Using $h = .5, .25, .1$, and $.05$, relative change is less than 10 percent for $h = .05$. Maximum error occurs at $x = .6$ and it is 12.07 percent. The Euler difference formula was used in the form, $u_{i+1} = (1 - 3h)u_i + hx_i^2$

3. Using $h = .5, .25, .1$ and $.05$, the relative change is less than 3 percent for $h = .1$.

section 15.4

1. (a)

x	$u(x, .5)$	$u(x, .25)$	$y_R(x)$	$y(x)$
.5	−.50	.08	.66	.25290
1.0	.38	.16	−.06	.23128

2. (a)

x	$u(x, .5)$	$u(x, .25)$	$y_R(x)$	$y(x)$
.5	0.0	.06	.12	.10419
1.0	.24	.29	.34	.33452

section 15.5

1. (a) $h = .5$, $y' + 3y = x^2$

x	$u(x)$	$y(x)$	$E(x)$
0.0	1.0	1.0	0.0%
.5	.69	.25290	−172.8
1.0	.65	.23128	−181.0

$h = .25$

x	$u(x)$	$y(x)$	$E(x)$
.25	.54	.47673	−13.3%
.5	.32	.25290	−26.5
.75	.25	.19250	−29.9
1.0	.28	.23128	−21.1

(c) $h = .5,\ y' + 2y(y-1) = 0$

x	$u(x)$	$y(x)$	$E(x)$
.5	.72	.7311	1.5%
1.0	.85	.8808	3.5

$h = .25$

x	$u(x)$	$y(x)$	$E(x)$
.25	.62	.62246	.4%
.50	.73	.73106	.1
.75	.81	.81757	.9
1.00	.87	.88080	1.2

2. (a)

x	$u(x, .5)$	$u(x, .25)$	$y_R(x)$	$E_R(x)$
.5	.69	.32	−.05	119.8%
1.0	.65	.28	−.09	138.9

(c)

x	$u(x, .5)$	$u(x, .25)$	$y_R(x)$	$E_R(x)$
.5	.72	.73	.74	−1.22%
1.0	.85	.87	.89	−1.04

3. (a)

x	$u(x, .5)$	$u(x, .25)$	$y_R(x)$	$y(x)$	$E(x, .5)$	$E(x, .25)$	$E_R(x)$
0.0	1.0	1.0	1.0	1.0	0.0%	0.0%	0.0%
.50	.687	.320	− .047	.25290	−171.6	−26.5	118.6
1.00	.648	.274	− .100	.23128	−180.2	−18.5	143.2

(c)

x	$u(x, .5)$	$u(x, .25)$	$y_R(x)$	$y(x)$	$E(x, .5)$	$E(x, .25)$	$E_R(x)$
0.0	.5	.5	.5	.5	0.0%	0.0	
.50	.719	.728	.737	.73106	1.7	.4	− .8%
1.0	.856	.876	.896	.88080	2.8	.5	−1.73

5. (a)

x	$u(x, .5)$	$u(x, .25)$	$y_R(x)$	$y(x)$	$E(x, .5)$	$E(x, .25)$	$E_R(x)$
0.0	1.0	1.0	1.0	1.0	0.0%	0.0%	0.0%
.50	.69	.22	− .25	.25290	−172.8	13.0	198.9
1.0	− .82	.01	.84	.23128	454.5	95.7	−263.2

(c)

x	$u(x, .5)$	$u(x, .25)$	$y_R(x)$	$y(x)$	$E(x, .5)$	$E(x, .25)$	$E_R(x)$
0.0	.5	.5	.5	.5	0.0%	0.0%	0.0%
.50	.72	.74	.76	.73106	1.51	−1.22	−3.96
1.0	.90	.89	.88	.88080	−2.18	−1.04	.09

section 15.6

1. (a)

x	u_i
1	.3430
5	1.0132
10	1.0343

2. (a)

x	u_i
1	.7751
5	1.0303
10	1.0343

(c)

x	u_i
1	1.2375
5	1.0364
10	1.0344

3. (a)

x	$u_i\ (a_0 = .3)$	$u_i\ (a_0 = .7)$	$u_i\ (a_0 = .9)$
1	.6037	.8961	.9860
5	1.0260	1.0325	1.0338
10	1.0343	1.0344	1.0344

section 16.2

1. (a) $\dfrac{a}{s^2 + a^2}$, $s > 0$ **(c)** $\dfrac{b}{(s-a)^2 + b^2}$, $s > a$

7. (a) $\dfrac{A(e^{-s} - 1)^2}{s(1 - e^{-2s})}$ **(c)** $\dfrac{A}{s^2}\left[1 - \dfrac{sTe^{-sT}}{1 - e^{-sT}}\right]$ **(e)** $\dfrac{A}{s^2+1}\,\dfrac{(e^{-s\pi} + 1)^2}{(1 - e^{-2s\pi})}$

section 16.3

1. (a) $6e^{4x}$ **(c)** $2e^{2x}(e^x - 1)$ **(e)** x^2 **(g)** $\frac{1}{3}(\sin x + \sin 2x)$

section 16.4

1. (a) $y = 3e^x - 2e^{2x}$ **(c)** $y = \frac{1}{2}(\cos x + \cosh x)$
(e) $y = \frac{1}{50}(80e^{-2x} - 74e^{-3x} - 3\sin 4x - 6\cos 4x)$
(g) $y_1 = \frac{1}{5}(6e^{-4x} - e^x)$, $y_2 = \frac{1}{5}(9e^{-4x} + e^x)$
(i) $y = e^{4x}(\cos 3x - \frac{4}{3}\sin 3x)$
(k) $y = \frac{1}{10}[e^{-x}(\cos x - 5\sin x) + 2\sin 2x - \cos 2x]$
2. (a) $y = \frac{1}{4}(2x - 1 - e^{2(1-x)})$
(c) $y = \frac{1}{9}(18e^{-(x+1)} - 11e^{-3(x+1)} + 3x - 4)$

section 16.5

1. (a) $\dfrac{1}{s}(e^{-s} + 1)$

2. (a) $y = \frac{1}{5}(e^{2x} - e^{-3x}) + \frac{H(x-4)}{60}[6e^{2(x-4)} + 4e^{-3(x-4)} - 10]$

(c) $y = \frac{H(x-\pi)}{8}[-2e^{-2(x-\pi)} + e^{-4(x-\pi)} + 1]$

(e) $y = H(x-2)[1 - \cos(x-2)] - H(x-1)[1 - \cos(x-1)] + \sin x - \cos x + 1$
4b. (i) 1

5. (a) $\theta = \frac{F_0}{6}[e^{-3t} - 3e^{-t} + 2 - H(t-1)[e^{-3(t-1)} - 3e^{-(t-1)} + 2]]$

6. (a) $\theta = F_0\sqrt{\frac{l}{g}}\left[t - \sqrt[4]{\frac{l}{g}}\sin\sqrt[4]{\frac{g}{l}}\,t + H(t-\pi)\left(\cos\sqrt[4]{\frac{g}{l}}\,t\right.\right.$

$\left.\left.\sqrt[4]{\frac{l}{g}}\sin\sqrt[4]{\frac{g}{l}}\,t + t + 1\right)\right]$

7. (a) $q = CE_0[H(t-1)(1 - e^{-(t-1)/RC}) - H(t-2)(1 - e^{-(t-2)/RC})]$
10. (a) $y = \frac{3}{8}(\cos x - \cos 3x)$

(c) $y = \cos x + \int_0^x \sin u f(x-u)\,du$

section 16.6
1. (a) yes. $M = 1$, A any nonnegative number, $a = 3$
(c) yes. $M = 1$, A any nonnegative number, $a > 6$ **(e)** no.

appendix I section I.1
1. (a) $v' + kv/m = q$, $v(0) = 0$ **(b)** $y' = (mg/k)(1 - e^{-kt/m})$, $y(0) = 0$
(c) $y' = (mg/k)(1 - e^{-kt/m})$, $y(0) = h$ **3. (b)** $y = \frac{1}{2}gt^2$
5. $y'' + ky'/m = g$, $y(0) = 0$, $y'(0) = v_0$

section I.2
1. $\theta'' + g\theta/l = 0$, $\theta(0) = 0$, $\theta'(0) = \theta$ **3.** $\pi/8$

section I.3
1. $Ry' + y/C = E$, $y(t_0) = a_0$; $Ri' + i/C = E'$
5. $(R_0 + 2\alpha i)i' + i/C = E'$; no

appendix II
1. (a) $x > 1$ **(b)** $x < 1$ **(c)** $x \neq (2n+1)\pi/2$, $n = 0, \pm 1, \pm 2, \ldots$
(d) $x > 1$
4. $i = e^{i\pi/2}$, $-i = e^{-i\pi/2}$, $1 + i = \sqrt{2}e^{i\pi/4}$, $1 - i = \sqrt{2}e^{-i\pi/4}$, $i^2 = e^{i\pi}$

7. $i^2 = e^{i\pi} = -1$, $(1 + i)^3 = 2^{3/2}e^{i3\pi/4}$, $(3 + i2)^2 = 13e^{i2\tan^{-1}(2/3)}$,
$(i)^{-2} = -1 = e^{-i\pi}, (1 + i)^{-3} = 2^{-3/2}e^{-i3\pi/4}$
8. $\sqrt{i} = e^{i\pi/4}, e^{i5\pi/4}$; $(1 + i)^{1/3} = 2^{1/6}e^{i\pi/12}, 2^{1/6}e^{i3\pi/4}, 2^{1/6}e^{i17\pi/12}$;
$\sqrt[3]{1} = 1, e^{i2\pi/3}, e^{i4\pi/3}$; $\sqrt[4]{4} = \sqrt{2}(e^{in\pi/2})$ where $n = 0,1,2,3$
$(3 + i7)^{1/4} = (58)^{1/8}(e^{i(\theta/4+n\pi/2)})$ where $n = 0,1,2,3$
$\theta = \tan^{-1}(7/3)$
9. **(a)** $f' = \alpha e^{\alpha x} \cos \beta x - \beta e^{\alpha x} \sin \beta x + i\alpha e^{\alpha x} \sin \alpha + ie^{\alpha x} \cos x$
(b) $f' = (1/x) + i\beta \cos \beta x$

appendix III

1. **(a)** $\frac{2}{3}$ **(c)** $\frac{\pi}{2}$ **(e)** $\frac{1}{2}$

REFERENCES

Specific references are included in the text. More generally we list the following.

Two advanced undergraduate texts on ordinary differential equations which extend the treatment of this text are:

1. Coddington, E. A., *An Introduction to Ordinary Differential Equations*, Prentice-Hall, Inc., Englewood Cliffs, N. J., 1961.

2. Hurewicz, W., *Lectures on Ordinary Differential Equations*, The M.I.T. Press, Cambridge, Mass., 1958.

Several books that contain a wide selection of the applications of differential equations are:

3. Batschelet, E., *Introduction to Mathematics for Life Scientists*, Springer-Verlag, New York, 1973.

4. Betz, H.; Burcham, P. B.; and Ewing, G. M., *Differential Equations with Applications*, 2nd Edition, Harper & Row, Publishers, New York, 1964.

5. Lotka, A. J., *Elements of Mathematical Biology*, Dover Publications, Inc., New York, 1956.

6. Noble, B., *Applications of Undergraduate Mathematics in Engineering*, Mathematical Association of America, New York, 1971.

7. Pain, H. J., *The Physics of Vibrations and Waves*, John Wiley & Sons, Ltd., London, 1968.

8. Rescigno, A. and Segre, G., *Drug and Tracer Kinetics*, Blaidell Publishing Company, Waltham, Mass., 1966.

9. Salvadori, M. G. and Schwarz, R. J., *Differential Equations in Engineering Problems*, Prentice-Hall, Inc., Englewood Cliffs, N. J., 1954.

10. Kiselev, A. I.; Krasnov, M. L.; and Makarenko, G. I., *Ordinary Differential Equations*, English translation published by Frederick Ungar Publishing Co., New York, 1967.

INDEX